Molecular Techniques in Taxonomy

NATO ASI Series

Advanced Science Institutes Series

A series presenting the results of activities sponsored by the NATO Science Committee, which aims at the dissemination of advanced scientific and technological knowledge, with a view to strengthening links between scientific communities.

The Series is published by an international board of publishers in conjunction with the NATO Scientific Affairs Division

A Life Sciences B Physics	Plenum Publishing Corporation London and New York
C Mathematical and Physical Sciences D Behavioural and Social Sciences E Applied Sciences	Kluwer Academic Publishers Dordrecht, Boston and London
F Computer and Systems Sciences G Ecological Sciences H Cell Biology I Global Environmental Change	Springer-Verlag Berlin Heidelberg New York London Paris Tokyo Hong Kong Barcelona Budapest

NATO-PCO DATABASE

The electronic index to the NATO ASI Series provides full bibliographical references (with keywords and/or abstracts) to more than 30000 contributions from international scientists published in all sections of the NATO ASI Series. Access to the NATO-PCO DATABASE is possible in two ways:

– via online FILE 128 (NATO-PCO DATABASE) hosted by ESRIN, Via Galileo Galilei, I-00044 Frascati, Italy.

– via **CD-ROM** "NATO-PCO DATABASE" with user-friendly retrieval software in English, French and German (© WTV GmbH and DATAWARE Technologies Inc. 1989).

The CD-ROM can be ordered through any member of the Board of Publishers or through NATO-PCO, Overijse, Belgium.

Series H: Cell Biology Vol. 57

Molecular Techniques in Taxonomy

Edited by

Godfrey M. Hewitt
Andrew W. B. Johnston

University of East Anglia
School of Biological Sciences
Norwich, NR4 7TY
United Kingdom

and

J. Peter W. Young

John Innes Institute
Colney Lane
Norwich NR4 7UH
United Kingdom

Springer-Verlag
Berlin Heidelberg New York London Paris Tokyo
Hong Kong Barcelona Budapest
Published in cooperation with NATO Scientific Affairs Division

Proceedings of the NATO Advanced Study Institute on Molecular Techniques in Taxonomy held in Norwich (England), July 9–20, 1990.

ISBN 3-540-51764-2 Springer-Verlag Berlin Heidelberg New York
ISBN 0-387-51764-2 Springer-Verlag New York Berlin Heidelberg

Library of Congress Cataloging-in-Publication Data
NATO Advanced Study Institute on Molecular Techniques in Taxonomy (1990 : Norwich, England) Molecular techniques in taxonomy / edited by Godfrey M. Hewitt, Andrew Johnston, and J. Peter W. Young. (NATO ASI series. Series H, Cell biology ; vol. 57) "Proceedings of the NATO Advanced Study Institute on Molecular Techniques in Taxonomy held in Norwich (England), July 9–20, 1990"—T.p. verso.
ISBN 0-387-51764-2 (U.S.)
1. Biology—Classification—Molecular aspects—Congresses. I. Hewitt, Godfrey M. II. Johnston, A. III. Young, J. Peter W. IV. Title. V. Series. QH83.N32 1990 574'.012—dc20 91-17381

Printed in Germany

Typesetting: camera ready by authors

31/3140-543210 – Printed on acid-free-paper

Preface

The NATO Advanced Study Institute on Molecular Techniques in Taxonomy was held on 9-20 July 1990 at the University of East Anglia, Norwich, England. In recent years, molecular techniques have begun to have a major impact in taxonomic and evolutionary studies, and we felt that many scientists were now ready to apply the new techniques and would appreciate the chance to learn more about them at first hand. We were clearly right, because the meeting was heavily oversubscribed and we had to turn away many worthy applicants while attempting to maintain a balance of interests and experience. The result was that about a hundred very diverse, very talented, and very enthusiastic people gathered in Norwich; they were so full of ideas and newly-discovered mutual interests that the success of the meeting was ensured from that point on.

As organizers, however, we felt obliged to impose some structure on these interactions, so we had asked our invited lectures to give overviews of major areas, or presentations of their own work as case histories, or practical demonstrations of the critical stages of common laboratory techniques, or in some cases all three of the above! These formed the essential backbone of the course (and of this book), but a still wider range of approaches was represented at the meeting because every participant brought a poster describing their recent work, and some also gave short oral presentations in convivial evening workshops.

This display of diverse talents was essential to the success of the "hands-on" phase of the meeting, in which we aimed to facilitate ("organize" is too strong a word!) a free market in laboratory techniques. After hearing the lectures, each participant was asked which techniques they would most like to try out. Then we matched them in small groups with a suitable tutor, and tried to provide the necessary materials and laboratory facilities. This approach was astonishingly successful, considering that the tutors were operating in a foreign environment with untested materials, and in some cases without prior warning. Under expert supervision, our groups of novices successfully prepared DNA from everything from grass to grasshoppers, and demonstrated RFLPs using digoxigenin-labelled probes. Others experienced at first hand the frustrations, and occasionally the joys, of PCR, and everyone had the opportunity to discuss their future plans with others interested in the same techniques.

Indeed, one of the great delights of the meeting was the free flow of ideas and expertise among all the participants. We invited "teachers" and we offered places to "learners", but many participants of both kinds were world-class experts in their field, and in practice everyone found they had something to learn, and almost everyone had something to teach. For practical reasons, though, this book has to be limited to the contributions of the "official" lecturers. We have arranged them in three main sections: overviews of important topics in molecular taxonomy, case studies of the successful application of molecular methods to taxonomic and evolutionary questions, and protocols for a range of generally applicable methods. It would be artificial to keep these categories entirely isolated and pure, though, and the reader will certainly find some interesting examples among the overviews, some useful methods in the case histories, and even some general reviews enhancing the protocols.

There is one important element missing from this book, and that is the computational analysis of taxonomic data and the reconstruction of phylogenetic trees. This was very ably covered at the meeting by Dave Swofford and Gary Olsen, who provided us with a two-day minicourse in the principles, as well as hands-on computer tuition. However, Dave and Gary had just written a major review of the subject, and felt that they could not better it just yet. Accordingly, we recommend very strongly their chapter on phylogeny reconstruction in *Molecular Systematics* (edited by D.M. Hillis & C. Moritz, published by Sinauer, 1990), and that book is also a very valuable and comprehensive source of protocols and practical advice on all the main laboratory techniques.

Our task as organizers would have been impossible without the help of many others. Our impromptu approach to laboratory practicals would certainly have degenerated into chaos but for the exceptional professional skill and dedication of the technicians of the School of Biological Sciences, in particular Alan Cavill, Derek Fulton and Ian Twaite. Deborah Clemitshaw provided a level of secretarial support normally reserved for top executives and helped smooth the paths of many participants. The participants were invited to a civic reception in Norwich Castle by the Lord Mayor, and the Vice-Chancellor of the University gave a reception before the "Norfolk Dinner" in the Sainsbury Centre on the last night. Several companies assisted with displays and products. To these and all the others - our sincere thanks. Most of all we thank the lecturers and participants for following the example of Chaucer's Scholar of Oxford "and gladly wolde he lerne, and gladly teche".

We believe that the Advanced Study Institute was a success, and fulfilled a real need. Indeed, some of the participants asked, as they left, to be invited to the next meeting! There will not be another one, at least not in exactly the same format, but we hope that this book will be helpful to all those who were not able to come, as well as a a reminder for those who were here.

Godfrey Hewitt, Andy Johnston, Peter Young,
NORWICH

Acknowledgements

In addition to the major financial support from NATO for this ASI, we are also grateful to Natural Environmental Research Council UK, Aldrich Chemical Co. Ltd, Anachem, Applied Biosystems Ltd, Boehringer Corp Ltd, BDH Chemicals, BIO-Rad Laboratories Ltd, Denley Instruments, Fisons, Genetic Research Instrumentation Ltd, Gibco, LEEC Ltd, Labsystems, Laser Labsystems Ltd, Kodak Ltd, Mailbox International Ltd, Milligen Biosearch, Sigma, Stratec Scientific Ltd, UVP, Whatman and finally the City of Norwich and University of East Anglia for their assistance and support.

Contents

Protocols

A PROLOGUE

PAST AND FUTURE OF TAXONOMY

Jean R David
Laboratoire de Biologie et Genetique Evolutives CNRS
91198 Gif-sur-Yvette Cedex
France

In his most influential book"The Growth of Biological Thought", Ernst Mayr used three words in the subtitle, "Diversity, Evolution, and Inheritance". These words clearly reflect the historical changes, and progression in the scientific interests of mankind.

A consideration of the diversity of living beings first led to the description of "taxa", mostly characterised by their external appearance. These descriptions were formally defined by the Swedish naturalist K von Linne, in the middle of the 1 8th century, and his rules of nomenclature are valid still. Living beings were clustered, mainly on the basis of morphology, but also by their capacity to breed together, into groups called species. Even today, the typological definition of the species is used in at least 95% of the new descriptions.

The concept of Evolution was first developed by Lamarck in early 1 9th century. It was suggested that similar taxa were so, not because they performed similar roles in nature, but because they shared a common ancestry in a recent past. This idea of progressive transformations, initially very controversial, became prevalent among biologists, especially after Darwin proposed that natural selection was the main cause of the evolutionary process. Taxonomists became involved, not only in the description of species, but also in the establishment of the evolutionary relatedness between taxa, i.e. their phylogeny.

The beginning of the 20th century saw a major advance when Mendel's laws were rediscovered and began to be actively investigated. Genetic analyses of phenotypic variants have led us to progressively unravel the chemistry of the genetic program and to understand the molecular basis of life itself. The results and concepts of molecular biology are now pervading all fields of Biology, just as Evolution did a century ago.

In the present world, taxonomy has two major functions: to keep and improve the inventory of extant species, and establish the phylogeny of taxa. Work on the first task was begun many centuries ago. So it is no wonder that by comparison with the lightning growth of some modern disciplines, such as Genetics on Immunology, taxonomists are sometimes considered as old-fashioned. Even worse, the fierce competition now existing for research money sometimes leads some people to suggest that taxonomy has achieved its goals, that it is no longer an active research field, and that except for museum conservation and species identifications it does not need or deserve much money.

This NATO-ASI meeting was organized to show precisely the contrary. Although an ancient discipline, taxonomy remains an active field with an important future. I would like to remind you that modern taxonomy has integrated all the technical progress provided by the growth of other disciplines, and continues to do so. For example, electron microscopy has permitted the morphological analysis of very small organisms. The use of molecular techniques, such as allozyme analysis, has revealed many cases of cryptic, morphologically

NATO ASI Series, Vol. H 57
Molecular Techniques in Taxonomy
Edited by G. M. Hewitt et al.

identical species. The tremendous development of computers gives us hope that automatic identification methods can be worked out. Not only have the techniques evolved recently but also the ideas. The biological species concept of Mayr and the cladistic methods of classification formalized by Hennig have brought about conceptual revolutions. Whilst not all taxonomists are cladists, cladistics has promoted new ways of reasoning. For example the recognition of the fact that ancestral traits are not informative for establishing relatedness. A monophyletic group is identified when its members share several derived traits i.e. synapomorphy In the living world extant:: species should be considered, as "sisters", not as "mother and daughter". Similarly the concept of "living fossils" is also difficult to defend.

In the last decade, the achievements of molecular genetics have resulted in a technical revolution in biology, as well as a huge accumulation of DNA sequence data. Sequencing the same gene in different species, geneticists have discovered that they could analyze their results in many ways, giving birth to a new and active discipline, molecular evolution. The capacity to analyze and compare the genomes without the technical constraints of making crosses and of studying several generations, is a major advance. Molecular sequences provide the ultimate information for calculating distances and establishing dendrograms and phylogenies; it is no surpise that these techniques and results have generated great excitement in the scientific community.

Spreading the utilization of these techniques was the goal, and the achievement, of this ASI conference. A broad array of the presently available possibilities were presented and are illustrated by the articles in this book. These include RFLP polymorphism, DNA minisatellites and fingerprints, DNA-DNA hybridization, mtDNA, satellite DNA, ribosomal RNA and, of course, the use of PCR to obtain such DNA sequences. The talks clearly established which technique should be preferred, depending on the question asked and the problem investigated. Thus DNA fingerprints permit the analysis of the mating structure of a population, and a knowledge of the relatednesses over two or three generations. While at the opposite end of the time scale, groups which have diverged for hundreds of million years are better characterised by certain ribosomal RNA sequences.

Almost all of these techniques were demonstrated practically. Detailed, concrete protocols have been provided and, most importantly, there were pertinent discussions led by expert people which pointed out the many practical difficulties which are so often encountered and which are so difficult to explain in written protocols. The benefit of this to the participants forms the cornerstone of the conference.

It was pointed out several times that the techniques have grown very rapidly, and are still improving so quickly that technical constraints in molecular studies are almost nil. On the other hand, theoretical constraints are many, not only for the choice of the methods but also for the interpretation of data. In this respect, the use of powerful computers is now obligatory, and the establishment of efficient and adequate programs is a very active and exciting field of scientific endeavour. The talks clearly explained the various assumptions and problems which are encountered when producing such programs. When DNA sequences are compared, it seems inevitable that we use parsimony methods, even if we believe that nature was not parsimonious. The diversity and complexity of the problems raised by the analysis of sequences has produced a new brand of investigators, "molecular evolutionists". Such a reductionist approach is modern, stimulating and rewarding. There is however a clear danger of loosing the real scope of biology, which is the study of organisms. Knowing in great detail how a molecule has evolved is fascinating information. We should not forget, however that this molecule was embedded within different organisms all along its evolutionary path, and for the moment, molecular evolution does not tell us how and why organisms evolved.

The fact that practising evolutionary biologists and taxonomists are now interested in molecular techniques should mean that they will ask more pertinent questions while keeping in sight important unresolved problems. The talks and posters in this conference addressed such scientific questions, including population polymorphism, spatial and temporal variations, the relationships between morphology and molecular differences, the mechanisms of speciation, the rates of change in different traits and taxa, and of course numerous aspects of phylogenies. Great progress can be made by a joint utilization of various available techniques, and also by the choice of favourable organisms as models. In this latter respect, the drosophilid family appears to offer numerous and diverse opportunities. It can be seen as an evolutionary paradigm and will likely remain so in the foreseeable future.

Evolutionary biology, including taxonomy, population genetics and several other fields, is now a major discipline. Furthermore it should grow since, the more we know, the more we become aware of what remains to be discoverd and understood. In spite of their great analytical power, been pointed out many times, is that during 3 billion years of Evolution, Nature has dealt with phenotypes, and the existence of genotypes and of genetic programs are very recent discoveries. The genotypic molecules which are accessible to analysis, such as allozymes, structural proteins, ribosomal RNA, or satellite DNA, appear to be products, not the causes, of Evolution. Understanding the causes will be to understand the mechanisms of speciation and the genetic bases of adaptive changes, including morphological, physiological and behavioural variations. This will be the formidable task of all biological disciplines in the next century since, as pointed out by Dobzhansky, "nothing in biology makes sense except in the light of evolution".

In this general context, taxonomists will make a significant contribution, for beyond the simple goal of establishing true phylogenies they will ask important questions and bring in valuable biological knowledge. They know that, over time, some traits have been surprisingly conserved; others have shown very rapid, unpredictable changes. What are the genetic bases of traits, ancestral or derived, which are used in morphological cladistics? To take an example, in *Drosophila,* the subgenus *Scaptodrosophila* is characterised by the presence of a very tiny propleural bristle, apparently devoid of any important function, and which does not exist in other subgenera. What are the genes which are responsible for the formation of this bristle? Why did they keep functioning in one phylum and lose their function in others? Such questions must be addressed initially to very simple problems, but even so are beyond the capacities of genetic investigation for the present.

In conclusion, taxonomy, as all other biological disciplines, has to integrate new techniques and new concepts. It looks well set to do so. Currently the most exciting and rapidly progressing field is the establishment of molecular phylogenies which, we hope and believe, reflects the true historical process of Evolution. Fortunately since taxonomists will keep in touch with morphological studies, they should also contribute to the solution of a major problem in biological research, i.e. to bridge the gap between genotypes and phenotypes. Taxonomy should now see a renewal of intellectual interest, as a basic discipline and, hopefully, better funding from research agencies.

TAXONOMY: AN ESSENTIAL KEY TO EVOLUTIONARY BIOLOGY

R J Berry

Department of Biology
University College London
London WC1E 6BT

Variation and Biology

The core of biology is variation - of genes, phenotypes, species, ecosystems. This is a major difference from the so-called exact sciences. Carl Pantin (1968) has written of the latter, "physics and chemistry have been able to become exact and mature just because much of the wealth of natural phenomena is excluded from their study". He goes on, "There is no need for the physicist as such to go to biology for data until in the last resort he has to take into account the fact that the observer is a living creature. I would call such sciences 'restricted'. In contrast, biology and geology are 'unrestricted'. Men of science devoted to these fields must be prepared to follow the analysis of their problems into every other kind of science. If they wish to advance their subject they cannot possibly say, "I will not burden my mind with chemistry, physics, or anything but my special interest".

Variation - or biodiversity as it now tends to be called - may occur at the individual, intraspecific, specific or community/ecosystem level, or in DNA, cell, phenotype or population. It is inseparable from the effective study of biology. In past centuries, advances in biology were limited by the assumption that organisms were analogous to physical machines, and this restriction persisted until the thrall of Plato's essences were lifted and the invariant species of Carl Linnaeus became first the polytypic species of the late nineteenth century, and thence the clinally varying, highly polymorphic metapopulation of the mid twentieth century (Mayr,

NATO ASI Series, Vol. H 57
Molecular Techniques in Taxonomy
Edited by G. M. Hewitt et al.

1982; Gilpin, 1987). Molecular biology is built upon techniques to recognize variation at the chemical level, but from the point of view of evolution, conceptually it is no more than that practised by the old-time naturalists in their study of natural systems.

Attempts to remove variation from biology must be resisted in the interests of seeking truth on the one hand and advancing science on the other. It is ironic that T H Huxley, "Darwin's bulldog", the man who did so much to challenge the negative reactions to evolutionary ideas following the publication of the Origin of Species, stimulated the unhealthy practice of teaching by "types" and hence avoiding having to cope with variation. He confessed, "I am afraid there is very little of the genuine naturalist in me. I never collected anything, and species work was always a burden to me; what I cared for was the architectural and engineering part of the business, the working out of the wonderful unity of plan in the thousands and thousands of diverse living constructions, and the modification of similar apparatuses to serve diverse ends" (Huxley, 1890).

Do not be seduced by the common misapprehensions about reductionism, on the grounds that science is properly reductionist, in contrast to the confused woolliness of "holism". We must distinguish between methodological reductionism which is a necessary part of scientific inference and experiment, and which clearly recognizes its simplifying premises; and ontological reduction, which involves a dogmatic assumption that complex wholes are 'nothing but' their component parts (Ayala, 1974; Barnett, 1988). Biology cannot be reduced to molecules.

Evolutionary Synthesis and Re-Synthesis

Evolutionary thinking is impossible when variation is ignored; it was Darwin's genius to show that linking inherited variation to a struggle for existence leads to natural selection. Ospovat (1981) has pointed out the pre-1859 assumption of a perfect adaptation of organisms to their environment led to a belief in a "natural law" of nature functioning as a well-adjusted mechanism, enforcing harmony among organisms, and between them and their environment. There was thus no need to challenge the idea that the laws of nature were devised by God to achieve his ends; Darwin's particular accomplishment was to destroy this mechanical view of natural law.

Darwin recognized that variation is the glory of biology, not merely an inconvenient complication. His recognition came from his openness to ideas from a wide range of disciplines - agriculture, horticulture, animal husbandry, sociology, palaeontology and biogeography, as well as biology in the narrow sense. He was a naturalist in the original and noble meaning of that name; concerned with understanding the whole, and not bound by theory.

The importance of this eclectic outlook is evident to anyone with a knowledge of the history of biology: advances have taken place whenever variation has been taken seriously, and stagnation has resulted whenever typology has replaced 'population thinking'. The confusion in evolutionary biology in the 1920s can be attributed to the inability of geneticists to appreciate the causes of palaeontological variation and conversely, of palaeontologists to appreciate genetical variation (Provine, 1971; Berry, 1982). The problem still exists: a more recent conflict has arisen from the same root - the failure of some biologists to realize that genetical mechanisms exist which will produce 'punctuational' events in the fossil record (Berry, 1985).

Biological Education

If variation is the core to a proper understanding of biology, any neglect of variation must lead to a defective understanding of the subject. There is a widespread acceptance that formal taxonomic teaching has declined in degree courses in Britain over recent years. A review group in taxonomy set up by the Advisory Board for the Research Councils (ABRC) circulated university and polytechnic biology departments in the mid-1970s and found: "Until the [1939-45] war the undergraduate training of botanists, zoologists and palaeontologists had a strong content of anatomy and systematics. Graduates with this background could and did move without difficulty into taxomonic posts, or posts in which basic skill in taxonomy were an asset. This situation has changed radically in the last 30 years. The scope of first degrees in biological subjects has broadened, and the branches in which spectacular advances have been made emphasised. Consequently taxonomic teaching in undergraduate courses has decreased or disappeared, more so in zoology than in botany" (Advisory Board for the Research Councils 1979, para. 807).

Other inquiries have come to depressingly similar conclusions. A survey of university Zoology Departments found that the number of staff between 1975 and 1985 had remained approximately constant, but their interests, teaching responsibilities and research activities had widened considerably (Simkiss 1986). In other words, the available expertise was spread much more thinly in a greatly enlarged subject. A Natural Environment Research Council (1976) study of the Role of Taxonomy in Ecological Research stated: "Skilled teaching of taxonomy is declining and the number of dedicated taxonomists is insufficient to keep pace with the needs of ecology, palaeontology and perhaps other subjects as well". The European Science Foundation (1977) identified the need for collaboration and joint action so as to make the best use of a limited resource of both manpower and other resources. Ride (1978) described the perception of taxonomy in Australia recently as "a fourth rate game of little significance, played by fourth-rate scientists in fourth-rate institutions". The Association of Systematic Collections (of the USA) concluded that "many of the problems affecting institutions in the United States are not only common to institutions in other countries but also their resolution is dependent upon co-operation at the international level" (Edwards, 1986).

The undergraduate situation has not been compensated by any increase in postgraduate teaching. The ABRC (1979, para 829) working party believed: "that the numbers of potential taxonomists who annually pass through our universities and polytechnics are adequate to maintain taxonomy in a viable condition", but this is increasingly perceived as over-optimistic.

There is a need for professional taxonomists, but they will always be a small minority of biologists. The NERC inquiry into taxonomy concluded that: "the decline of taxonomic activity and expertise... has reached the point of hampering the development of disciplines which rely on an adequate level of taxonomic support.. Ecological research is one such discipline... Very frequently ecological study has lost impetus owing to the unforeseen need to identify specimens or to unravel the taxonomy of a difficult group" (Natural Environment Research Council 1976). Tilling (1987) commented that the effects of taxonomic incompetence has produced a dearth of school teachers with "the ability, confidence or inclination to tackle the teaching of ecology"; this has been documented in a survey of pupil's attitudes by Gayford (1985).

However, it is not only ecological research which is hindered by taxonomic ignorance. Nature conservation depends on a knowledge of the distribution and changes therein of plants and animals, but the Nature Conservancy Council (1984) accepts that even in Britain the basic data relating to this are incompletely known. Stubbs (1986) expressly linked the inadequacy of museum resources (especially trained staff) with a hampering of conservation efforts.

Ecology is the most obvious discipline affected by the inability to correctly identify organisms, but obviously all field studies (behaviour, genetics, parasitology, demography, etc.) depend on a capacity to recognize taxa; in other words, to be able to practice taxonomy, at least at a basic level Evolutionary studies are also involved. Thus the adaptationist debate of the 1860s, the biometrical debate of the 1900s, the palaeontological debate of the 1920s and 1930s, and the neutralist debate of the 1960s were all the fruit of misunderstanding variation. The same condemnation can be levelled over the cladism, punctuationalism, and creationist disputes of the 1970s and 1980s. Failure to appreciate the range and relevance of variation has been a recurring stumbling-block to biological progress, and variation in this sense is really a synonym for taxonomy. It is manifested by the commonness of the four modern myths identified by Cain (1983): the neutrality of enzyme variants, the constancy of the 'biological clock', the de-coupling of macro-evolution from microevolution, and the unthinking assumption of parsimony by taxonomists. This not to claim that a well-taught biologist will never err, but to assert the reasonable proposition that a basic understanding of taxonomy is less likely to lead to a wrong judgment.

But there is an aspect to biological education which is larger than the place of systematics or taxonomy in it, and that is the value of biology (which necessarily includes taxonomy) in the education of the whole person. More than any other academic discipline (apart possibly from medicine, which is after all only applied biology), biology relates individuals to their environment, physical, organic and cultural (Ashby, 1978). The Nature Conservancy Council (1984) has recognized this in defining its own purpose as primarily cultural: "that is the conservation of wild flora and fauna, geological and physiographic features for their scientific, educational, recreational, aesthetic, and inspirational value. The term cultural should not be misconstrued: it is used here in the broadest sense as referring to the whole mental life of a nation. This cultural purpose shades imperceptibly into that which is clearly economic, that is, dealing with aspects of resource utilisation providing the commodities for material existence and regulated by commercial factors. It is, perhaps, undesirable to distinguish sharply between the two, for both are necessary to quality of life, and many nature conservation activities serve both purposes".

Interestingly, in the context of the present essay, the statement continues: "The conservation of genetic variability also spans the two aspects".

Conservation, of course, is only one application of biology, but it is a topic which brings out the necessity of taking in the whole of biology, and not diminishing the subject by omitting essential features. Biology without variation should not properly be called biology; systematics is the formalization of variation, and hence an integral and necessary part of biology. The diminution of the systematic element in biology courses is not a conscious reductionism, but it is none the less a dangerous restriction on the subject. Hill's (1954, in Pantin 1968) aphorism that "Physics and chemistry will dominate biology only by becoming biology" can be turned around: the danger is that biology will become nothing more than physics and chemistry if it neglects variation.

Conclusion

The one feature which distinguishes biology from virtually all other sciences is the existence of variation; indeed, Darwin's revolution was based on incorporating inherited variation with the ecological observation of a widespread struggle for existence among living organisms. Progress in biology has hesitated whenever variation has been neglected; the history of evolutionary understanding is full of warning parables showing this. It follows that taxonomy (which is the formal study of variation) is an essential tool for all who aspire to be proper biologists.

Acknowlegement

This essay is based on a paper given at the Jubilee Meeting of the Systematics Association in 1977, and published by Oxford University Press in Prospects in Systematics, ed. D.L. Hawkesworth in 1978.

REFERENCES

Advisory Board for the Research Councils (1979) Taxonomy in Britain. Her Majesty's Stationery Office, London.

Anonymous (1974) Trends, priorities and needs in systematic and evolutionary biology. Syst Zool 24:416-39.

Ashby E (1978) Reconciling Man with the Environment. Oxford University Press, London.

Ayala R J (1972) Darwinian versus non-Darwinian evolution in natural populations of Drosophila Proc Sixth Berkeley Symp Math Stat Prob 5:211-36.

Barnett S A (1988) Biology and Freedom. Cambridge University Press, Cambridge.

Berry R J (1982) Neo-Darwinism. Edward Arnold, London.

Berry R J (1985) Natura non facit saltum. Biol J Linn Soc 26:301-5.

Cain A J (1983) Concluding remarks. In Protein Polymorphism: Adaptive and Taxonomic Significance (ed G S Oxford and D Rollinson) pp 391-7. Academic Press, London and New York.

Edwards S R (1986) The Association of Systematics Collections. In A National Plan for Systematic Collections? (ed P J Morgan) pp136-155. National Museum of Wales, Cardiff.

European Science Foundation (1977) ESRC Review No 13 Taxonomy in Europe. European Science Foundation, Strasbourg.

Gayford C G (1985) Biological fieldwork - a sample of the attitudes of sixth-form pupils in schools in England and Wales. J Biol Educ 19:207-12.

Gilpin M E (1987) Spatial structure and population vulnerability. In Viable Populations for Conservation (ed M E Soulé) pp125-139. Cambridge University Press, Cambridge.

Huxley T H (1890) Autobiography [Oxford Paperback Edition (ed G.R. De Beer) 1983]. Oxford University Press, Oxford.

Mayr E (1982) The Growth of Biological Thought. Belknap Press, Cambridge, Mass.

Natural Environment Research Council (1976) The Role of Taxonomy in Ecological Research. Working Party Report. Her Majesty's Stationery Office, London.

Nature Conservancy Council (1984) Nature Conservation in Great Britain. Nature Conservancy Council, London.

Ospovat D (1981) The Development of Darwin's Theory. Natural History, Natural Theology and Natural Selection 1838-1859. Cambridge University Press, Cambridge.

Pantin C F A (1968) Relations between the Sciences. Cambridge University Press, Cambridge.

Provine W B (1971) The Origins of Theoretical Population Genetics. Chicago University Press, Chicago and London.

Ride W D L (1978) Towards a national biological survey. Search, 9:73-82.

Simkiss K (1986) Zoo-technology, molecular zoology? Biologist 33: 206-7.

Stubbs A E (1986) The role of biological collections in relation to the Nature Conservancy Council. In A National Plan for Systematic Collections? (ed P J Morgan) pp106-110. National Museum of Wales, Cardiff.

Tilling S M (1987) Education and taxonomy: the role of the Field Studies Council and AIDGAP. Biol J Linn Soc 32, 87-96.

VARIATION AT THE DNA LEVEL: SOMETHING FOR EVERYONE

Martin Kreitman
Department of Ecology and Evolution
Princeton University
Princeton, N.J. 08544
USA

Introduction: Pretentiousness notwithstanding, if there is such a thing as a central paradox in biology it is that all organisms resemble one another but at the same time are different. Darwin, of course, recognized the universality of this paradox, and it led him to a theory of descent with modification, the foundation of our modern theory of evolution.

Nowhere is this paradox more apparent than in the study of gene function and gene evolution. Over evolutionary time all genes accumulate mutations, but only certain sites are "accessible" to change - certain nucleotide sites or codon triplets can be indefinitely preserved by purifying selection. Other nucleotide sites diverge rapidly in evolution. Curiously, the two great traditions in biology today, organismal and evolutionary biology forming one tradition and cell and molecular biology forming the other, rely on the same molecular comparisons for insights but have staked their claims on different sides of the central paradox. Molecular biologists focus on the conservation of DNA sequences to understand what is essential to the normal expression and function of a gene; evolutionists rely on variation and change to reveal pattern and process in evolution. Molecular biology views variation, whether it occurs within or between a species, as functionless molecular noise, while evolutionists view the absence of change as unrevealing. (While many would like to view the genetic material as a common meeting ground for the two scientific traditions, viewed this way, I'm afraid it is little more than a point of common departure.)

To the evolutionist it is variation that reveals the tempo and mode of the evolutionary process. This is true whether ones interest is in assigning parentage,

NATO ASI Series, Vol. H 57
Molecular Techniques in Taxonomy
Edited by G. M. Hewitt et al.

measuring gene flow across populations, inferring patterns of migration, colonization or a species expansion, understanding the evolutionary causes of polymorphism and molecular evolution, or determining phylogenetic relationships among species. Variation, in the broad sense, is as much the raw material of evolutionary study as it is the raw material of evolution.

The purpose of this paper is to provide a loosely technical overview of the kinds of variation that exist at the level of DNA and an equally loose overview of techniques allowing it to be revealed. I will not pay much attention to the problem of determining what kind of DNA variation is most appropriate for answering which kind of evolutionary question. That is left for the reader (just use a little common sense). My goal here is to make the reader aware of the many types of DNA variation available to them, the main strategies for its identification, and the techniques available for carrying out a study. However, this is not a comprehensive review of the subject. Instead, it is a rather idiosyncratic overview of the approaches we have been using in my lab, or approaches I would like to see exploited more intensively.

What I mean by variation: I will focus on techniques for revealing variation within species - classically called polymorphisms - and also for revealing variation between reasonably closely related species - so-called substitutions. Whereas the species boundary is the classic dividing line between population genetics and organismal evolution, between what can be studied genetically and what can only be studied comparatively, there is no need to make any such distinction, either theoretically or empirically, for DNA. For the purposes of this chapter, indeed for the purpose of understanding molecular evolution, there is no need to view variation within a species any differently than variation between species.

To make this point more concrete, consider the neutral theory of molecular evolution. As Motoo Kimura rightly points out, under this theory, "Polymorphism (within a species) is just a transient phase of molecular evolution" (Kimura, 1983). Indeed, the neutral theory takes no special accounting of speciation: allelic variants within a species are genealogically related to common ancestral sequences (which may or may not be present in the same species) just as allelic variants in different species also have common ancestors further back in time. The topology of an allele genealogy does not change across species boundaries, only the time depth of the tree.

The neutral theory makes specific predictions about how to relate variation within and between species. Selection theory does likewise. The ability to quantify DNA evolution on more than one time scale, and the opportunities it affords for testing evolutionary theories, is in my view one of the most attractive features of DNA as a tool for studying evolution. This being the case, it is important to have techniques and methodologies for revealing variation of homologous segments of DNA, whether come from individuals within a species or come from different species. This review, although focusing primarily on how to find polymorphism within species, will also apply equally well to interspecific variation among closely related species.

What type of variation to study: DNA can vary in many different ways - including (but certainly not limited to) single nucleotide change, insertion and deletion, simple-repeat length variation, and hypervariable minisatellite tandem repeat sequence variation - and each kind of polymorphism can be of use to the evolutionist. Variation is not even confined to one genome: all organisms possess a nuclear and a mitochondrial genome (plants also possess a chloroplast genome).

Different kinds of mutations have different rates of occurrence. Each class of mutation is not suitable for studying all evolutionary processes. In particular, one should choose a class of variation whose mutational rate matches the time-constant of the evolutionary process being investigated. For example, immensely polymorphic tandem repeat sequences, which can have mutation rates of 5% or more per generation, allow accurate parentage analysis, but are rather useless for determining the relationships of species or even for determining the genetic distances between populations within a species. Divergence simply occurs too quickly. At the other extreme, the exceedingly slow-evolving 18S ribosomal genes are useful for delineating the deepest branches of life, but are uninformative about recent evolution.

The choice of which part of the genome, or indeed, which genome (mitochondrial or nuclear) to survey will depend on both strategic and tactical (or technical) considerations. In this section I will deal with the key strategic issue, knowing which of the many kinds of variation to exploit. First I will describe several major categories of DNA polymorphism and will indicate some of the advantages and disadvantages of each type. I will then discuss the merits of studying mitochondrial vs. nuclear gene variation. (I wonder, for example, how many of the investigations of variation in mitochondrial DNA chose this molecule without consideration of alternative sources and types of variation?) I leave for the next section a consideration of some tactical issues, namely deciding on the appropriate method.

As a first step in deciding what polymorphism to study, I present a brief overview of types of variability likely to be encountered in a eukaryotic genome. The major types of polymorphism are summarized in Table 1.

Base Substitutions: Most nucleotide base substitutions occur at low rates; for the vast majority of species no more than two nucleotides (or alleles) will be found in a sample at any particular site. In general, the number of mutations found to be segregating within a species or population sample will be a function of the product of population size and mutation rate. It is possible that some bacterial species or marine invertebrate with immense population sizes will have three or even all four nucleotides present at a site, but at least for *Drosophila* having more than two nucleotides at a site is rare.

The proportion of nucleotide sites that are polymorphic in a sample of genes will depend on certain species-specific population parameters, population size and mutation rate in particular, as well as on the region of DNA being considered. In general, the level of polymorphism (or the rate of evolution) is expected to be inversely correlated with the constraints on the functionality of the DNA. Coding regions of genes, for example, tend to have little polymorphism compared to introns and noncoding regions. It is likely to be rare for noncoding regions to have higher levels of polymorphism (or evolve at a faster rate) than pseudogenes, which themselves are thought to evolve only by genetic drift. By this accounting of variation, it is widely thought that the highest levels of polymorphism will be found in regions of DNA having no function; lower-than-neutral levels of polymorphism is thought to indicate the presence of functional (or selective) constraint (elaborated in Kimura 1983). The most likely place to find polymorphism, therefore, is in noncoding or so called "junk" DNA. Introns, for example, are generally relatively polymorphic.

How frequent are single-base nucleotide polymorphisms in noncoding DNA? One simple measure of polymorphism frequency is π, the probability that two alleles differ at a nucleotide site. In most instances π will be approximately equal to the average nucleotide site heterozygosity, or H_n. As tabulated in Table 2 *Drosophila* values (indicated as θ) range between 0.001 and 0.02. This means that in two randomly chosen genes from one species, between 1 in 1000 and 20 in

Table 1
Summary of DNA Variability

Number of Sites	Type of Variation	Alleles/ Locus	Methods for Scoring*	Reference#
Single Locus	Base Substitution	2	ASO AS-PCR RFLP: 6-Cutter DNA Seq.	1, 2 3, 4
	Ins/Del (Nontandem) 1-50bp	2	PCR DNA Seq. RFLP: 4-Cutter	5, 6 7
	Transposable Element	2	Southern	8
	Tandem Repeat Homo-nuc.	≥ 2	PCR DNA Seq. RFLP: 4-Cutter	 9, 10
	Tandem Repeat (Di-/Tri-nuc.)	≥ 2	PCR DNA Seq. RFLP: 4-Cutter	11
	Tandem Repeat Minisatellite	2 - >100	LS-PCR Southern Internal RFLP	12 13
Multilocus	Base Substitution	Many Haplotypes	DNA Seq. Southern	
	Tandem Repeat Minisatellite	Hugh Number of Haplotypes	Southern	14

* ASO: Allele Specific Oligonucleotide; AS: Allele-Specific; PCR: Polymerase Chain Reaction; RFLP: Restriction Fragment Length Polymorphism; Seq.: dideoxy DNA sequencing; LS: Locus-Specific; Southern: Southern blot analysis. # 1: Alves et al. 1988; 2: Erlich and Bugawan 1989; 3: Kwok et al. 1990; 4: Ruano and Kidd 1989; 5: Erlich 1989; 6: Saiki et al. 1988; 7: Kreitman and Aguadé 1986; 8: Southern 1975; 9: Kreitman 1983; 10: Jones and Kafatos 1982; 11: Weber and May 1989; 12: Jeffreys et al. 1988; 13: Jeffries et al. 1990; 14: Jeffreys 1987.

Table 2.
Summary of RFLP polymorphism in *Drosophila*

Species	Locus	Type of Study	Sample Size	Length of Region	Est. No. Sites*	No. Seg. Sites	$\hat{\theta}$**	Reference
melanogaster	Adh	6-cutter	18	12kb	23	4	0.006	Langley, et al. (1983)
	Adh	6-cutter	48	13kb	30	8	0.006	Aquadro, et al. (1986)
	Adh	4-cutter	87	2.7kb	526bp	17	0.006	Kreitman & Aguade (1986)
	87A heat shock	6-cutter	29	25kb	25	2	0.002	Leigh-Brown (1983)
	Amy	6-cutter	85	15kb	26	3	0.006	Langley, et al. (1988)
	G-6PD	6-cutter	122	13kb	28	5	0.003	Eanes, et al, (1989)
	Rosy	6-cutter	60	40kb	41	7	0.003	Aquadro, et al. (1988)
	Yellow-	6-cutter	64	106kb	176	9	0.001	Aguade, et al. (1989)
	Achaete-	6-cutter	109	30.5kb	63	9	0.003	Eanes, et al. (1989)
	Scute	6-cutter	49	120kb	67	10	0.003	Beech & Leigh Brown (1989)
	Zeste-tko	6-cutter	64	20kb	42	10	0.004	Aguade, et al. (1989)
	White	6- & 4-cutter	64	45kb	327	54	0.013	Miyashita & Langley (1988)
	Notch	6-cutter	37	60kb	58	15	0.005	Schaeffer, et al. (1988)
simulans	Rosy	6-cutter	30	40kb	56	28	0.019	Aquadro, et al. (1988)
pseudo-obscura	Adh	6-cutter	20	32kb	43	27	0.021	Schaeffer, et al. (1987)
	Rosy	4-cutter	58	5.2kb	147	66	0.092	Riley, et al. (1989)

*The number of sites is the number of restriction sites scored (not the number of nucleotide sites).

**Estimates based on methods of Nei & Tajima (1981), Engels (1981) or Hudson (1982). The methods generally give similar values.

1000 sites are expected to differ (that is to say are polymorphic). In humans, approximately 2 nucleotide sites in 1000 sites are polymorphic.

There are a number of instances in which an investigator might want to score only a single-site nucleotide polymorphism. In the case of risk assessment, for example, a particular polymorphism in linkage disequilibrium with a genetic disease gene would be useful as a molecular marker for the disease gene. In most formal genetic analyses, in genetic mapping experiments or in artificial selection experiments, following the segregation of single nucleotide polymorphisms may be sufficient.

For many population studies, though, it is desirable to identify as much polymorphism as is feasible. For single-site nucleotide polymorphisms, since there are in general only two alleles at each polymorphic site, the only way to increase the amount of genetic variability is to score more polymorphic sites. This is, of course, not difficult to do. A single Southern blot can reveal many polymorphic sites simultaneous, as can DNA sequencing. However, surveying more than one linked polymorphic site raises a rather thorny and intractable problem. The question is, should two polymorphisms be considered statistically independent of one another if they are closely linked physically on the DNA? The answer, in general, is no: the two sites may have correlated evolutionary histories if they are not on different linkage groups. For many kinds of problems, such as estimating the genetic distance between populations, or determining phylogenetic relationships of closely related species, it would be more informative to have data for unlinked polymorphic sites than the same number of linked sites. Unfortunately, I know of no simple way to obtain information about a large number of unlinked polymorphisms.

Given the relatively low level of single-base polymorphism per nucleotide site, a more expeditious way to quantify polymorphism is to identify alleles from the joint distribution of nucleotides at more than one polymorphic site. That joint distribution, I refer to as the haplotype of the allele. For n polymorphic sites with two nucleotides there are potentially 2^n haplotypes if recombination is available for reshuffling the polymorphic nucleotides among the alleles. Therefore, if a large number of alleles are required for a particular study, then it is always advantageous to score haplotypes rather than individual polymorphic sites.

Unfortunately, desirability by itself does not get results. For diploid organisms it is not always possible to identify the haplotype of a single allele. Consider the simplest case, determining the haplotype of alleles when there are two linked polymorphic sites. If each site is polymorphic for two alleles, "+" and "-", there are four possible haplotypes, ++, +-, -+ and --. Consider now an individual who is a heterozygote at both sites. There are two possible haplotype combinations yielding a double heterozygote: ++/-- and +-/-+. Unless the two alleles can be separated before molecular analysis, it is impossible to deduce the haplotype of doubly heterozygous individuals. The problem only gets worse for more than two polymorphic sites. Unfortunately, most simple molecular techniques for scoring polymorphism do not allow separation of the two alleles.

There are several ways around the problem. The simplest is to not work with diploid organisms. When this is not possible, mitochondrial DNA might be a good alternative: it is effectively a haploid genome. If possible, work with haplo-diploid animals. They offer the best of both worlds. Another approach is to do some genetics. For species that can be bred in the lab it may be possible to make homozygous lines by inbreeding or by following standard genetic crossing schemes. If a single homozygous line is available, it may be possible to cross wild individuals to the standard and then score the f_1 hybrids. If genetic crosses are not possible, another approach is to try to statistically infer the haplotypes of multiply-heterozygous individuals by recomposing them from known haplotypes. The latter are immediately known for all homozygous or singly-polymorphic individuals. Such a statistical approach has been developed by Clark (1990).

Several molecular approaches can be used to separate alleles of heterozygous individuals. One method is by cloning, an approach I do not recommend. One ingenious method for separating two alleles in PCR amplified DNA is by denaturing gradient-gel-electrophoresis (Myers, Sheffield and Cox 1989). This technique takes advantage of the fact that fragments of DNA differing only slightly in DNA sequence can often be forced to denature at slightly different conditions, or at slightly different points along the denaturing gradient. This technique can electorphoretically separate two alleles differing by single base substitutions. An application of this technique to the study of polymorphism at the major histocompatibility locus of humans is presented by Gyllensten (1989).

Some human geneticists are using an absurdly simple PCR-based technique for separating alleles of heterozygous individuals - simply serially diluting genomic DNA prepared from single individuals until there is on average only one DNA molecule representing any region (Ruano and Kidd 1990). Because it is sometimes difficult to amplify DNA from a single molecule and because contamination becomes an important issue, I am not confident about recommending this approach. It certainly uses Taq polymerase inefficiently.

Length polymorphism: Length polymorphism comes in essentially four varieties, which for convenience I call i) small unique length variants (1 - 30 bp); ii) large insertions - usually transposable elements; iii) length variants of simple di- or tri-nucleotide repeat sequences, and iv) minisatellite tandem repeat length variation.

Small unique length variants result from single-base insertions, small duplications and mismatch repair errors. This kind of variation forms an abundant class of polymorphism in noncoding DNA. For example, using a 4-cutter RFLP technique for identifying length variants, we have discovered 20 independent length variants in approximately 1500 bp of noncoding DNA in the alcohol dehydrogenase gene region of *D. melanogaster*, of which 12 are small-length variants. An excellent review of small-length variation of *Drosophila* is presented by Jones and Kafatos (1982) for the chorion gene region and a very revealing study of length polymorphism is presented by Aquadro et al. (1986). Simple repeat sequences and homonucleotide runs (poly-T, for example) appear to be highly susceptible to slippage during DNA replication. These kinds of polymorphism are easy to identify or score, either by 4-cutter RFLP analysis, or by PCR analysis.

Length polymorphism for simple dinucleotide or trinucleotide tandem repeats form a second class of small-length polymorphism that likely arises from slippage during replication. In humans (and other mammals) these simple repeat sequences are very abundant and a single "locus" can have many alleles. It is estimated that there are 50,000 - 100,000 (dC-dA)n/$(dG\text{-}dT)_n$ blocks in the human genome, with the number of tandem repeats/locus, n, ranging from 15 - 30 (Weber and May 1989). The function of these repeats are unknown, although they have been proposed to be hotspots for recombination, sites of gene regulation, or may be involved in chromosome structure (reviewed in Weber and May 1989). As many as eleven alleles have been identified at a locus in humans, with 80% of individuals being heterozygotes. Length variants can be identified by PCR amplification, followed by electrophoresis on polyacrylamide gels. Weber and May present a procedure for cloning new loci.

Simple tandem repeat loci are attractive alternatives, in my view, to protein gel electrophoresis for measuring polymorphism relevant to many population genetic problems. Their main advantages are their abundance, high levels of polymorphism, and the ability to resolve single-locus genotypes. In addition, they are believed to be widely dispersed through the genome. Simple repeat sequences have been identified in broad range of animals (Levinson et al. 1985) and plants (Condit et al. 1990).

Length polymorphism arising from transposable elements is of little value for population analysis because, although an element may occur many times within a single genome, their frequencies at particular sites are generally very low (reviewed in Charlesworth and Langley 1990). Other kinds of middle repetitive DNA, such as Alu sequences in humans and primates, which are not transposable elements at all but processed 7SL RNA genes (Ullu and Tschudi 1984), are also of little value to population genetics because their frequencies at particular sites are too high. Although Alu sequences occur approximately 100,000 times in the human genome, there is little polymorphism within a site. So, the vast majority of transposable elements, and a (still largely unknown) fraction of middle repetitive DNA, may not be variable enough to be of use in evolutionary genetics.

The same certainly cannot be said for minisatellite tandem repeat sequence DNA. Several different satellite sequence families have now been discovered, some of which are highly conserved in evolution but which have highly variable numbers of tandem repeats (VNTR) at a locus. A subset of these minisatellites share a common core sequence (approximately 15 bp) which is similar to the "Chi" sequence of *E. coli*, known to specify a recombination signal (see Jeffreys 1987). But it is not known whether the minisatellites have the same function in eukaryotes. Human VNTR probes have been successfully used for studying paternity in a number of organisms, especially birds (Gyllensten et al. 1989; Burke and Bruford 1987); a M13 bacteriophage also contains a sequence that can detect a minisatellite hypervariable locus (Vassart 1987). The mutation rate to new alleles has been studied at several hypervariable loci in humans and is highly variable among loci but can be at least as high as 7% (Jeffreys et al. 1988). Some highly variable loci have hundreds of length variants and have heterozygosities approaching 100%.

Nuclear vs. mitochondrial DNA: Despite its small size mitochondrial DNA has been, by far, the most popular genetic material for use in evolutionary molecular study. The primary advantages of mitochondrial DNA over nuclear DNA are four-fold: i) it is relatively easy to extract and purify; ii) it does not require cloning; iii) it it highly polymorphic and evolves quickly; iv) it is haploid. I have already discussed the advantages of studying haploid material - the haplotype of each allele can be unambiguously identified. Related to this point is the fact that mitochondrial DNA does not recombine. This means that phylogenies of alleles can be constructed using standard tree-building algorithms. The same can definitely not be said of nuclear DNA genes. Within-species mitochondrial phylogenies have been put to especially good use in the analysis of population structure and patterns of migration (reviewed in Avise et al. 1987). For example, Cann, Stoneking and Wilson (1987) have used human mitochondrial phylogenies to infer the direction of gene flow in early modern humans by showing that African populations have the most diverse and most ancestrally related mitochondrial DNAs.

Mitochondrial DNA has also been popular for determining absolutes dates of certain events, such as speciation, or in the case of humans the date of most recent common ancestry of the major races. Cann, Stoneking and Wilson, for example, estimate the mitochondrial "Eve" (which really just means the common ancestor) lived approximately 200,000 years ago. A word of caution is warranted for anyone having similar intentions: intraspecific dating based on mitochondrial DNA, no matter how good the calibration of the molecular clock rate, is fraught with problems. The most serious one is the huge expected variance around any estimated time. This is because the clock itself is driven by a stochastic mutational process. Without going into any detail, I would not be surprised if a reasonable confidence interval for the 200,000 year estimate for humans is between 50,000 and 1 million years (see Kreitman and Hudson 1991 for a method for constructing confidence intervals for polymorphism data; Ewens 1979 for basic theory). For this reason alone I would advise against absolute dating of within-species events using mitochondrial DNA time estimates.

One additional cautionary point should be raised about mitochondrial DNA: it doesn't always evolve quickly and it isn't always polymorphic within species. The conventional wisdom is that mitochondrial DNA in vertebrates, especially mammals, evolves 5 - 10 times faster than nuclear DNA (reviewed in Vawter and Brown 1986). But this is certainly not the case for *Drosophila*. Studies now clearly show single-copy nuclear DNA and mitochondrial DNA to be evolving at roughly similar rates (Solignac, Monnerot and Mounolou 1986; Caccone, Amato and Powell 1988). This means that mitochondrial DNA offers no inherent advantage in *Drosophila* for determining the phylogenetic relationship of species. It also means there should be **less** polymorphism within species. Theoretically, there should be 1/4 as much selectively neutral variation in mitochondrial DNA as in a diploid nuclear gene.

In addition, because the whole mitochondrial genome is one linkage group and does not recombine, it is very susceptible to a reduction in polymorphism resulting from positive selection (e.g., adaptive substitutions). After the selective fixation of an adaptive mutation and barring new mutation, only one mitochondrial allele will be present in the species - every copy will be identical by descent to the originally selected allele. Selective substitutions, therefore, sweep away ancestral variation and reduce levels of polymorphism. Selection does not actually have to be on a mutation in the mitochondrial DNA itself. Selection acting on any maternally inherited trait will also cause a concommittant change in mitochondrial gene frequencies!

In summary, mitochondrial DNA can offer an excellent entree into molecular evolutionary genetics and systematics. It is relatively easy to work with and to analyze. But some thought must be given to the possibility that this genome will not be evolving any faster than the nuclear genome, and the analysis of mitochondrial variation must proceed with some caution.

A quick survey of techniques: There are essentially only two approaches, indirect and direct, for surveying variation within and between (closely related) species. The indirect method reveals variation by filter hybridization with specific probes. The classical filter hybridization technique is the Southern blot (Southern 1975). DNA is restricted enzymatically, generally with six-base-recognizing enzymes, fragments are then electrophoretically separated by length on an agarose gel, and the DNA is finally transferred to a membrane (now usually charged or uncharged nylon). A specific radioactive (or nonradioactive) probe is prepared, generally a cloned piece of DNA, and then hybridized to the filter. The probe may also be an allele-specific oligonucleotide, which will recognize only one specific allele (Alves et al. 1988). After sufficiently stringent wash, the filter is exposed to X-ray film to reveal restriction fragment length polymorphism.

Southern blots using four-base-recognizing restriction enzymes offer approximately a 16-fold increase in the density of detectable polymorphism (Kreitman and Aguadé 1986). Because 4-cutter enzymes produce small fragments (generally less than 1000 bp) the digested DNA can be run on a denaturing DNA sequencing-type gels, and electrophoretically transferred to a membrane (Church and Gilbert 1984). The advantage of this approach is the ability of DNA sequencing gels to distinguish fragments differing in length by as little as a single nucleotide.

More recently, DNA amplification by the polymerase chain reaction with thermostable Taq polymerase (Saiki et al. 1988) has replaced cloning to provide substrate for direct analysis of variation. Once a segment of DNA is amplified, variation can be revealed by restriction enzyme digestion (either 4- or 6-cutter) followed by gel electrophoresis and ethidium bromide staining, or by DNA sequencing.

DNA sequencing of PCR amplified DNA has turned out to be something of a challenge. The ideal substrate for dideoxy sequencing is single-stranded DNA. PCR amplified DNA is, of course, double stranded. Several approaches for obtaining single-stranded DNA have been developed. First, it is possible to take a previously amplified DNA product and reamplify it with a single oligonucleotide primer. Often this will generate a certain excess of one DNA strand. A related approach performs the PCR amplification with a limiting amount of one primer, which is exhausted during the course of amplification (Gyllensten et al. 1989; Kreitman and Landweber 1990). Eventually the one strand, primed by the remaining primer, is produced in excess. When conditions are right this method can produce large amounts of single-stranded DNA for sequencing. But sometimes it is difficult to find appropriate conditions.

A clever alternative for producing single-stranded DNA is to digest away only one strand of PCR amplified DNA (Ochman and Higuchi 1990). Using one phosphorylated oligonucleotide primer (a "kinased" primer) in the amplification, followed by treatment with Lambda exonuclease (sold by BRL) - a 5' - 3' exonuclease that requires a terminal 5' phosphate to initiate digestion - leaves a single intact strand of DNA. My lab currently uses this method for most of our PCR-based DNA sequencing.

Finally, it is possible to directly sequence double-stranded PCR products. Sequencing DNA from low-melt agarose gels (Kretz, Carson and O'Brian 1989) or adding nonionic detergents to the double-stranded DNA prior to sequencing, improves the quality of double-stranded sequencing (R.M. Feldman, this conference).

PCR has become a popular technique for detecting length polymorphisms of all kinds. Oligonucleotide primers are constructed to amplify across a length polymorphism, be it a unique insertion/deletion, a simple tandem repeat, or a single-locus hypervariable tandem repeat sequence. The polymorphism is visualized simply by running the PCR products on an appropriate gel and staining with ethidium bromide (Weber and May 1989; Jeffreys et al. 1988).

Jeffreys, Neumann and Wilson (1990) have recently described a clever and simple method for revealing polymorphism within a PCR-amplified tandem-repeat sequence. Partially digesting a hypervariable single-locus PCR product with a restriction enzyme that cuts within the repeat sequence and running the products on an appropriate gel, can often reveal substantial polymorphism within the repeat locus. This approach has great potential for allowing rapid characterization of many genotypes of a single locus.

Which method to use: I have little advice about which method or approach to use. Certainly such a decision must depend on the biological problem at hand, and the resources available to the investigator. Because of the tremendous differences among species in levels of polymorphism, both nuclear and mitochondrial, it may be useful to do a quick survey, using any of the appropriate techniques before committing to a particular approach. The same is true for molecular studies of closely related species - a quick survey to determine the minimum and maximum extent of divergence between species might be useful in formulating a final strategy.

Finally, recognize in DNA the opportunity for a completely accurate assessment of variation. But with this opportunity comes the more likely possibility of errors. DNA sequence errors, unlike allozymes, are likely to be found out by other researchers. Save yourself some future embarrassment by keeping this in mind. As my thesis advisor would constantly remind us about DNA sequencing, "God is in the detail". To this I add, "So is an appropriate sampling strategy".

References

Aguadé, M., N. Miyashita and C.H. Langley 1989 Reduced variation in the *yellow-achaete-scute* region in natural populations of *Drosophila melanogaster. Genetics* **122:** 607-615.

Aguadé, M., N. Miyashita and C.H. Langley 1989 Restriction-map variation at the *Zeste-tko* region in natural populations of *Drosophila melanogaster. Mol. Biol. Evol.* **6:** 123-130.

Alves, A.M., D. Holland, M.D. Edge and F.J. Carr 1988 Hybridisation detection of single nucleotide changes with enzyme labelled oligonucleotides. *Nucl. Acids Res.* **16:** 8722.

Aquadro. C.F., S.F. Desse, M.M. Bland, C.H. Langley & C.C. Laurie-Ahlberg 1986 Molecular population genetics of the alcohol dehydrogenase gene region of *Drosophila melanogaster. Genetics* **114:** 1165-1190.

Aquadro, C.F., K.M. Lado and W.A. Noon 1988 The *Rosy* region of *Drosophila melanogaster* and *Drosophila simulans.* I. Contrasting levels of naturally occurring DNA restriction map variation and divergence. *Genetics.* **119:** 875-888.

Avise J.C. 1986 Mitochondrial DNA and the evolutionary genetics of higher animals. *Phil. Trans. Royal Soc. London. Series B:* **312:** 325-42.

Beech, R.N. and A.J. Leigh Brown 1989 Insertion-deletion variation at the *yellow-achaete-scute* region in two natural populations of *Drosophila melanogaster. Genet. Res., Camb.* **53:** 7-15.

Burke, T. and M.W. Bruford 1987 DNA fingerprinting in birds. *Nature* **327:** 149-52.

Caccone, A., G.D. Amato and J.R. Powell 1988 Rates and patterns of scnDNA and mtDNA divergence within the *Drosophila melanogaster* subgroup. *Genetics* **118:** 671-683.

Cann, R.L., M. Stoneking and A.C. Wilson 1987 Mitochondrial DNA and human evolution. *Nature* **325:** 31-36.

Charlesworth, B and C.H. Langley 1990 The Population Genetics of Transposable Elements in Drosophila. In, *Evolution at the molecular level,* edited by R.K. Selander, A.G. Clark and T.S. Whittam, Sinaur Associates, Inc., Sunderland MA.

Church, GM and W. Gilbert 1984 Genomic sequencing. *Proc. Natl. Acad Sci. USA* **81:** 1991-1995.

Clark AG. 1990 Inference of haplotypes from PCR-amplified samples of diploid populations. *Mol. Biol. Evol.* **7:**111-22.

Condit, R. and S.P. Hubbell 1991 Abundance and DNA sequence of two-base repeat regions in tropical tree genomes. *Genome* **34:** in press.

Eanes, W.F., J.W. Ajioka, J. Hey and C. Wesley 1989 Restriction-map variation associated with the G6PD polymorphism in natural populations. *Mol. Biol. Evol.* **6:** 384-397.

Eanes, W.F., J. Labate and J.W. Ajioka 1989 Restriction-map variation with the *yellow-achaete-scute* region in five populations of *Drosophila melanogaster. Mol. Biol. Evol.* **6:** 492:502.

Engels, W.R. 1981 Estimating genetic divergence and genetic variability with restriction endonucleases. *Proc. Natl. Acad. Sci. USA*. **78:** 6329-6333.

Erlich, H.A. and T.L. Bugawan 1989 HLA Class II gene polymorphism: DNA typing, evolution, and relationship to disease susceptibility. In, *PCR Technology*, H.A. Erlich, Editor. Stockton Press, N.Y.

Erlich, H. 1989 *PCR Technology: Principles and applications for PCR amplification.* Stockton Press, N.Y.

Ewens, W.J. 1979 *Mathematical population genetics.* Springer-Verlag, N.Y.

Gyllensten, U. 1989 Direct sequencing of *in vitro* amplified DNA. In, *PCR Technology*, H.A. Erlich, Editor. Stockton Press, N.Y.

Gyllensten, U.B. and H.A. Erlich 1989 Generation of single stranded DNA by the polymerase chain reaction and its application to direct sequencing of the HLA Dq locus. *Proc. Natl. Acad. Sci. USA* **85:** 7652-7656.

Gyllensten, U.B. S. Jakobsson, H. Temrin and A.C. Wilson 1989 Nucleotide sequence and genomic organization of bird minisatellites. *Nucl. Acids Res.* **17:** 2203-2214.

Higuchi, R. and H. Ochman 1989 Production of single-stranded DNA templates by exonuclease digestion following the polymerase chain reaction. *Nucl. Acids Res.* **17:** 5856.

Hudson, R.R. 1982 Estimating genetic variability with restriction endonucleases. *Genetics* **100:** 711-719.

Jeffreys, A.J. 1987 Highly variable minisatellites and DNA fingerprints. *Biochem. Soc. Trans.* **15:** 309-317.

Jeffreys, A.J., V. Wilson, R. Neumann and J. Keyte 1988 Amplification of human minisatellites by the polymerase chain reaction: towards DNA fingerprinting of single cells. *Nucl Acids Res.* **16** 10953-10971.

Jeffreys, A.J., R. Neumann and V. Wilson 1990 Repeat unit sequence variation in minisatellites: a novel source of DNA polymorphism for studying variation and mutation by single molecule analysis. *Cell* **60:** 473-485.

Jeffreys, A.J., N.J. Royle, V. Wilson and Z. Wong 1988 Spontaneous mutation rates to new length alleles at tendem-repetitive hypervariable loci in human DNA. *Nature* **332:** 278-281.

Jones, C.W. and F.C. Kafatos 1982 Accepted mutations in a gene family: evolutionary diversification of duplicated DNA. *J. Mol. Evol.* **19:**87-103.

Kimura, M. 1983 *The neutral theory of evolution.* Cambridge Univ. Press, Cambridge

Kreitman, M. 1983 Nucleotide polymorphism at the *Adh* locus of *Drosophila melanogaster. Nature* **304:** 412-417.

Kreitman, M. and M Aguadé 1986 Genetic uniformity in two populations of *Drosophila melanogaster* as revealed by filter hybridization of four-nucleotide-recognizing restriction enzyme digests. *Proc. Natl. Acad. Sci. USA* **83:** 3562-3566.

Kreitman, M. and L.F. Landweber 1989 A strategy for producing single-stranded DNA in the polymerase chain reaction. *Gene Anal. techn.* **6:** 84-88.

Kreitman, M. and R.R. Hudson 1991 Inferring the evolutionary histories of the *Adh* and *Adh-dup* loci in *Drosophila melanogaster* from patterns of polymorphism and divergence. *Genetics*, in press.

Kretz, K.A., G.S. Carson and J.S. O'Brien 1989 Direct sequencing from low melt agarose with Sequenase. *Nucl. Acids Res.* **17:** 5864.

Kwok, S., D.E. Kellog, N. McKinney, D. Spasic, L. Goda, C. Levenson and J.J. Sninsky. 1990 Effects of primer-template mismatches on the polymerase chain reaction: Human immunodeficiency virus type 1 model studies. *Nucl. Acids Res.* **18:** 999-1005.

Langley, C.H., E. Montgomery and W.F. Quattlebaum 1983 Restriction map variation in the *Adh* region of *Drosophila. Proc. Natl. Acad. Sci. USA* **79:** 5631-5635.

Langley, C.H., A.E. Shrimpton, T. Yamazaki, N. Miyashita, Y. Matsuo & C.F. Aquadro 1988 Naturally occurring variation in the restriction map of the *Amy* region of *Drosophila melanogaster. Genetics* **119:** 619-629.

Leigh Brown, A.J. 1983 Variation at the 87A heat shock locus in *Drosophila melanogaster. Proc. Natl. Acad. Sci. USA* **80:** 5350-5354.

Levinson, G. J.L. Marsh, J.T. Epplen and G.A. Gutman 1985 Cross-hybridizing snake satellite, Drosophila, and mouse DNA sequences may have arisen independently. *Mol. Biol Evol.* **2:** 494-504.

Miyashita, N. and C.H. Langley 1988 Molecular and phenotypic variation of the *white* locus region in *Drosophila melanogaster. Genetics* **120:** 199-212.

Myers, R.M., V.C. Sheffield and D.R. Cox 1989 Mutation detection by PCR, GC-clamps, and denaturing gradient gel electrophoresis. In, *PCR Technology,* H.A. Erlich, Editor. Stockton Press, N.Y.

Nei, M. and F. Tajima 1981 DNA polymorphism detectable by restriction endonucleases. *Genetics* **97:** 145-163.

Riley, M.A., M.E. Hallas and R.C. Lewontin 1989 Distinguishing the forces controlling genetic variation at the *Xdh* locus in *Drosophila pseudoobscura. Genetics* **123:** 359-369.

Ruano G, K.K. Kidd and J.C. Stephens 1990 Haplotype of multiple polymorphisms resolved by enzymatic amplification of single DNA molecules. *Proc. Natl. Acad. Sci USA* **87:** 6296-300.

Ruano G and K.K. Kidd 1989 Direct haplotyping of chromosomal segments from multiple heterozygotes via allele-specific PCR amplification. *Nucl. Acids Res.* **17:** 8392.

Saiki, R.K., D.H. Gelfand, S. Stoffel, S.J. Scharf, R. Higuchi, G.T. Horn, K.B. Mullis and H.A. Erlich 1988 Primer-directed enzymatic amplification of DNA with a thermostable DNA polymerase. *Science* **239:** 487

Shaeffer, S.W., C.F. Aquadro and W.W. Anderson 1987 Restriction-map variation in the alcohol dehydrogenase region of *Drosophila pseudoobscura. Mol. Biol. Evol.* **4:** 254-265.

Shaeffer, S.W., C.F. Aquadro and C.H. Langley 1988 Restriction-map variation in the *Notch* region of *Drosophila melanogaster. Mol. Biol. Evol.* **5:** 30-40.

Solignac, M., M. Monnerot and J.C. Mounolou 1986 Mitochondrial DNA evolution in the *melanogaster* species subgroup of *Drosophila. J. Mol. Evol.* **23:** 31-40.

Southern, E.M. 1975 Detection of specific sequences among DNA fragments separated by gel electrophoresis. *J. Mol. Biol.* **98:** 503-517.

Ullu, E. and C. Tschudi 1984 Alu sequences are processed 7SL RNA genes. *Nature* **312:** 171-172.

Vassart, G., M. Georges, R. Monsieur, H. brocas, A.S. Lequarre and D. Christophe 1987 A sequence in M13 phage detects hypervariable minisatellites in human and animal DNA. *Science* **235:** 683-684.

Vawter, L. and W.M. Brown 1986 Nuclear and mitochondrial DNA comparisons reveal extreme rate variation in the molecular clock. *Science* **234:** 194-196.
Weber, J.L. and P.E. May 1989 Abundant Class of Human DNA Polymorphisms which can be Typed using the Polymerase Chain Reaction. *Am. J. Hum. Genet.* **44:** 388-396.

MOLECULAR SYSTEMATICS AT THE SPECIES BOUNDARY: EXPLOITING CONSERVED AND VARIABLE REGIONS OF THE MITOCHONDRIAL GENOME OF ANIMALS VIA DIRECT SEQUENCING FROM AMPLIFIED DNA

Chris Simon[1]
Department of Ecology & Evolutionary Biology
University of Connecticut
Storrs, Connecticut 06269

Traditional morphological characters have provided a wealth of systematic information. This information has been used to establish phylogenetic relationships which we refine today using a wide variety of techniques. A common result of traditional morphological analyses has been identification of groups of related taxa. Relationships within and/or among these groups were often uncertain (e.g. orders of flowering plants, Heywood 1978; species groups of kangaroo rats, Hall 1981). Very closely related species were particularly difficult to connect phylogenetically using morphological characters (e.g. Hawaiian *Drosophila*, Hardy 1965; cichlid fishes, Fryer and Isles 1972). The addition of detailed morphometric analyses in the 1970's and 1980's increased our ability to identify taxa and discriminate among them but did not prove useful for deciphering relationships among taxa (Rohlf and Bookstein 1990).

In the late 1960's and throughout the 1970's allozyme characters provided simple genetic information which was useful for phylogenetic analysis (Buth 1984, Swofford and Berlocher 1987). This information reflected nucleotide substitutions in charged amino acids. The work of Coyne and others with varying pH and gel concentrations showed us that a large proportion of hidden allozyme variants could be visualized but that a complete picture of variation would require nucleotide sequences (reviewed in Coyne 1982).

At the same time that electrophoretic analyses were proliferating, chromosomal banding patterns (e.g. Carson et al. 1970), immunological distances (e.g. Maxon et al. 1979) and amino acid sequencing (reviewed in Goodman 1981) were providing additional phylogenetic information. In the late 1970's and early 1980's DNA restriction mapping began to be used for

---

[1] Address until 1/91: Zoology Program, Univ. Hawaii, Honolulu HI 96822

NATO ASI Series, Vol. H 57
Molecular Techniques in Taxonomy
Edited by G. M. Hewitt et al.

population level studies (reviewed by Avise et al. 1987) and in a few years, the techniques were perfected to the stage where small organisms such as insects could be examined (reviewed by Simon 1988).

In the mid 1970's techniques for the rapid sequencing of DNA were developed but it was not until 1985 when DNA amplification by the polymerase chain reaction (PCR) was devised (Saiki et al. 1985, Mullis et al. 1986) that sequence analysis of nucleotides became feasible on a large scale. PCR allows the rapid selection, isolation, amplification and sequencing of DNA regions of interest from small amounts of tissue (Simon et al., this volume) providing data on mutations at the nucleotide level which are extremely valuable for systematic studies.

This paper discusses the relative usefulness of nucleotide sequence data from various mitochondrial gene regions for studying phylogenetic relationships among animal species at a variety of taxonomic levels. Important considerations which are discussed include an understanding of patterns of evolutionary rate variation within and among taxa, structural and functional constraints, multiple substitutions at sites (saturation) and the accumulation of nucleotide bias, the effects of saturation on distance phylogenies versus character-based phylogenies, and the importance of comparative studies of closely related species. The appropriateness of various molecular techniques for systematics applications are discussed in light of the above considerations.

ADVANTAGES OF MITOCHONDRIAL DNA FOR PHYLOGENETIC SYSTEMATICS

The advantages of using mitochondrial DNA for systematic studies have been pointed out numerous times (e.g. Lansman et al. 1981, Brown 1985, Wilson et al. 1985, Simon 1988). The list of useful characteristics includes ease of extraction and manipulation, the simplicity of the molecule, lack of recombination, and high mutation rate. The observed rapid average rate of evolution of the mtDNA molecule is what gives these data the potential for use in species and population level studies, yet this view of mtDNA obscures probably the most useful aspect of the molecule. That is, it contains slowly evolving "highly conserved" regions as well as rapidly evolving "highly variable" regions (Cann et al. 1987, Carr et al. 1987, Clary and Wolstenholme 1987, Uhlenbusch et al. 1987, Haucke and Gellissen 1988, Neefs et al. 1990).

The rate of animal mtDNA evolution varies among lineages, among genes, and within genes. Molecular systematists can use this evolutionary rate variation to their advantage. The divergence rate of a particular segment of DNA will determine the systematic level at which it is appropriately used for phylogenetic study. Rapidly evolving DNA is useful for studies of recently diverged taxa, but may be too different to be useful in differentiating genera or species belonging to particularly ancient genera. If taxon divergence time is long, many sites are likely to become saturated (change more than once). In such saturated regions, homologies will be difficult to determine and sequences will be difficult if not impossible to align (Brown et al. 1982, Nei and Gojobori 1986, Simon et al. 1990, Swofford and Olsen 1990).

Conserved regions are useful for studies of distantly related taxa. It is only in conserved regions that distantly related taxa can be aligned. In protein coding regions, identification of amino acid codons, in which second positions are highly conserved, aids in alignment. In ribosomal DNA regions, structurally or functionally important conserved sequence blocks can serve as markers between which the more variable regions can be aligned with differing degrees of confidence (Olsen 1988).

Knowledge of conserved regions, has one more very important utility, it allows us to choose so-called "universal" oligonucleotide primers for amplification via PCR (Simon et al., this volume). This is of critical importance in the study of species which have not yet been sequenced.

RATE OF MITOCHONDRIAL DNA EVOLUTION AND ITS RELATION TO PHYLOGENY CONSTRUCTION: PATTERNS AND PROCESSES

The Average Rate of Evolution of Mitochondrial DNA is Greater Than That of Nuclear DNA

Brown et al. (1979, 1982) pointed out that in primates the average rate of evolution of the mitochondrial molecule estimated by restriction analysis and measured by sequencing is 5-10 times higher than the average rate of evolution of primate nuclear DNA. Miyata et al. (1982) found this same pattern in other mammals. In contrast, Powell et al. (1986) and Solignac et al. (1986a) found that in Drosophila the average rate of evolution of nuclear and mitochondrial DNAs were similar. Vawter and Brown (1986) reached the same conclusion by analyzing DNA data from sea urchins. Both

Powell et al. (1986) and Vawter and Brown (1986) suggested that the average rate of evolution of nuclear DNA in invertebrates is higher than the average rate of evolution of nuclear DNA in mammals.

Recently, Sharp and Li (1989) using published sequence data showed that Drosophila nuclear genes do evolve faster than mammalian nuclear genes and that Drosophila mitochondrial genes evolve faster than some nuclear genes and at approximately the same rate as others. They calculated the rate of substitution at silent sites in Drosophila nuclear genes and found it to be at least three times higher than the rate of substitution at silent sites in mammalian nuclear genes. They pointed out that in Drosophila, silent sites of nuclear genes with strong codon bias evolve half as fast as those with weak codon bias and that mitochondrial DNA silent sites evolve at about the same rate as nuclear DNA silent sites with weak codon bias and two times faster than nuclear DNA silent sites with strong codon bias.

Rate comparisons between silent positions of nuclear (calmodulin) and mitochodrial (COII) genes among 11 strains of Caenorhabditis elegans (Thomas and Wilson, submitted), revealed a silent substitution rate which is 10 fold greater in the mitochondrial gene. Because these strains are recently diverged, it is unlikely that multiple substitutions have occurred. All but one of the 26 substitutions observed was silent. Thus, highly divergent rates of evolution between nuclear and mitochondrial genes is not a phenomenon restricted to vertebrates.

The average rate of evolution of the mitochondrial genome, while interesting from a theoretical standpoint, is less relevant for systematic studies than the rate of evolution of individual nucleotide positions. When selecting characters for phylogenetic analysis, it is necessary to calculate or estimate site specific or region specific substitution rates so that appropriate characters can be excluded or down-weighted according to their observed level of variability in the taxa under investigation. Weighting is discussed in a later section.

Different Mitochondrial Lineages Have Different Evolutionary Rates

The rate of mtDNA evolution is not necessarily constant even between closely related taxa. DeSalle and Templeton (1988) studied two lineages of Hawaiian Drosophila, one that was postulated to have undergone many founder events and another that had not. They found a 3-fold difference in the rate of evolution of the two lineages. As they predicted, the lineage with the history of many founder events had the faster rate. Crozier et al. (1989) using parsimony analysis of insect mtDNA sequence data with the mouse and *Xenopus* as outgroups found different substitution rates on branches leading to *Drosophila* and honeybees, respectively. In periodical cicadas (Simon, Franke, Martin, McIntosh, and Pääbo, unpublished data), Domain 3 of the 12S rRNA gene evolves at a faster rate in one species than in the other two.

In vertebrates, Kocher et al. (1989) found evidence which suggests that there may be a five-fold higher rate of amino acid substitution in mammals and birds relative to fishes. Because they examined only a small segment of the genome (300 bp), they cautioned that more data is needed. Thomas and Beckenbach (1989) sequenced on the order of 2000 bp in six salmonid fish species and found high amino acid homology with the corresponding genes of the frog *Xenopus*. Data from *Xenopus* and salmonid fish taken together provide evidence for a slow-down in the rate of evolution in cold-blooded vertebrates compared to mammals. A similar result has been obtained for sharks (A. Martin, pers. comm.). The implication of the above findings for phylogenetic analysis is that methods which assume homogeneous rates of evolution in all lineages (e.g. UPGMA) should be avoided.

The Rate of mtDNA Evolution Varies Among and Within Gene Regions and is Related to Functional Constraints

When nucleotide sequence information became available for the mitochondrial genome of more than one animal, it was immediately obvious that different genes were evolving at different rates. Bibb et al. (1981) compared mouse mtDNA to that of humans and Anderson et al. (1982) compared cow mtDNA to human mtDNA. In these examples, protein coding genes varied in sequence similarity between approximately 60% and 80%. Variation in the rate of mtDNA evolution within genes was also striking. More recent sequence information has strengthened and clarified these conclusions (e.g. Clary and Wolstenholme 1985, Roe et al. 1985, Jacobs et al. 1988).

Control Region. The most variable region of the mtDNA is the so-called A+T-rich region in insects (Fauron and Wolstenholme 1980, Clary and Wolstenholme 1987) and the D-Loop region in vertebrates (Upholt and David 1977, Crews et al. 1978, Aquadro and Greenberg 1983). This region is called the control region and surrounds the origin of replication of the molecule.

With few exceptions (Cann and Wilson 1983; Jacobs et al. 1988; Cantatore et al. 1989), it is the only region of the mitochondrial DNA of animals that does not code for any gene products. As in nuclear non-coding regions (introns, spacer regions, pseudogenes) this lack of coding suggests a lack of functional constraints and would explain the hypervariability (Li et al. 1984, Hillis and Davis 1986). This level of variability has made the control region useful for studies of relationships below the species level (Cann et al. 1987, Thomas et al. 1990).

In lower vertebrates and invertebrates, the control region is sometimes characterized by extreme length variability (Moritz et al. 1987, Rand and Harrison 1989, Boyce et al. 1989, Martin and Simon 1990). The largest and most variable so far recorded is that of bark weevils which varies between 9 and 13 Kb (Boyce et al. 1989).

<u>Protein Coding Genes</u>. Animal mitochondria contain 13 protein coding genes (Attardi and Schatz 1988). In some aspects, the evolution of these genes is similar to that of nuclear protein genes, in other aspects, it is quite different. For two mitochondrial protein coding genes in primates (NADH dehydrogenase 4 & 5), Brown et al. (1982) found the silent substitution rate to be 80-86% times higher than the rate of substitution in codon positions which result in amino acid replacements. For 6 protein coding genes in <u>Drosophila yakuba</u> and <u>D. melanogaster</u> (Wolstenholme and Clary 1985) found 86% of the substitutions to be silent. Within the mitochondrial cytochrome b coding region for a variety of animals, silent codon positions are more variable (Kocher et al. 1989, Irwin et al. in press). These results are similar to those found for nuclear DNA where silent codon positions have been documented to evolve faster than replacement positions because they lack functional constraints (Li et al. 1985, Lewontin 1989).

In addition to the predictable variation in rate among the three codon positions there is also variation in the rate of evolution from codon to codon. Thomas and Beckenbach (1989) described spatial heterogeneity in the distributions of mutations from four mitochondrial protein genes in fishes. Spatial heterogeneity in the mitochondrial cytochrome b genes of a wide variety of vertebrates fits predictions based on previous structure/function models (Kocher et al. 1989, Irwin et al. in press).

In nuclear protein coding genes, mutations in third positions of codons of 4-codon families are not entirely silent. There is often a preference for usage of one codon over another. The strength of the codon bias varies among nuclear encoded proteins and is inversely related to rate of evolution of these genes (Sharp and Li 1989).

Proteins in animal mitochondrial DNA are less influenced by codon bias than proteins in nuclear DNA because an expanded codon recognition pattern allows a single tRNA species to decode as many as four codons, making these positions effectively silent (Gray 1989). The prevalence of A and T nucleotides in silent codon positions in *Drosophila* (Wolstenholme and Clary 1985) would therefore not reflect codon bias but rather the strange preponderance of A's and T's in insect mitochondrial DNA in general (DeSalle et al. 1987, Simon et al. 1990). However, the fact that the frequency of T <—> C transitions in the 3rd position of two codon families in *Drosophila* was twice the corresponding value for 4 codon families suggested to Wolstenholme and Clary (1985) that T <—> C transitions may be selected against in silent positions of 4-codon families; similar analysis of more closely related *Drosophila* species might shed light on this problem.

tRNA Genes. In comparisons of mitochondrial tRNAs from five vertebrate species (rat, mouse, cow, and *Xenopus*), Gadaleta et al. (1989) showed that the degree of conservation of different functional regions within tRNAs varied, with the anticodon loop being the most conserved region, followed by the dihydrouridine (DHU) stem. The DHU loop, the ribothymidine pseudouridine cytosine (TΨC) loop and the TΨC stem were the most variable regions. The degree of conservation of the 5' half was higher than that of the 3' half. In closely related salmonid species, for $tRNA^{ARG}$ and $tRNA^{GLY}$, variation is concentrated in the DHU loop (Thomas and Beckenbach 1989). In nematodes, *Ascaris suum* and *Caenorhabditis elegans* (Wolstenholme et al. 1987), the TΨC loop is absent.

Rate of evolution varies among as well as within mitochondrial tRNAs. In the rat, mouse, cow, and frog comparison, above, the tRNAs that coded for the most used amino acids were most conserved. Comparisons among humans showed that tRNA genes that recognize four-fold degenerate codons are generally more variable than tRNA genes that recognize two-fold degenerate codons (Cann et al. 1984). This observation does not hold for *Drosophila* yakuba compared to Drosophila melanogaster (Wolstenholme and Clary 1985), but, again, the effect may be hidden by substitutional saturation and more recently diverged species should be analyzed.

The 22 mitochondrial tRNA genes are less constrained by structure and function than their highly conserved nuclear counterparts; they evolve at a higher rate (Wilson et al. 1985). Nevertheless, mitochondrial tRNAs evolve slower than mitochondrial protein coding genes (e.g. Bibb et al. 1981, Wolstenholme and Clary 1985), an indication that they are structurally or functionally more constrained than these protein genes. An excellent demonstration of the effect of structure and function on the rate of evolution of mitochondrial tRNAs is the study of Thomas et al. (1989) who showed that in sea urchins, the tRNA leucine CUN gene appears to have lost its function, been incorporated into the ND5 protein subunit gene, and to have diverged at the typically higher rate of the ND5 gene.

Ribosomal Genes. Ribosomal RNA genes in general have been characterized as highly conserved (Noeller and Woese 1981, Hixson and Brown 1986, Moritz et al. 1987). As a consequence of this nucleotide conservation, Dams et al. (1988) were able to align ribosomal DNA sequences (small subunit) for eukaryotes (vertebrates, invertebrates, plants, protists and fungi), archebacteria, eubacteria, plastids (from higher and lower plants and protists) and mitochondria (vertebrates, invertebrates, higher and lower plants, protists and fungi). Nevertheless, as has been pointed out many times, the entire ribosomal gene is not highly conserved, rather, it is a mosaic of alternating conserved and variable segments.

As in all genes, the level of variability of particular rRNA gene segments is related to structural and functional constraints. For example, one of the most conserved segments in the small ribosomal RNA gene codes for the site of tRNA attachment (Simon et al. 1990). Other conserved regions are important for maintaining the characteristic secondary and tertiary structures of the rRNA molecules (Noller and Woese 1981). Ribosomal DNA genes code for single stranded rRNA which folds and bonds to itself to form helical paired stems and unpaired loops (secondary structure) (Figure 1). In the ribosome, the folded rRNA is further bent into a three dimensional (tertiary) structure held together by proteins.

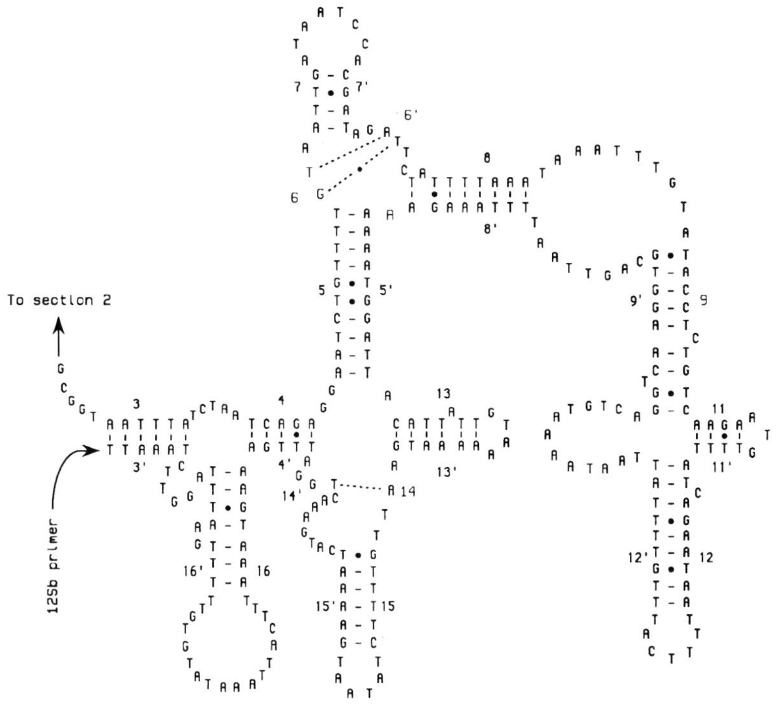

Figure 1. The secondary structure of domain III of the mitochondrial small subunit (12S) RNA from the 13-year periodical cicada Magicicada tredecim, illustrating short- and long-range stems and unpaired regions. Data published in Simon et al. 1990.

In the mitochondrial genome there are two ribosomal subunits, the small or 12S subunit and the large or 16S subunit (Attardi and Schatz 1988). Each subunit is divided into three or four sections or domains. The stems that hold these major sections together are particularly well conserved (Simon et al. 1990) and are called long-range stems because in the gene which codes for this RNA, the bases which code for one side of the helix are located far away from the bases which code for the other side (e.g. Figure 1, stem 3). There are other stems where the bases which code for one side of the helix are located near the bases which code for the other side of the helix, separated by only three to a dozen or so nucleotides. These are called short range stems and many of these are poorly conserved (e.g. stem 16, Figure 1). However, not all short range stems are poorly conserved (e.g. stem 7 Figure 1). Other functional constraints, for example roles related to binding of ribosomal proteins, could cause short range stems to be highly conserved (Simon et al. 1990).

The systematic implications of the level of conservation of helical stems versus unpaired loops has been discussed for 5S and 5.8S rRNA (found in nuclear ribosomes but not in mitochondria) by Wheeler and Honeycutt (1988). In their analysis of selected arthropods and gastropods, they suggested that trees constructed from unpaired bases were more concordant with trees based on morphological data than those constructed from paired bases. However, this conclusion should be taken with a grain of salt because the tree based on paired regions was only 1% longer than the tree based on unpaired regions (304 steps versus 300 steps). Furthermore, the 5S and 5.8S subunit genes are small--between 100 and 200 base pairs--and do not code for stems with long distance interactions such as those found in the small and large rRNAs. Thus, even if their conclusion was well founded, it could not be generalized to more complex rRNA genes such as the mitochondrial 12S and 16S where level of constraint is a more or less continuous variable and has a large range.

The selection of nucleotide characters for phylogenetic studies is not a simple problem. In general, for a given taxonomic level the best characters are those whose states are constant within taxa and variable among taxa, but not so variable that noise is introduced into the analysis. For molecular data, noise is created by multiple substitutions at a site.

Within Species Variation in Evolutionary Rate

When collecting nucleotide sequence data for phylogenetic analysis it is important to estimate within species variation. Estimates of within species variability are necessary for correcting estimates of genetic distance among species (Wilson et al. 1985). They are also useful for determining whether more than one state is present for each nucleotide site and how those states are shared among populations and among species. Knowledge of character variability is necessary for designing sampling strategies (Archie et al. 1989).

Another important reason to examine within species mtDNA sequence variation is to look for potentially misleading interspecific hybridization (Harrison 1989, Aubert and Solignac 1990) or multiple ancestral lineages which might be shared among species (Neigel and Avise 1986). The latter is important because gene trees are not necessarily equivalent to species trees especially for recently separated species (See, Wilson et al. [1985] and Avise et al. [1987] for discussions of this very important topic.)

One study has examined within population and among population variation in mtDNA nucleotide sequence data. Thomas et al. (1990) studied three populations (subspecies) of kangaroo rats each sampled at two points in time (PCR allowed sampling of turn-of-the-century museum skins). Sample sizes for the populations examined (old museum skins vs. fresh samples) were: (14, 20); (21, 19); (14, 24). For the 106 individuals, 225 base pairs were sequenced from the D-Loop and two adjacent tRNAs. Within-population sequence divergence ranged from .2 to 1.7%. Levels of variability were remarkably similar in old and new populations. Among-population sequence divergence, corrected for within-population variation, ranged from 1.5 to 2.0%. One individual of a second kangaroo rat species was sampled and found to differ in sequence by 13-16%; the lack of a transition bias suggested to the authors that this value was a severe underestimate (see below).

Within- and among-population mtDNA sequence divergence has been estimated using restriction site data. Avise et al. (1987) summarized studies of vertebrate mtDNA. Most included an average of 6 or fewer individuals per locality (range 1.5 - 21) . The number of localities sampled ranged from 4 - 40. Within species variability was generally low (less than 1% sequence divergence). The exceptions (ranging from 2-8%) were all cases in which a geographic discontinuity subdivided the species into two or more isolated or semi-isolated units.

Martin and Simon (1990) examined restriction site variation in 118 periodical cicadas (2 species, 16 populations) and compared these figures to values for other invertebrate (and vertebrate) taxa. For invertebrates, within species variability ranged from .01% sequence divergence (in 17-year cicadas) to 4.8% (in Hawaiian <u>Drosophila</u>, DeSalle et al. 1986b). Both periodical cicada species examined exhibited low variability relative to other taxa. This may be attributable to population bottlenecks occurring during the Pleistocene, recent colonization following glacial retreat, and/or high among-family variance in reproductive success (Martin and Simon 1990). The unusually high degree of within species genetic divergence in Hawaiian <u>Drosophila</u> could have resulted from migration, from strong local population subdivision, and/or from survivorship of more than one ancestral lineage following a recent speciation event (DeSalle et al. 1986a).

MULTIPLE NUCLEOTIDE SUBSTITUTIONS AT A SITE AND PHYLOGENETIC ANALYSES

As nucleotide changes accumulate over time, it is inevitable that some substitutions will be masked by subsequent substitutions at the same site (Jukes and Cantor 1969). The lower the constraint on a nucleotide position, the higher the probability for multiple substitutions. Regions with high levels of multiple substitutions are described as saturated. The result of saturation is that the nucleotide divergence between two taxa will be underestimated and patterns of nucleotide substitution will be altered (Brown et al. 1982, DeSalle et al. 1987). Corrections for multiple substitutions at single sites are necessary in order to obtain accurate estimates of evolutionary divergence among taxa.

Altered Patterns of Nucleotide Substitution

Transitions and Transversions. Brown et al. (1982) found transitions to outnumber transversions 9:1 in a comparison of an 896 bp segment containing genes for three transfer RNAs and parts of two proteins (NADH 4 & 5) for human and chimps. In comparisons of mitochondrial DNA of the small ribosomal subunit in these same primates, Hixson and Brown (1986) found similar transition/transversion ratios.

On straight probabilistic grounds--considering all possible base changes--transversions should outnumber transitions 2:1. Brown et al. (1982) suggested that the high proportion of transitions in their primate comparisons was due chiefly to a bias in the mutation process, selection at the molecular level, rather than to selection at the level of gene products. Wilson et al. (1985) reiterated this point and suggested that transitions are selected for by tautomeric base pairing that allows purine-pyrimidine mismatches and that these mismatches are rarely if ever corrected due to deficient mtDNA repair systems. Purine-pyrimidine mismatches in turn lead to transitions in the next round of replication (Topal and Fresco 1976).

Transition Bias, Saturation, and Structural and Functional Constraints. By their nature, transversions accumulate over time and obscure the record of transitions; this phenomenon was observed by Derancourt et al. (1967), Brown et al. (1982), DeSalle et al. (1987) and modeled by Holmquist (1983). For example, Brown et al. (1982) found 92% transitions in an 896 bp region (three tRNAs and parts of NADH 4 & 5) of human versus chimp but the percentage of transitions for pairs of species with increasingly longer divergence times fell until in comparisons of primate with non-primate species it reached 45%. Miyamoto et. al. (1989) found a similar pattern in the 12S gene for closely versus more distantly related bovids; this pattern was accelerated in the rapidly evolving D-loop region of these same species. For strains of Drosophila melanogaster, transitions accounted for 83% of changes in silent codon positions (4 protein coding genes); for the more distantly related pair Drosophila melanogaster and Drosophila yakuba, transitions accounted for only 18% of such changes (6 protein coding genes) (Wolstenholme and Clary 1985). In comparisons of nematode strains for the COII gene, transitions outnumbered transversions 12/1. For comparisons among species, the observed substitutional pattern showed transitions only slightly more common that transversions (1.3 : 1). In comparisons of two nematode genera, transversions outnumber transitions 1.5 : 1 (Thomas and Wilson, submitted).

Aquadro, et al. (1984) modeled the dynamics of the substitution process. Their findings support the hypothesis that the decrease in the proportion of transitions observed as divergence increases is a consequence of transversions obscuring previous transitions and further suggest that although a portion of the mtDNA molecule evolves at an extremely rapid rate, a significant portion is under strong selective constraints. This latter point is discussed in general and in more detail by Fitch and Markowitz (1970), Fitch (1986), Takahata (1987) and Palumbi (1989).

As discussed above, variable selective constraints can be imposed by functional considerations (e.g. amino acid codon specificity, active sites of proteins, tRNA attachment sites in rRNA) or structural considerations (e.g. secondary and tertiary structure in rRNAs, tRNAs and proteins) and can affect the rate of evolution of the region under consideration. The degree of constraint varies among genes. A clear demonstration of variability of constraint among different mitochondrial protein coding regions can be seen by comparing rates of evolution calculated for the "slowly evolving" COI gene versus the "rapidly evolving" ATPase 8 gene. The table below illustrates that the COI gene is so constrained that it changes little in amino acid composition even in distant comparisons; the Atpase 8 gene has many more positions that are free to change.

COMPARISON	% AMINO ACID SIMILARITY COI	ATP8	REFERENCE
Human vs. Mouse	90	46	Bibb et al. 1981
Mammals* vs. Chicken	86	27	Desjardins & Morais 1990
Human vs. Urchin	75	21	Jacobs et al. 1988

* Mammals = Average [mouse, cow, human]

The result of structural and functional constraints is that, comparing slowly evolving nucleotide positions for distantly related species is similar to comparing rapidly evolving positions for more closely related species. This phenomenon is what has allowed molecular evolutionists to use nuclear ribosomal genes to construct phylogenetic trees for animals in different phyla and arrive at an hypothesis that agrees to some extent with traditional expectations (Field et al. 1988). Thus, high transition/transversion ratios can be found in comparisons of mtDNA nucleotide sequences from closely related species in the most variable positions (with few substitutions of any kind elsewhere) and in comparisons of distantly related species only in the most conserved positions. High transition/transversion ratios in mtDNA have been found in: D-Loops compared among humans (Aquadro and Greenberg 1983); D-loop and tRNA genes from populations of 3 kangaroo rat subspecies (Thomas et al. 1990); rapidly evolving proteins and 16S genes in Hawaiian Drosophila species that have diverged by less than 2% (DeSalle et al. 1987); proteins and tRNA genes in the sibling species D. simulans and D. mauritiana (Satta et al. 1987); silent positions of protein genes in six salmonid fishes (Thomas and Beckenbach 1989); silent positions of codons from the cytochrome b gene of sheep versus goats (Irwin et al. in press); 12S genes from five closely related primate species (Hixson and Brown 1986); 12S genes from Bison and Cow (Miyamoto et al. 1989) and conserved regions of human, Drosophila, and cicada 12S genes (Simon et al. 1990).

With reference to transition bias, there are exceptions to the generalization that comparing slowly evolving positions for distantly related species is similar to comparing rapidly evolving positions for closely related species. For example, second positions of codons are highly constrained because substitutions in these positions, whether transitions or transversions, cause amino acid replacements. Functionally unimportant amino acid replacements--those not involved in protein secondary or tertiary structure or active sites--would be expected to show second position transition bias early in evolution. But, functionally important amino acids may only be able to change if the replacement amino acid has similar biochemical properties. In this case, there would be no reason to expect second position substitutions to be biased towards transitions early in evolution.

Transition Bias and Rate Variation Within and Among Lineages. Although all evidence suggests that transition rate in mtDNA initially exceeds transversion rate, it is possible that transition rate may not be constant within and among lineages. Variation in observed transition rate could be due to an actual difference in the percent of transitions or due to a difference in the rate of evolution (speed of saturation). *Drosophila melanogaster* has a lower transition rate than its sibling species which themselves are more similar in rate to primates. Within primates, chimpanzees were 5.0% divergent in nucleotide sequence from humans and exhibited 95% transitions in silent positions (averaged over 2 protein coding genes) (Brown et al. 1982). Strains of *D. melanogaster* which were only 0.5% divergent in nucleotide sequence (therefore unlikely to have experienced saturation) had only 83% transitions in silent codon positions (averaged over 4 protein coding genes) (Wolstenholme and Clary 1985, deBruijn 1983). However, this lower rate is not characteristic of all *Drosophila*. The closely related species pair *D. simulans* and *D. mauritiana* were found to be 2.1% divergent in nucleotide sequence in two protein coding genes and transitions accounted for 93% of substitutions in silent sites (Satta et al. 1987).

Theoretical models in a maximum likelihood framework have been developed to estimate variation in evolutionary rates among lineages where the absence of close relatives makes the empirical calculation of exact transition and transversion rates impossible. Hasegawa and Kishino (1989) compared the complete mitochondrial genomes of human, mouse and cow and concluded that the transversion rate was lower in cow than in human or mouse. Transition rate could not be compared due to the high degree of saturation. Kishino and Hasegawa (1990) analyzed the 896bp primate data set of Brown et al. (1982) combined with homologous sequences from mouse and cow and suggested that within hominoids, the transition rate is higher in the great-apes + human clade than in the gibbon. When data for these same 896 base pairs became available for seven additional primate species (Hayasaka et al. 1988), Hasegawa et al. (1990) estimated that the rate of evolution of lemurs was much lower than that of other primate lines for first and second positions of codons and all tRNA sites. Third positions of codons were not analyzed due to their rapid rate of evolution resulting in saturation (transitional difference have nearly reached saturation even between humans and chimps, Hasegawa et al. 1985). Transition rate in lemurs appeared to be only about 1/10 of that in other primates examined. An important test of this result would be to make direct nucleotide substitution comparisons among representatives of the 15 species and five genera of lemurs and to compare these findings to similar data collected for other prosimian taxa. It is only by comparing closely related species that these various effects can be sorted out.

A serious limitation of Kishino and Hasegawa's theoretical rate estimation procedure is the inability to take into account possible rate variation within lineages. In the example above, only one lemur lineage was examined. But, the definition of a lineage is a matter of scale. In mitochondrial inheritance, the occurrence of one mutation can define a new lineage. Lemurs, at present, consist of at least 15 nuclear gene lineages (species) and many more mitochondrial lineages. If any of these vary in rate, they would invalidate Hasegawa's model which assumes no variation within "the lemur lineage."

Another potentially serious limitation of Hasegawa's studies is the assumption of constant nucleotide composition among lineages. Saccone et al. (1989) demonstrated that apparent rate variation can be due to a change in nucleotide composition within and among genomes. They developed a model of gene evolution which takes into account variation in nucleotide composition. Their results cast doubt not only on the lemur study, but also on the claim of Li et al. (1987) that their has been a large acceleration of nuclear gene evolution in rodents and the claim by Bishop and Friday (1987) that nuclear DNA indicates a closer relationship between birds and mammals than between either group and reptiles or amphibians.

<u>Nucleotide Composition Bias</u>. DeSalle et al. (1987) found that multiple substitutions at mtDNA sites obscured transitional changes at a faster rate in flies than in primates due to a preponderance of A's and T's in the fly

sequences. All Drosophila examined so far have mitochondrial genomes that are 77-80% A+T nucleotides (Clary and Wolstenholme 1985, DeSalle et al. 1987, Satta et al. 1987, Garesse 1988). This bias would impose constraints on possible base changes. Therefore, two species of Drosophila could evolve rapidly (as rapidly as primates for example) and still not become very different over time (Solignac et al. 1986a, DeSalle et al. 1987).

High A+T bias may be a general phenomenon in insects. It has been noted in all other groups examined: mosquito (HsuChen et al 1984), locust (Uhlenbusch et al. 1987), cricket (Rand and Harrison 1989), bee (Crozier et al. 1989), and cicadas (Simon et al. 1990). A+T bias may characterize the mitochondrial genomes of protostome invertebrates in general. Extreme A+T-richness has also been found in nematodes (Wolstenholme et al. 1987, Thomas and Wilson, submitted) and in five genera of spiders (12S gene; C. Hayashi, pers. comm., H. Croom pers. comm.). Deuterostome invertebrates do not show this extreme bias (e.g. Strongylocentrotus purpuratus complete mtDNA sequence has only 59% A+T; Jacobs et al. 1988). Nucleotide bias changes the dynamics of the evolution of the molecule and would have to be taken into account in evolutionary rate corrections which attempt to estimate the number of multiple substitutions at sites.

A+T nucleotide bias, when present, tends to accumulate in hypervariable sites. Simon et al. (1990) noted a negative correlation in the degree of A+T bias and the degree of conservation of bases in the mitochondrial small ribosomal subunit gene of cicadas. In Drosophila yakuba, A+T bias in silent positions (94%) is greater than in the mitochondrial molecule in general (77%); this bias is not present in human mtDNA which has an A+T content of 56% with silent sites being 49% A+T (DeSalle et al. 1987). Bernardi and Bernardi (1986) compiled data which showed that the nucleotide bias of genomes in general tends to be reflected in the third positions of codons. Muto and Osawa (1987) found the same phenomenon.

The cause of nucleotide bias is unknown. DeSalle et al. (1987) suggested that for some unknown reason the Drosophila mitochondrial genome is functionally constrained to remain high in A's and T's such that, "at any given time the number of silent sites able to fix a base substitution...is small because most substitutions (which would be transitions) would lower the content of adenine and thymine." Perhaps the random acquisition of a high A+T genome in invertebrate evolution resulted in the fixation of biochemical systems adapted to this bias (Wolstenholme and Clary 1985). More research is needed to determine cause and effect.

Effects of Saturation on Distance Phylogenies versus Character-based Phylogenies

Phylogenetic inference methods can be broken into two categories, those that create trees based on genetic distances among taxa and those that create trees based on presence of shared character states (Felsenstein 1988, Swofford and Olsen 1990). Distance-based methods could be subdivided into those that assume homogeneous rates of evolution in all branches (e.g. UPGMA) and those that do not (e.g. neighbor joining). Character-based methods can be broken down into maximum likelihood methods, character-based parsimony methods, and the method of invariants. A discussion of methods is not within the scope of this paper but the effect of molecular evolutionary processes on the two general categories of methods is relevant. Comments below are directed toward distance-based phylogenetic methods and character-based parsimony although many points pertain to other methods as well. Maximum likelihood methods and invariants-based methods although intuitively very appealing are computationally difficult for large numbers of taxa.

Alignment. Correct alignment is the most important step in nucleotide sequence data preparation for phylogenetic analysis regardless of method; it is equivalent to assigning character homologies. High levels of multiple substitutions between two taxa coupled with any degree of insertions or deletions will make DNA sequences impossible to align. Areas of uncertain alignment should not be used for phylogenetic analysis (Simon et al. 1990, Swofford and Olsen 1990). Between areas of certain and uncertain alignment there are gray areas. The challenge which faces molecular biologists at present is to determine where in a sequence an alignment stops being reliable. Toward this end, several methods of phylogenetic analysis for DNA sequences include alignment evaluation simultaneously with tree construction (Sankoff and Cedergren 1983, Feng and Doolittle 1984, Hogeweg and Hesper 1984, Hein 1989ab, Nanney et al. 1989, Cheeseman and Kanefsky 1990).

Even when it is possible to align sequences based on conservation of amino acids or based on ribosomal RNA secondary structure, unconstrained positions can be so saturated that they contain phylogenetically random information. Tests for phylogenetic randomness of nucleotide sequence data have been devised (Archie 1989ab, Faith 1989, Faith and Cranston 1991). Undetected random changes would have a dramatic effect on estimates of genetic distance and corresponding phylogenetic inferences and, if not down-weighted, could distort estimates of relationship created by character-based parsimony.

Effects of Saturation on Distance-based Phylogenetic Trees. Because multiple substitutions at individual sites, nucleotide bias, and structural/functional constraints limit our ability to calculate accurate genetic distances among taxa, considerable effort has gone into designing appropriate corrections. Corrections for multiple substitutions have become increasingly sophisticated over time. Jukes and Cantor (1969) devised a method based on a model in which all changes at all sites are equally likely and in which one nucleotide will change to any of the other three with equal probability. In later models, transition bias is taken into account (Kimura 1980, Brown et al. 1982). One class of correction methods uses nucleotide substitution data to create substitution matrices from which past events can be back calculated (reviewed in Nei 1987). Recent models incorporate functional constraints such as amino-acid-specifying codon positions (Li et al. 1985, Nei and Gojobori 1986, Lewontin 1989), ribosomal secondary structure (Manske and Chapman 1987, Olsen pers. comm.), variable rates of evolution among lineages (Kishino and Hasegawa 1990) and nucleotide bias (Saccone et al. 1989). All of these methods depend on simplifying assumptions and provide only approximations.

Martin et al. (1990) investigated the accuracy of genetic distance estimates and discussed phylogenetic inferences made from them. They proceeded by randomly subsampling known mtDNA sequences and then determining the variance of genetic distance estimates among subsamples. They showed that patterns of non-random distribution of substitutions (variation in rate of evolution within and among genes) will affect sample sizes needed to estimate accurate distances and that necessary sample size will often exceed the typical 300-500 bp sampled by a single PCR reaction. They concluded that "gross errors in estimation of evolutionary relatedness can occur when calculations of genetic distance are based on few base pairs." They recommended sequencing larger sections of DNA or sampling numerous individuals/species that have diverged at the same time. However, there are two issues which are confused here. An accurate estimate of phylogenetic relationships does not require an accurate estimate of the average genetic distance among lineages for the whole genome. In order to assess the effect of small samples of bases on the determination of relationships, phylogenies would need to be calculated from the subsampled data and compared (as in Archie et al. 1989), or a randomization test applied (Archie 1989ab). As long as there is no systematic bias in the data, samples of 250-500 bp could provide accurate phylogenetic trees. Systematic errors can be minimized for both distance and character data by removing long branches from a tree and by eliminating or down-weighting unreliable (saturated) data (Swofford and Olsen 1990). Irwin et al. (in press) found that incorporating 3rd positions in a neighbor-joining tree of 20 mammals (based on the complete cytochrome b gene) produced results that were significantly different from the tree based on first and second positions. The tree based on first and second positions was more similar to classical expectations and to parsimony trees based on any combination of characters and position weights.

There are several important philosophical differences between distance-based phylogenetic methods and character-based parsimony which should not be overlooked. First is the unappealing property of distance-based analyses that all information on evolutionary change is averaged into one number for each pair of taxa. Second, to paraphrase Swofford and Olsen (1990), the assumptions involved in distance methods (such as additivity and clock-like evolution) are rarely evident or discussed and the justification of the algorithm itself often seems to be the objective of the study. On the contrary, in the case of methods will a well-defined optimality criterion such as parsimony, the objective is usually related to a (more or less) concrete set of assumptions.

Effects of Saturation on Character-based Parsimony Trees. Multiple substitutions at a site can cause evolutionary reversals of character states and convergent evolution, collectively known as homoplasy. Parsimony methods search for the shortest length tree for a given data set in an attempt to minimize homoplasy and to maximize the number of shared derived characters defining groups of taxa. Theoretically, parsimony can find an evolutionary pattern of relationships even in the presence of considerable homoplasy as long as there are not many homoplasious characters which all suggest the same wrong pattern (Swofford and Olsen 1990). Character-based parsimony is intuitively appealing because it uses information from many characters some of which may be informative at lower taxonomic levels, others of which may define relationships at higher levels. Cautions to be observed when using this method are that character-based parsimony is adversely affected by: 1) extreme rate variation among lineages especially when the total amount of change is high (Felsenstein 1988); and 2) outlier taxa which are distantly related to all other taxa--the phenomenon of "long branches attracting" (Hendy and Penny 1989).

In saturated regions it is impossible to know what base changes have occurred; shared-derived character states have been obliterated. Such regions should be discarded from analysis. This can be accomplished, as noted above, by an evaluation of the ability to align specific regions, or by down-weighting characters known to be variable such as silent positions of codons in species that are diverged by roughly 10% or more or positions in tRNA and rRNA stems that are known to be variable. In control regions, there are few patterns of constraints which can suggest a priori weighting schemes.

For any gene region, even when some pattern of constraint is known, that pattern may not be uniform. For example: not all short-range rRNA stems are variable; not all third positions of codons are silent, only those that are four-fold degenerate; not all four-fold degenerate third positions are silent, only those with no codon bias.

It is important to realize that assigning no weights to individual characters is an extreme form of weighting in which all characters are assumed to evolve at the same rate (Swofford and Olsen 1990). An assumption which is certainly not true. The term "unweighted" should be replaced with the term "uniformly weighted."

Manske and Chapman (1987) suggest weighting nucleotides in ribosomal genes using site specific character variability calculated from the observed sequences via an information theory index similar to the Shannon Weaver formula. My laboratory is experimenting with procedures to weight 12S ribosomal regions according to variability observed in closely versus distantly related taxa. By dividing the molecule into segments belonging to one of three or four rate classes, each rate class can be corrected individually. Wheeler and Honeycut (1988) suggested assigning half-weights to paired rRNA gene stem positions because a substitution on one side of a stem would require a compensatory change on the opposite side.

Unfortunately, the situation is not so simple. Often ribosomal stems contain G·T or A·C base pairs which are allowed by tautomeric base pairing (Topal and Fresco 1976) and represent easy transitions from a G-C or A-T pair (Simon et al. 1990). Whatever weighting is used, it is instructive to compare alternative weightings and evaluate their effects.

Williams and Fitch (1989) present a method called dynamic weighting where character weights are assigned based on comparisons of observed nucleotide sequences. Noting that positions that change frequently are unreliable indicators of relationship, they suggest weighting positions inversely to their frequency of substitutions. This philosophy is similar to the information theory method of Manske and Chapman (1987) discussed above except that in dynamic weighting, character variability is assessed in reference to an initial tree while in the information theory method this is not the case. Dynamic weighting is also philosophically and methodologically similar to Farris' (1969) successive approximations approach in which characters with high consistency (equated with low homoplasy; Kluge and Farris 1969) on an initial tree are weighted more heavily. In both procedures successive iterations usually converge in a few cycles on one tree and this tree is not necessarily identical to the seed tree (suggesting that the weighting is not trivial). A drawback of dynamic weighting is that the solution is sensitive to the initial tree, the initial matrix of substitutional values, and to the inclusion of autapomorphies.

Farris (1990) pointed out that his program Hennig-86 uses an updated successive approximations approach which includes a corrected homoplasy index. The original index was shown to be biased by the number of taxa and number of characters used in a study (Archie, 1989a, Farris 1989). The revised homoplasy measure used, called the "retention index", is the same as Archie's "homoplasy excess ratio maximum" (Archie 1990, Farris 1990), which is an estimator of the "homoplasy excess ratio" used in his test of character state randomness (Archie 1989ab). The consistency index <u>for individual characters</u> can be rescaled to vary between zero and one by multiplying it by the retention index and this value can be used for character weighting (Farris 1990, Archie 1990).

Kocher et al. (1989) suggest that variable positions should be used for comparing closely related species and conserved positions for comparing more distantly related species. Of course, it is impossible to know which species are closely related and which distantly related without reference to a phylogeny. Luckily, as the result of classical systematic studies (based on morphology, chromosomes, allozymes, etc.), for most taxa, we have some idea of species-group or familial relationships. In these cases, subsets of taxa can be analyzed separately using characters most appropriate for that taxonomic level. It is difficult to simultaneously optimize estimates of relationships among very closely related taxa and very distantly related taxa especially since the absence of many intermediate groups will create long branches.

Swofford and Olsen (1990) suggest that weighting which favors transversions will have the same effect as using variable characters to group closely related taxa and conserved characters to group more distantly related taxa. The reasoning behind this is that transversions occur rarely and are more reliable; transitions are likely to be indicators of highly variable sites. By this weighting scheme, distantly related species will be grouped based on transversions while closely related species, between which few transversions exist, will be grouped based on transitions. This method will be valid as long as saturated regions and/or positions have been removed from the data. In saturated regions/positions transversions accumulate over evolutionary time and erase the record of transitions. If many of these saturated positions were present in the data, favoring transversions would be the wrong strategy.

It should be pointed out here, that none of the weighting schemes discussed above takes into account the masking effect of highly biased nucleotide composition. High A+T or G+C bias will reduce the observed substitutional variability (Saccone et al. 1989). The more closely related the taxa under consideration, the less the problems associated with recognizing biased substitution processes.

Several phylogenetic analysis programs use a generalized parsimony method in which the user (Swofford 1990, Maddison and Maddison 1990) or the program (Williams and Fitch 1989) defines a matrix which describes the cost associated with a transition from state *i* to state *j*. Modification of the cost matrix allows experimentation with different kinds of weighting.

Character-based parsimony methods search for the shortest length tree for the data as a best estimate of tree topology i.e. patterns of evolutionary relationships among taxa. Branch lengths produced in the process represent one of many possible solutions under a specified optimization procedure (e.g. favoring parallelisms over reversals [DELTRAN] or vice versa [ACCTRAN]; Swofford and Maddison 1987) and a specified model of evolution (e.g. Wagner model, Dollo model, Camin-Sokal model; Swofford 1990). These branch lengths are not corrected for multiple substitutions and therefore do not represent actual amounts of evolution for those characters. Several methods have been suggested to solve this problem for character-based parsimony methods (Goodman et al. 1974, Langley and Fitch 1974, Fitch and Bruschi 1987, Hendy and Penny 1989).

One group of branch-length correcting methods is based on the reasoning that the number of nucleotide substitutions observed between a sequence and its remote ancestor is an increasing function of the number of branching events between them. The simplest of this type of branch-correction methods was developed by Fitch and Bruschi (1987). It is philosophically similar to the earlier methods of Goodman et al. (1974) and Langley and Fitch (1974) and operates by increasing the lengths of lineages that have fewer nodes. The correction for each interval is proportional to the number of substitutions already present in the most-parsimonious tree. Lineages with dense branching may not need to be corrected (Fitch and Beintema 1990). Although this class of methods makes the relative rates of evolution among lineages more realistic, none of them is expected to correct for all superimposed substitutions and the rates will be underestimated seriously if many long branches are present on the tree (Fitch and Bruschi 1987).

Hendy and Penny (1989; Hendy 1989) have developed a modification of character-based parsimony which they call the "closest tree" method. This method uses the number of observed changes in each branch of the tree to reconstruct the estimated actual number of changes based on the evolutionary branching process model of Cavender (1978). It currently handles cases of up to 20 taxa using a branch and bound estimation procedure. The closest tree method will produce consistent trees even when ordinary parsimony does not; it avoids the problem of long branches attracting. It directly provides branch lengths, corrected for multiple substitutions, estimated from the data (as long as sufficient data exist to get good estimates). Hendy and Penny point out that under the molecular clock hypothesis, these branch lengths should be linearly proportional to time intervals such that deviations from the clock can be easily recognized.

The closest tree method is still in the process of refinement. To better fit nucleotide sequence data Hendy and Penny plan to go beyond Cavender's model which uses two-state characters rather than four-state characters and assumes equal probabilities of change at every site and equal rates of change between the states. Hendy and Penny's closest tree method is conceptually similar to the Manske and Chapman (1987) idea (which was developed for distance methods and modified distance methods) in that the original data is used to correct branch lengths following the construction of an initial tree.

Once a parsimony tree is calculated, statistical tests such as bootstrapping (for individual branches) (Felsenstein 1988) or randomization tests (Archie 1989ab, Faith 1990) can be used to test for statistically significant patterns.

USEFULNESS OF RESTRICTION MAPPING VERSUS SEQUENCING FOR PHYLOGENETIC ANALYSIS

The first methods used for constructing evolutionary hypotheses from restriction data employed presence of shared sites or fragments to calculate genetic distances among taxa from which trees could be built (Nei and Li 1979, Engels 1981, Kaplan and Risko 1981). Templeton (1983) developed a method of reconstructing evolutionary trees from restriction data which took into account probabilities of convergent losses and gains of enzyme cutting sites. His approach combined parsimomy with the philosophy of character compatibility analysis. Weaknesses of his method include the need to treat all sites recognized by one enzyme as a single character and the use of a Wagner parsimony optimization which assumes that all changes are equally likely (DeBry and Slade 1985).

Picking up where Templeton left off, DeBry and Slade (1985) develop an explicit probability model for the evolution of restriction sites and show that Dollo parsimony rather than Wagner parsimony is a better estimator of evolutionary trees because it is based on the assumption that any particular character is rarely gained but often lost. This assumption will only be true if: 1) the rate of substitution is constant across lineages and within and among gene regions; 2) the average substitution rate is the same for all base changes; 3) fragments with identical mobilities result from changes at homologous sites; and 4) the base composition of the DNA is 1:1:1:1. As

DeBry and Slade acknowledge, and as this paper has discussed, we know that none of the above assumptions is uniformly true. Thus, although their method is an improvement over other methods and heuristically very valuable for understanding restriction site evolution, it appears virtually impossible to construct a realistic probability model for the evolution of restriction sites except in cases of very closely related taxa where only a few substitutions have occurred (as in the population studies reviewed by Avise et al. 1987).

Nucleotide sequence information has an advantage over restriction site data in its higher resolving power (Kocher et al. 1989). Thomas and Beckenbach (1989) found that in all pairwise comparisons among six salmonid species, the sequence divergence observed directly was higher than previously predicted based on restriction analysis. Furthermore, for restriction sites, it is difficult to know exactly which gene region is being studied, whether the change observed is silent or selected, how many nucleotide substitutions have occurred inside a site, the exact degree of nucleotide bias, and the number of transitional versus transversional substitutions. Furthermore, restriction analysis cannot distinguish between mobility differences due to conformational mutations and those due to length variation (Vigilant, et al. 1988). Restriction sites are like black boxes.

COMBINING SEQUENCING WITH OTHER MOLECULAR TECHNIQUES

Although sequencing is faster and easier with PCR, it is still expensive. One way to optimize costs and save time is to use sequence information in combination with other techniques.

Reference Sequences

In order to know whether restriction fragments with identical mobilities are truly homologous, it is necessary to map their location with reference to each other. Mapping requires multiple double digests and can be very time consuming. An easier way to map restriction sites is to first sequence the region of interest or if already sequenced retrieve that sequence from Genbank or EMBL and to locate specific endonuclease recognition sequences a priori or a posteriori. This technique was used successfully by Cann et al. (1984) to map 441 cleavage sites in 112 humans. Computer programs now exist which search for restriction sites by identifying their characteristic palindromic arrays.

High Resolution Restriction Mapping

Once a gene region has been sequenced for a number of taxa, comparisons can be made and diagnostic substitutions can be identified. For population biological and systematic studies it will then be necessary to survey large numbers of individuals to describe variation in the substitutions of interest. If these substitutions lie within restriction sites, high resolution restriction mapping can be used to detect them. Kreitman and Agaude (1986) used high resolution restriction mapping to study population level variation in Drosophila melanogaster. This method uses probes of approximately 3 kilobases to survey a restricted portion of the genome and thereby simplify band identification. (Plant geneticists use similar techniques [reviewed in Palmer et al. 1988].) In addition, Kreitman and Aguade used a Drosophila melanogaster reference sequence to calculate expected fragment sizes. Combining these techniques, they were able to reliably score fragments as small as 60-70 base pairs and identify insertion/deletion differences of only one base pair. This procedure can be simplified via PCR.

Once small (200-500 bp) pieces of mtDNA from the region of interest have been amplified via PCR, they can be cut using 4-base-recognizing restriction enzymes. Because PCR produces large quantities of DNA, cut fragments can be identified via ethidium bromide staining. This alleviates the necessity to use radioactivity and to wait hours for autoradiographs to develop. My laboratory has found that PCR provides sufficient quantities of DNA to allow EtBr detection of digested pieces as small as 50 base pairs on a 2-4% agarose minigel.

Allele Specific Dot-blot Probing

Recently, techniques have been developed for detecting single base substitutions using dot blots of DNA probed with synthetic oligonucleotide primers (Miller and Barnes 1986, Saiki et al. 1986, Verlaan-de Vries et al. 1986, Oste 1988, Erlich and Bugawan 1990, Kogan and Gitschier 1990). This technique is useful if mutational differences to be studied do not lie within restriction sites. For this procedure, hybridization stringencies are adjusted such that each probe will attach only to the exact sequence of bases to be screened. Alternative probes are constructed for each substitution possibility using a reference sequence as a guide. For example, suppose two reference individuals were sequenced--one from population 1 and one from population 2--and found to differ at only one site, site Z. Suppose that in population 1, site Z contained an adenine (A), and in population 2, site Z contained a guanine (G). Two probes would then be constructed: one with an A at site Z and one with a G at site Z. All members of populations 1 and 2 would then be screened using both probes to discover which possessed and an A and which possessed a G at site Z. With proper positive and negative controls, these techniques have been successfully used for medical screening.

Allele Specific PCR Primers

Allele specific primers are similar to allele specific probes in that multiple primers are constructed which match alternative substitutions at a single site. The site of interest is located at the 3' end of the primer to help ensure that amplification will not take place if the primer does not match the substitution. Stringency of the PCR reaction (annealing temperature) is adjusted carefully such that each primer will only anneal to and amplify the individuals possessing the specific substitution that primer is designed to match (Wu et al. 1989, Erlich and Bugawan 1990).

Allele specific PCR primers have been used successfully to detect lactate dehydrogenase (LDH) polymorphisms in the fish, *Fundulus heteroclitus* (Powell et al., submitted). Two mutations thought to be selectively important were found by sequencing the LDH gene. Only one of these mutations caused a replacement of a charged amino acid and could be detected by starch gel electrophoresis. In order to survey populations for both mutations, allele specific primers were designed for the segment of DNA bounded by the two mutations. Two sets of two opposing primers were constructed; each had one of the potentially substituted bases at its 3' end. To achieve equal reaction stringencies, binding strength for the four oligonucleotide primers was adjusted by varying their length (12 versus 14 bp) (M. Powell, pers. comm.) By this procedure, all possible substitutions could be detected.

CONCLUDING REMARKS

Molecular sequencing techniques, augmented by PCR, have revolutionized systematics studies and provided a wealth of phylogenetic information. They will continue to do so. When analyzing DNA sequence data it is necessary to understand the processes governing its evolution and be aware of potential biases. This paper has used animal mitochondria as a focal point for the discussion of molecular processes and their phylogenetic implications.

A major point which has been reiterated in many sections of this paper is that the comparison of closely related species is critical to the understanding of the evolution of nucleotide sequences. In such comparisons it is immediately obvious what changes occur first and what biases are present. It is only then, through comparisons of taxa from a hierarchy of divergence levels, that the masking effect of multiple substitutions can be sorted out and more accurate evolutionary trees can be constructed.

ACKNOWLEDGEMENTS

This paper benefited from discussions with J. Archie, R. Cann, D. Irwin, T. Kocher, M. Mickevich, S. Pääbo, D. Swofford, K. Thomas, A.C. Wilson and many participants in the NATO Advanced Studies Institute in Molecular Taxonomy. I thank J. Archie, A. Martin and D. Swofford for comments on the manuscript and C. McIntosh for help with the figure. Financial support for data collection and for the author during the writing of this paper were provided by NSF BSR 88-22710.

REFERENCES

Anderson S, De Bruijn MHL, Coulson AR, Eperon IC, Sanger F Young IG (1982) Complete sequence of bovine mitochondrial DNA. Conserved features of the mammalian mitochondrial genome. J Mol Biol 156:683-717

Aquadro CF, Greenberg BD (1983) Human mitochondrial DNA variation and evolution: analysis of nucleotide sequences from seven individuals. Genet 103:287-312

Aquadro CF, Kaplan N, Risko KJ (1984) An analysis of the dynamics of mammalian mitochondrial dna sequence evolution. Mol Biol Evol 1:423-434

Archie J (1989a) A randomization test for phylogenetic information in systematic data. Syst Zool 38:239-252

Archie J (1989b) Homoplasy excess ratios: New indices for measuring levels of homoplasy in phylogenetic systematics and a critique of the consistency index. Systematic Zoology 38:253-269

Archie J (1990) Homoplasy excess statistics and retention indices: A reply to Farris. Systematic Zoology 39:169-174

Archie J (1991) Tests to distinguish between phylogenetic information and random noise in nucleotide sequence data. In: Dudley E (ed) Proceed IV Intnl Congress Syst Evol Biol, Dioscorides Press, in press.

Archie JW, Simon C, Martin A (1989) Small sample size does decrease the stability of dendrograms calculated from allozyme-frequency data. Evolution 43:678-683

Attardi G, Schatz G (1988) Biogenesis of mitochondria. Annu Rev Cell Biol 4:289-333

Aubert J, Solignac M (1990) Experimental evidence for mitochondrial DNA introgression between Drosophila species. Evolution, in press

Avise JC, Arnold J, Ball RM, Bermingham E, Lamb T, Neigel JE, Reeb CA, Saunders NC (1987) Intraspecific phylogeography: The mitochondrial bridge between population genetics and systematics. Annu Rev Ecol Syst 18:489-522

Bernardi G, Bernardi G (1986) Compositional constraints and genome evolution. J Mol Evol 24:1-11

Bibb MJ, Van Etten RA, Wright CT, Walberg MW, Clayton DA (1981) Sequence and gene organization of mouse mitochondrial DNA. Cell 26:167-180

Bishop MJ, Friday AE (1987) Tetrapod relationships: the molecular evidence. In: Patterson C (ed) Molecules and morphology in evolution: conflict or compromise? Cambridge University Press

Boyce TM, Zwick ME, Aquadro CF (1989) Mitochondrial DNA in the bark weevils: Size, structure and heteroplasmy. Genetics 123:825-836

Brown WM (1985) The Mitochondrial Genome of Animals. In: MacIntyre RJ (ed), Molecular Evolutionary Genetics, Plenum, NY, p 95

Brown WM, George M Jr, Wilson AC (1979) Rapid evolution of animal mitochondrial DNA. Proc Natl Acad Sci 76:1967-1971

Brown WM, Prager EM, Wang A, Wilson AC (1982) Mitochondrial DNA sequences of primates: tempo and mode of evolution. J Mol Evol 18: 225-239

Buth DG (1984) The application of electrophoretic data in systematic studies. Annu Rev Ecol Syst 15: 501-522

Cann RL, Wilson AC (1983) Length mutations in human mitochondrial DNA. Genetics 104:699-711

Cann RL, Brown WM, Wilson AC (1984) Polymorphic sites and the mechanism of evolution in human mitochondrial DNA. Genetics 106:479-499

Cann RL, Stoneking M, Wilson AC (1987) Mitochondrial DNA and human evolution. Nature 325: 31-36

Carr SM, Brothers AJ, Wilson AC (1987) Evolutionary inferences from restriction maps of mitochodrial DNA from nine taxa of Xenopus frogs. Evolution 41: 176-188

Carson HL, Hardy DE, Spieth HT, Stone WS (1970) The evolutionary biology of Hawaiian Drosophilidae. In: Hecht MK, Steere WC (eds) Essays in Evolution and Genetics in Honor of Theodosius Dobzhansky Appleton-Century-Crofts, New York, p 437

Cantatore P, Roberti M, Rainaldi G, Gadaleta MN, Saccone C (1989) The complete nucleotide sequence, gene organization, and genetic code of the mitochondrial genome of Paracentrotus lividus. J Biol Chem 264:10965-75

Cavender JA (1978) Taxonomy with confidence. Math Biosc 40:271-280

Cheeseman P, Kanefsky B (1990) Evolutionary tree reconstruction. In: AAAI String Symposium, Minimum Message Length and Coding. Stanford Univ Press

Clary DO, Wolstenholme DR (1985) The mitochondrial DNA molecule of Drosophila yakuba: Nucleotide sequence, gene organization, and genetic code. J Mol Evol 22:252-271

Clary DO, Wolstenholme DR (1987) Drosophila mitochondrial DNA: conserved sequence in the A+T rich region and supporting evidence for a secondary structure model of the small ribosomal RNA. J Mol Evol 25:116-125

Coyne J (1982) Gel electrophoresis and cryptic protein variation. In: Rattazzi M, Scandalios J, Whitt G (eds) Isozymes: Current Topics in Biological and Medical Research, vol 6. Alan R Liss, New York, p 1

Crews S, Ojala D, Posakony J, Nishiguchi J, Attardi G (1978) Nucleotide sequence of a region of human mitochondrial DNA containing the precisely identified origin of replication. Nature 277:192-198

Crozier RH, Crozier YC, Mackinlay AG (1989) The CO-I and CO-II region of honeybee mitochondrial DNA: Evidence for variation in insect mitochondrial evolutionary rates. Mol Biol Evol 6:399-411

Dams E, Hendriks L, Van de Peer Y, Neefs J-M, Smits G, Vandenbempt I, De Wachter R (1988) Compilation of small ribosomal subunit RNA sequences. Nucl Acids Res 16 supl:r87-r173

de Bruijn MHL (1983) Drosophila melanogaster mitochondrial DNA, a novel organization and genetic code. Nature 304:234-241

DeBry RW, Slade NA (1985) Cladistic analysis of restriction endonuclease cleavage maps within a maximum-likelihood framework. Syst Zool 34: 21-34

Derancourt J, Lebor AS, Zukerkandl E (1967) Sequence des nucleotides et evolution. Bull Soc Chim Biol 49: 577-603

DeSalle R, Freedman T, Prager EM, Wilson AC (1987) Tempo and mode of sequence evolution in mitochondrial DNA of Hawaiian Drosophila. Proc Natl Acad Sci USA 83:6902-6906

DeSalle R, Giddings LV, Templeton AR (1986a) Mitochondrial DNA variability in natural populations of Hawaiian Drosophila. I. Methods and levels of variability in D. silvestris and D. heteroneura populations. Heredity 56:75-85

DeSalle R, Giddings LV, Kaneshiro KY (1986b) Mitochondrial DNA variation in natural populations of Hawaiian Drosophila. II. Genetic and phylogenetic relationship of natural populations of D. silvestris and D. heteroneura. Heredity 56:87-96

DeSalle R, Templeton A (1988) Founder effects and the rate of mitochondrial DNA evolution in Hawaiian Drosophila. Evol 42:1076-1084

Desjardins P, Morais R (1990) Sequence and gene organization of the chicken mitochondrial genome. J Mol Biol 212:599-634

Engels WR (1981) Estimating genetic divergence and genetic variability with restriction endonucleases. Proc Natl Acad Sci USA 78:6329-6333

Erlich HA, Bugawan TL (1990) HLA DNA typing. In: Innis MA, Gelfand DH, Sninsky JJ, White TJ (eds) PCR Protocols: A Guide to Methods and Applications. Academic Press, San Diego, p 261

Faith DP (1990) Chance marsupial relationships? Nature 345:393-394

Faith DP, Cranston PS (1991) Could a cladogram this short have arisen by chance alone?: an index of cladistic character covariation. Cladistics, in press

Farris JS (1969) A successive approximations approach to character weighting. Syst Zool 18:374-385

Farris JS (1988) Hennig-86. A computer program for phylogenetic analysis available from the author. 41 Admiral Street, Port Jefferson Station, New York 11776

Farris JS (1989) The retention index and the rescaled consistency index. Cladistics 5:417-419

Farris JS (1990) The retention index and homoplasy excess. Systematic Zoology 38:406-407

Fauron C M-R, Wolstenholme DR (1980) Intraspecific diversity of nucleotide sequences within the adenine + thymine-rich region of mitochondrial DNA molecules of Drosophila mauritiana, Drosophila melanogaster and Drosophila simulans. Nucleic Acids Res 8:5391-5410

Felsenstein J (1988) Phylogenies from molecular sequences: inference and reliability. Ann Rev Genet 22:521-565

Feng D-F, Doolittle RF (1987) Progressive sequence alignment as a prerequisite to correct phylogenetic trees. J Mol Evol 25:351-360

Field KG, Olsen GJ, Lane DJ, Giovannoni SJ, Ghiselin MT et al (1988) Molecular phylogeny of the animal kingdom. Science 239:748-752

Fitch WM, Beintema JJ (1990) Correcting parsimonious trees for unseen nucleotide substitutions: the effect of dense branching as exemplified by ribonuclease. Mol Biol Evol 7:438-443

Fitch WM (1986) The estimate of total nucleotide substitutions from pairwise differences is biased. Phil Trans Roy Soc (Lond) B, 316:317-324

Fitch WM, Bruschi M (1987) The evolution of prokaryotic ferredoxins--with a general method correcting for unobserved substitutions in less branched lineages. Mol Biol Evol 4:381-394

Fitch WM, Markowitz (1970) An improved method for determining codon variability in a gene and its application to the rate of fixation of mutations in evolution. Biochem Genet 4: 579-593

Fryer G, Isles TD (1972) The Cichlid Fishes of the Great Lakes of Africa: Their Biology and Evolution. Oliver and Boyd, Edinburgh

Gadaleta G, Pepe G, DeCandia G, Quagliariello C, Sbisa E, Saccone C (1989) The complete nucleotide sequence of the Rattus norvegicus mitochondrial genome: cryptic signals revealed by comparative analysis between vertebrates. J Mol Evol 28:497-516

Garesse R (1988) Drosophila melanogaster mitochondrial DNA: Gene organization and evolutionary considerations. Genetics 118:649-663

Goodman M (1981) Decoding the pattern of protein evolution. Prog Biophys Mol Biol 37:105-164

Goodman M, Moore GW, Barnabas J, Matsuda G (1974) The phylogeny of human globin genes investigated by the maximum parsimony method. J Mol Evol 3:1-48

Gray MW (1989) Origin and evolution of mitochondrial DNA. Annu Rev Cell Biol 5:25-50

Hall ER (1981) The mammals of North America, ed 2. John Wiley and Sons, New York, p 573

Hardy DE (1965) Diptera: Cyclorrapha II, Series Schizophora, Section Acalypterae I, Family Drosophilidae. Insects of Hawaii, Vol 12. University of Hawaii Press, Honolulu

Harrison R (1989) Animal mitochondrial DNA as a genetic marker in population and evolutionary biology. Trends Ecol Evol 4:6-11

Hasegawa M, Kishino H (1989) Heterogeneity of tempo and mode of mitochondrial DNA evolution among mammalian orders. Jpn J Genet 64:243-258

Hasegawa M, Kishino H, Yano T (1985) Dating of the human-ape splitting by a molecular clock of mitochondrial DNA. J Mol Evol 26:132-147

Hasegawa M, Kishino H, Hayasaka K, Horai S (1990) Mitochondrial DNA evolution in primates: transition rate has been extremely low in the lemur. J Mol Evol 31:113-121

Haucke H-R, Gellissen G (1988) Different mitochondrial gene orders among insects: exchanged tRNA gene positions in the COII/COIII region between an orthopteran and a dipteran species. Curr Genet 14:471-476

HsuChen C-C, Kotin RM, Dubin DT (1984) Sequences of the coding and flanking regions of the large ribosomal subunit RNA gene of mosquito mitochondria. Nucl Acids Res 12:7771-7785

Hayasaka K, Gojobori T, Horai S (1988) Molecular phylogeny and evolution of primate mitochondrial DNA. Mol Biol Evol 5:626-644

Hendy MD (1989) The relationship between simple evolutionary tree models and observable sequence data. Syst Zool 38:310-321

Hendy MD, Penny D (1989) A framework for the quantitative study of evolutionary trees. Syst Zool 38:297-309

Hein J (1989a) A new method that simultaneously aligns and reconstructs ancestral sequences for any number of homologous sequences, when the phylogeny is given. Mol Biol Evol 6:649-668

Hein J (1989b) A tree reconstruction method that is economical in the number of pairwise comparisons used. Mol Biol Evol 6:669-684

Heywood VH (1978) Flowering Plants of the World. Mayflower Books Inc., New York

Hillis DM, Davis SK (1986) Evolution of ribosomal DNA: Fifty million years of recorded history in the frog genus Rana. Evolution 40:1275-1288

Hixson JE, Brown WM (1986) A comparison of the small ribosomal RNA genes from the mitochondrial DNA of the great apes and humans: sequence, structure, evolution, and phylogenetic implications. Mol Biol Evol 3:1-18

Hogeweg P, Hesper B (1984) The alignment of sets of sequences and the construction of phyletic trees: an integrated method. J Mol Evol 20:175-186

Holmquist R (1983) Transitions and transversions in evolutionary descent: an approach to understanding. J Mol Evol 19:134-144

HsuChen C-C, Kotin RM, Dubin DT (1984) Sequences of the coding and flanking regions of the large ribosomal subunit RNA gene of mosquito mitochondria. Nucl Acids Res 12:7771-7785

Irwin DM, Kocher TD, Wilson AC (1990) Evolution of the cytochrome b gene of mammals. J Mol Evol in press

Jacobs HT, Elliott DJ, Math VB, Farquharson A (1988) Nucleotide sequence and gene organization of sea urchin mitochondrial DNA. J Mol Biol 202:185-217

Jukes TH, Cantor CR (1969) Evolution of protein molecules. pp 21-120 in HW Munro, ed Mammalian protein metabolism. Academic Press, New York

Kaplan N, Risko K (1981) An improved method for estimating sequence divergence of DNA using restriction endonuclease mappings. J Mol Evol 17:156-162

Kimura M (1980) A simple method for estimating evolutionary rate of base substitutions through comparative studies of nucleotide sequences. J Mol Evol 16:111-120

Kishino H, Hasegawa M (1990) Converting distance to time: an application to human evolution. Methods Enzymol 183:550-570

Kluge AG, Farris JS (1969) Quantitative Phyletics and the Evolution of Anurans. Syst Zool 18:1-32

Kocher TD, Thomas WK, Meyer A, Edwards SV, Pääbo S, Villablanca FX, Wilson AC (1989) Dynamics of mitochondrial DNA evolution in animals: Amplification and sequencing with conserved primers. Proc Natnl Acad Sci USA 86:6196-6200

Kogan SC, Gitschier J (1990) Genetic prediction of Hemophilia A. In: Innis MA, Gelfand DH, Sninsky JJ, White TJ (eds) PCR Protocols: A Guide to Methods and Applications. Academic Press, San Diego, p 288

Kreitman M, Aguade M (1986) Genetic uniformity in two populations of Drosophila melanogaster as revealed by filter hybridization of four nucleotide-recognizing restriction enzyme digests. Proc Natl Acad Sci USA 83: 3562-3566

Langley CH, Fitch WM (1974) An examination of the constancy of the rate of molecular evolution . J Mol Evol 3:161-177

Lansman RA, Shade RO, Shapira JF, Avise JC (1981) The use of restriction endonucleases to measure mitochondrial DNA sequence relatedness in natural populations. III. techniques and potential applications. J Mol Evol 17: 214-226

Lewontin RC (1989) Inferring the number of evolutionary events from DNA coding sequence differences. Mol Biol Evol 6:15-32

Li W-H (1987) Models of nearly neutral mutations with particular implications for nonrandom usage of synonymous codons. J Mol Evol 24:337-345

Li W-H, Wu C-I, Luo C-C (1984) Nonrandomness of point mutation as reflected in nucleotide substitutions in pseudo genes and its evolutionary implications. J Mol Evol 21:58-71

Li W-H, Wu C-I, Luo C-C (1985) A new method for estimating synonymous and nonsynonymous rates of nucleotide substitutions considering the relative likelihood of nucleotide and codon changes. Mol Biol Evol 2:150-174

Li W-H, Tanimura M, Sharp PM (1987) An evaluation of the molecular clock hypothesis using mammalian DNA sequences. J Mol Evol 25:330-342

Maddison WP, Maddison DR (1990) MacClade. A program for the analysis of character evolution and the testing of phylogenetic hypotheses. Distributed by Sinauer Associates, Sunderland, Mass.

Manske CL, Chapman DJ (1987) Nonuniformity of nucleotide substitution rates in molecular evolution: Computer simulation and analysis of 5S ribosomal RNA sequences. J Mol Evol 26:226-251

Martin AP, Kessing BD, Palumbi SR (1990) Accuracy of estimating genetic distance between species from short sequences of mitochondrial DNA. Mol Biol Evol 7:485-488

Martin A, Simon C (1988) Anomalous distribution of nuclear and mitochondrial DNA markers in periodical cicadas resulting from life cycle plasticity. Nature (London) 336:247-249

Martin A, Simon C (1990) Differing levels of among-population divergence in the mitochondrial DNA of 13- versus 17-year periodical cicadas related to historical biogeography. Evolution 44:1066-1080

Maxson LR, Highton R, Wake DB (1979) Albumin evolution and its phylogenetic implications in the plethodontid salamander genera Plethadon and Ensatina. Copeia 1979:502-508

Miyamoto MM, Tanhauser SM, Laipis PJ (1989) Systematic relationships in the artiodactyl tribe Bovini (family Bovidae), as determined from mitochondrial DNA sequences. Syst Zool 38:342-349

Miyata T, Hayashida H, Kikuno R, Hasegawa M, Kobayashi M, Koike K (1982) Molecular clock of silent substitutions: At least six fold preponderance of silent changes in mitochondrial genes over those in nuclear genes. J Mol Evol 19:28-35

Miller JK, Barnes WM (1986) Colony probing as an alternative to standard sequencing as a means of direct analysis of chromosomal DNA to determine the spectrum of single-base changes in regions of known sequence. Proc Natl Acad Sci USA 83: 1026-1030

Moritz C, Dowling TE, Brown WM (1987) Evolution of animal mitochondrial DNA: Relevance for population biology and systematics. Ann Rev Ecol Syst 18: 269-92

Mullis K, Faloona F, Scharf S, Saiki R, Horn G, Erlich H (1986) Specific enzymatic amplification of DNA in vitro: the polymerase chain reaction. Cold Spring Harbor Symposium on Quantitative Biology 51: 263-273

Muto A, Osawa S (1987) The guanine and cytosine content of genomic DNA and bacterial evolution. Proc Natl Acad Sci USA 84:166-169

Nanney DL, Preparata RM, Preparata FP, Meyer EB, Simon EM (1989) Shifting ditypic site analysis: Heuristics for expanding the phylogenetic range of nucleotide sequences in Sankoff analysis. J Mol Evol 28:451-459

Neefs J-M, Van de Peer Y, Hendriks L, and DeWachter R (1990) Compilation of small ribosomal subunit RNA sequences. Nucl Acids Res 18 supplement:2237-2318

Nei M (1987) Molecular evolutionary genetics. Columbia Univ Press, NY

Nei M, Gojobori T (1987) Simple methods for estimating the numbers of synonymous and nonsynonymous nucleotide substitutions. Mol Biol Evol 3:418-426

Nei M, Li W-H (1979) Mathematical model for studying genetic variation in terms of restriction endonucleases. Proc Natl Acad Sci USA 76:5269-5273

Nei M, Tajima F (1987) Problems arising in phylogenetic inference from restriction-site data. Mol Biol Evol 4:320-323

Neigel JE, Avise JC (1986) Phylogenetic relationships of mitochondrial DNA under various demographic models of speciation. In: Karlin S, Nevo E (eds) Evolutionary Processes and Theory. Academic Press, p 515

Nicoghosian K, Bigras M, Sankoff D, Cedergren R (1987) Archetypical features in tRNA families. J Mol Evol 26:341-346

Noller HF, Woese CR (1981) Secondary structure of 16S ribosomal RNA. Science 212:403-411

Olsen GJ (1988) Phylogenetic analysis using ribosomal RNA. Methods Enzymology 164:793-812

Oste C (1988) Polymerase chain reaction. Biotechniques 6:162-166

Palmer Palmer JD, Jansen RK, Michaels HJ, Chase MW, Manhart JR (1988) Chloroplast variation and plant phylogeny. Ann Mo Bot Gard 75:1180-1206

Palumbi S (1989) Rates of molecular evolution and the fraction of nucleotide positions free to vary. J Mol Evol 29:180-188

Penny D (1982) Toward a basis for classification: the incompleteness of distance measures: incompatibility analysis and phenetic classification. J Theoret Biol 96: 129-142

Powell MA, Crawford DL, Lauerman T, Powers DA. Analysis of cryptic alleles of Fundulus heteroclitus lactate dehydrogenase by a novel allele specific polymerase chain reaction. Submitted to Mol Biol Evol

Powell JR, Caccone A, Amato GD, Yoon C (1986) Rates of nucleotide substitution in Drosophila mitochondrial DNA and nuclear DNA are similar. Proc Natl Acad Sci USA 83:9090-9093

Rand DM, Harrison RG (1989) Molecular population genetics of mitochondrial DNA size variation in crickets. Genetics 121:551-569

Roe BA, Ma DP, Wilson RK, Wong JFH (1985) The complete nucleotide sequence of the Xenopus laevis mitochondrial genome J Biol Chem 260(17):9759-74

Rohlf FJ Bookstein FL (1990) Introductions to methods for landmark data. In: Rohlf FJ Bookstein FL (eds) Proceedings of the Michigan Morphometrics Workshop. Museum of Zoology, University of Michigan Press, Ann Arbor, p 220

Saccone C, Pesole G, Preparata G (1989) DNA microenvironments and the molecular clock. J Mol Evol 29:407-411

Saiki RK, Scharf S, Faloona F, Mullis KB, Horn GT, Erlich HA, Arnheim N (1985) Enzymatic amplification of B-globin genomic sequences and restriction site analysis for diagnosis of sickle cell anemia. Science 230: 1350-1354

Saiki RK, Bugawan TL, Horn GT, Mullis KB, Erlich HA (1986) Analysis of enzymatically amplified B-globin and HLA-DQ DNA with allele-specific oligonucleotide probes. Nature 324:163-166

Sankoff DD, Cedergren RJ (1983) Simultaneous comparison of three or more sequences related by a tree. In: Sankoff D, Kruskal B (eds) Time Warps, String Edits, and Macromolecules: The Theory and Practice of Sequence Comparison. Addison-Wesley, Reading, MA, p 253

Satta Y, Ishiwa I, Chigusa SI (1987) Analysis of nucleotide substitutions of mitochondrial DNAs in Drosophila melanogaster and its sibling species. Mol Biol Evol 4:638-650

Sharp P, Li W-H (1989) On the rate of DNA sequence evolution in Drosophila. J Mol Evol 28:398-402

Simon C (1988) Evolution of 13- and 17-year Periodical Cicadas (Homoptera: Cicadidae: Magicicada). Bull Entomol Soc Amer 34:163-176

Simon C, Paabo S, Kocher T, Wilson AC (1990) Evolution of the mitochondrial ribosomal RNA in insects as shown by the polymerase chain reaction. In: Clegg M, O'Brien S (eds), Molecular Evolution. UCLA Symposia on Molecular and Cellular Biology, New Series, Vol. 122, Alan R. Liss, Inc., NY, p 235

Simon C, Franke A, Martin A (1991) The Polymerase chain reaction: DNA extraction and amplification. In: Hewitt GM (ed), Molecular Taxonomy, NATO Advanced Studies Institute, Springer Verlag, Berlin

Solignac M, Monnerot M, Mounolou J-C (1986a) Mitochondrial DNA evolution in the melanogaster species subgroup of Drosophila. J Mol Evol 23:31-40

Solignac M, Monnerot M, Mounolou J-C (1986b) Concerted evolution of sequence repeats in Drosophila mitochondrial DNA. J Mol Evol 24:53-60

Swofford DL (1990) PAUP: Phylogenetic Analysis Using Parsimony, Version 3.0. Illinois Natl Hist Surv, Champaign, IL

Swofford DL, Berlocher SH (1987) Inferring evolutionary trees from gene frequency data under the principle of maximum parsimony. Syst Zool 36:293-325

Swofford DL, Maddison WP (1987) Reconstructing ancestral character states under Wagner parsimony. Math Biosci 87:199-229

Swofford DL, Olsen GJ (1990) Phylogeny reconstruction. In: Hillis DM, Moritz C (eds) Molecular Systematics. Sinauer, Sunderland, Mass, p 411

Takahata N (1987) On the overdispersed molecular clock. Genet 116:169-179

Templeton A (1983) Phylogenetic inference from restriction endonuclease cleavage site maps with particular reference to the humans and apes. Evolution 37:221-244

Templeton A (1987) Nonparametric phylogenetic inference from restriction cleavage sites. Mol Biol Evol 4:315-319

Thomas WK, Beckenbach A (1989) Variation in salmonid mitochondrial DNA: evolutionary constraints and mechanisms of substitution. J Mol Evol 29:233-245

Thomas WK, Maa J, Wilson AC (1989) Shifting constraints on tRNA genes during mitochondrial DNA evolution in animals New Biologist 1:93-100

Thomas WK, Pääbo S, Villablanca FX, Wilson AC (1990) Spatial and temporal continuity of kangaroo rat populations shown by sequencing mitochondrial DNA from museum specimens. J Mol Evol 31:101-112

Thomas WK, Wilson AC. Mode and tempo of molecular evolution in the nematode Caenorhabditis. Submitted to Genetics

Topal MD, Fresco JR (1976) Complementary base pairing and the origin of substitution mutations. Nature 263:285-289

Uhlenbusch I, McCracken A, Gellissen G (1987) The gene for the large (16S) ribosomal RNA from the Locusta migratoria mitochondrial genome. Curr Genet 11:631-638

Upholt WB, David IB (1977) Mapping of mitochondrial DNA of individual sheep and goats: rapid evolution in the D loop regions of mitochondrial DNA. Cell 11:571-583

Vawter L, Brown WM (1986) Nuclear and miotochondrial DNA comparisons reveal extreme rate variation in the molecular clock. Science 234:194-196

Verlaan-de Vries M, Bogaard M, van den Elst H, van Boom J, van der Eb A, Bos J (1986) A dot-blot screening procedure for mutated ras oncogenes using synthetic oligodeoxynucleotides. Gene 50: 313-320

Vigilant L, Stoneking M, Wilson AC (1988) Conformational mutation in human mtDNA detected by direct sequencing of enzymatically amplified DNA. Nucleic Acids Res 16:5945-5955

Wheeler W, Honeycut R (1988) Paired sequence difference in ribosomal RNAs: Evolutionary and phylogenetic implications. Mol Biol Evol 5:90-96

Williams PL, Fitch WM (1989) Finding the weighted minimal change in a given tree. In: Fernholme B, Bremer K, Jornval H (eds), Nobel Symposium on The Hierarchy of Life, Elsevier, Cambridge p 453

Wilson AC, Cann RL, Carr SM, George M, Gyllensten UB, Helm-Bychowski KM, Higuchi RG, Palumbi SR, Prager EM, Sage RD, Stoneking M (1985) Mitochondrial DNA and two perspectives on evolutionary genetics. Biol J Linn Soc 26: 375-400

Wolstenholme DR, Clary DO (1985) Sequence Evolution of Drosophila Mitochondrial DNA. Genetics 109: 725-744

Wolstenholme DR, MacFarlane JL, Okimoto R, Clary DO, Wahleithner JA (1987) Bizarre tRNAs inferred from DNA sequences of mitochondrial genomes of nematode worms. Proc Natl Acad Sci USA 84:1324-1328

Wu DY, Ugozzoli L, Pal BK, Wallace RB (1989) Allele-specific enzymatic amplification of B-globin genomic DNA for diagnosis of sickle cell anemia. Proc Natl Acad Sci USA 86:2757-2760

RIBOSOMAL RNA PHYLOGENIES

M. Solignac, M. Pélandakis, F. Rousset, & A. Chenuil.

Laboratoire de Biologie et Génétique évolutives
CNRS, 91198 Gif-sur-Yvette Cedex, France

For a long time, cytochrome *c*, which is a slowly evolving protein, was the reference molecule to establish molecular phylogenies for very divergent eukaryotic lineages (Fitch and Margoliash 1967). Considering bacterial and chloroplastic *c*-type cytochomes together with cytochrome *c*, the first universal phylogenetic tree has been established. However, in this tree, the Eukaryote kingdom constituted nothing but a small internal branch of the bacterial tree (Dayhoff 1976): in fact, although encoded by the nucleus, the branching point of cytochrome *c* does not indicate the emergence of the nucleus of the Eukaryotes but that of their mitochondria: the cytochrome *c* gene was transferred to the nucleus from the genome of the primitive endosymbiote whose remnant cytoplasmic genes are to-day the mitochondrial DNA. In addition, *c*-type cytochromes are considered too short to work correctly for very deep nodes and they are absent in several bacterial species.

The role of a universal molecular chronometer is currently ensured mainly by ribosomal RNA (rRNA) molecules. Owing to their ubiquitous occurrence and to the high conservation of some regions of their sequences, rRNAs are ideal chronometers to trace back the origin of the major taxa. In addition, they include a mosaic of regions exhibiting various evolutionary rates and are apt to give reliable information whatever the level of divergence between the species considered. The interest of rRNA analysis is reinforced by the possibility of direct sequencing.

The principal characteristics of the rRNAs and of their corresponding genes, as well as the advantages to study rRNA sequences, are summarized below. Then are surveyed the most important results obtained from comparisons of rRNA sequences for the building of the so-called universal phylogenetic tree, but also for the determination of the relationships between less distantly related species.

NATO ASI Series, Vol. H 57
Molecular Techniques in Taxonomy
Edited by G. M. Hewitt et al.

Ribosomal RNA and DNA

The bacterial ribosome is a complex aggregation of about 50 ribosomal proteins and at least three separate classes of ribosomal RNAs identified by their sedimentation coefficient: 16S (or small subunit = SSU rRNA), 23S (large subunit = LSU rRNA) and 5S. The ribosomes of chloroplasts are very similar to those of Prokaryotes but they contain a 4.5S rRNA, homologous of the 3' terminal region of the 23S rRNA. Mitoribosomes have smaller rRNAs and less proteins than prokaryotic ribosomes; only those of plants contain a 5S rRNA. In Eukaryotes, four classes of rRNAs are present: 18S, 28S, 5S and 5.8S; the 18S and 28S are respectively equivalent to the prokaryotic 16S and 23S and the 5.8S rRNA is structurally homologous to the 5' terminal region of the prokaryotic 23S RNA. In several eukaryotic species, the large subunit is processed (de Lanversin and Jacq 1989). Increase in size of eukaryotic rRNAs is the result of the presence of a number of stretches absent in bacterial and organellar RNAs and variable in length for the different Eukaryotes. They are interspersed with regions of higher sequence and secondary structure conservation collectivelly named the conserved core. The fourfold difference in length between the shortest and the longest SSU or LSU rRNAs (Raué et al. 1988) is mainly attributable to the length variation of the divergent domains.

Within Prokaryotes, the 5S, 16S and 23S rRNA genes are grouped in one or generally several ribosomal operons (7 in *Escherichia coli*) interspersed with tRNA genes. In chloroplasts, the organisation of the genes is prokaryotic and the operon is located in a duplicated region of the genome with opposite orientations. The eukaryotic nucleus generally includes a high number of rRNA genes, virtually identical, and associated in one or several clusters on the chromosomes (nucleolar organizers). Generally 5S rRNA genes are clustered separately from the other subunits. In yeast, they are included in the unit of repetition but their sequence is specified by the opposite strand. These repeats are characterized by a concerted (or coincidental or horizontal) evolution of the different copies as seen through the intraspecific homogeneity of the genes and of their spacers contrasting with the interspecific variability (Gerbi 1985).

RNAs are characterized by their primary structure, i.e. the sequence of nucleotides, their secondary structure, i.e. the folding of the strand on itself in a plane, and their tertiary structure i.e. the tridimensional shape (Turner et al 1988). The primary structure contains the most accessible and perhaps useful information for phylogeny and taxonomy. However, secondary structures are of great help for sequence alignment because they are highly conserved. In addition, they are slowly evolving and can be used as characters per se for phylogenetic inferences. The complementarity of the bases which determines the secondary structure is maintained by a particular mode of evolution, the compensatory substitutions.

Ribosomal RNAs, the Rosetta stone of phylogenetics?

Although some precautions have to be kept in mind in the use of rRNAs to infer phylogenies (see Rothschild et al. 1986) the advantages of these molecules are numerous (Woese, 1967b).

Universality and homology. The rRNAs are found in Archaebacteria, Eubacteria, and in Eukaryotes (nucleus and organelles) and consequently are universal. The homology of rRNA genes whatever their origin is not questionable. The secondary structure of the core of the molecule is very conservative and the function of the rRNA is the same everywhere.

Number of residues. Their long size makes 16S and 23S rRNAs (and obviously the sum of both) more reliable chronometers than smaller molecules such as tRNAs (about 70 nucleotides) or 5S rRNAs (about 120 nucleotides). The length of the molecules is sufficient to avoid the undesirable effects of homoplasy on phylogenetic reconstitutions.

Sequence variability. Nucleotide sites of the rRNAs exhibit a wide range of evolutionary rates. For a given region, the rate of evolution is relatively constant but the difference in rate between conserved cores and divergent domains is over 100 fold . So the choice of the region to be sequenced can be adapted to the taxonomical distance of the organisms under comparison. The sequence does not need to be complete, providing that the partial sequences cover homologous regions, are long enough, and can be aligned.

Sequences available. The number of available sequences constitutes an additional advantage. For most of the major phyla, one or several reference sequences are already available and as many outgroups as wished can be chosen. The last compilation encompasses 71 LSU sequences (13 Eubacteria, 7 Archaebacteria, 16 Eucaryotes, 8 chloroplasts, 27 mitochondria, Gutell et al. 1990), 275 SSU sequences, not necessarily complete (42 Eubacteria, 16 Archaebacteria, 60 Eukaryotes, 11 chloroplasts, 36 mitochondria, Neefs et al. 1990) and 667 5S RNA sequences (267 Eubacteria, 44

Archaebacteria, 319 Eukaryotes, 11 eukaryotic pseudogenes, 20 chloroplasts, 6 plant mitochondria, Specht et al. 1990).

Technics of study. Comparison between species was first based on oligonucleotide catalogs, collections of sequences of oligonucleotides obtained by digestion of rRNA with ribonuclease T_1 which cleaves specifically at G residues. About 400 bacterial species have been studied in Woese's laboratory (Woese 1987a, b). Complete sequences were first obtained with the short 5S rRNA, using the partial chemical degradation of Peattie (1979). Sequences of the long rRNAs are now obtained by sequencing cloned genes. In addition, partial sequences can be quickly obtained through elongation of universal primer with the reverse transcriptase using rRNA as template (Qu et al. 1983, Lane et al. 1985). As for other genes, sequences obtained through PCR amplification of rDNA are in progress. In the past, several important results have been obtained with 5S rRNA (see for example Hori and Osawa 1979, Hendricks et al. 1986, Wolters and Erdman 1986); since more and more complete sequences of SSU and LSU rRNAs become available, studies are now preferentially focussed on these molecules.

The universal phylogenetic tree

The contribution of the rRNA to the elucidation of the general history of living forms and of the origin of organelles has been decisive. However, the results obtained on these two points are still debated in the litterature.

The three primary kingdoms. From the comparison of oligonucleotide catalogs Woese and Fox (1977) recognized three primary lines of evolutionary descent or primary kingdoms or urkingdoms: Eubacteria, Archaebacteria and Eukaryotes (Fig. 1). The primary kingdom Eukaryotes contains the organisms most familiar to us (i.e. multicellular organisms: animals, plants and fungi) but they occupy only a small domain within the eukaryotic line of descent where the most important branches belong to unicellular organisms. Bacteria evolution was essentially unknown until rRNA sequences were analyzed and have revealed the existence of two very divergent prokaryotic lineages: Eubacteria and Archaebacteria. The first group includes the classical Bacteria. They are splitted into about 10 phyla which have few relations with previous binary distinctions such as Gram positive - Gram negative bacteria, Cyanobacteria (blue-green algae) - other bacteria, photosynthetic - nonphotosynthetic or autotroph - heterotroph bacteria, Mycoplasmes - other bacteria and so on. Some of the

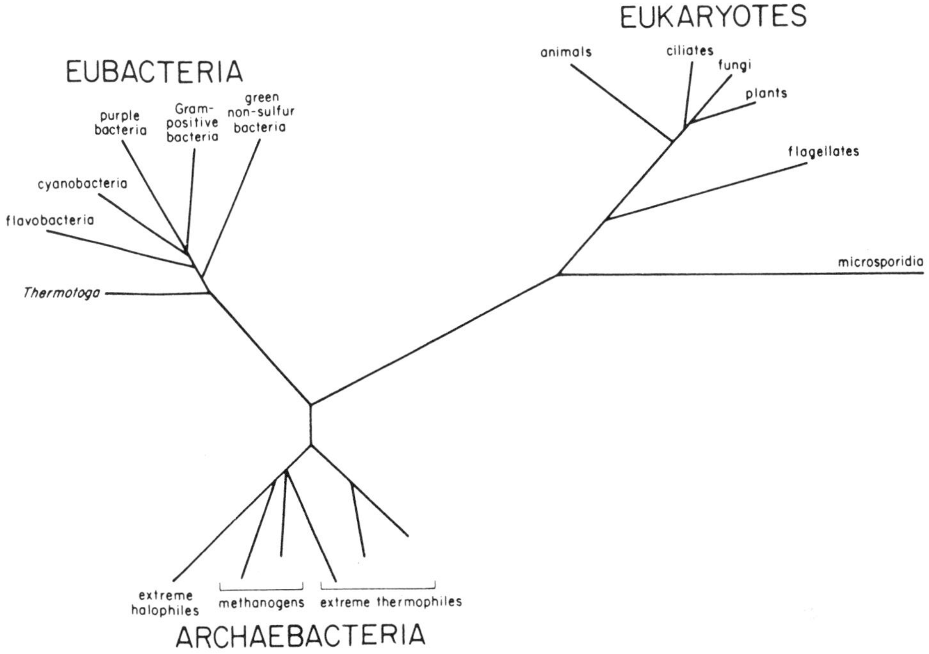

FIGURE 1. The universal unrooted phylogenetic tree showing three distinct primary kingdoms: Eubacteria, Archaebacteria and Eukaryotes (from Woese 1987a).

groups are conserved, most are not. Archaebacteria, the new primary kingdom, is comprized of three main lineages: Methanogens which produce methane from carbon dioxide; Halobacteria which thrive in hypersaline waters; Sulfobacteria (formerly thermoacidophilic) which require sulfur as an energy source and are extremely thermophile.

The Eocyte tree. The Woese and Fox basic trifurcation of the living forms has been disputed by Lake et al. (1984) who, arguing on the shape of ribosomes (Fig. 2A) found four primary groups: Eubacteria, Archaebacteria, Eukaryotes and Eocytes. Eocytes, a new appellation for the sulfur-dependant bacteria, are thought to be a lineage independant from the other Archaebacteria (i.e. Halobacteria and Methanogens).

The Lake's ribosome tree has been later supported by his analysis of 16S rRNA sequence data (Lake 1988), with a new method of clustering, the evolutionary parsimony (EP) method (Fig. 2B). The alternative topologies of the two four-branched networks (Fig. 2C) are respectively called the archaebacterial network (defended by Woese) and the Eocyte tree (defended by Lake). It has to be noticed that for a given

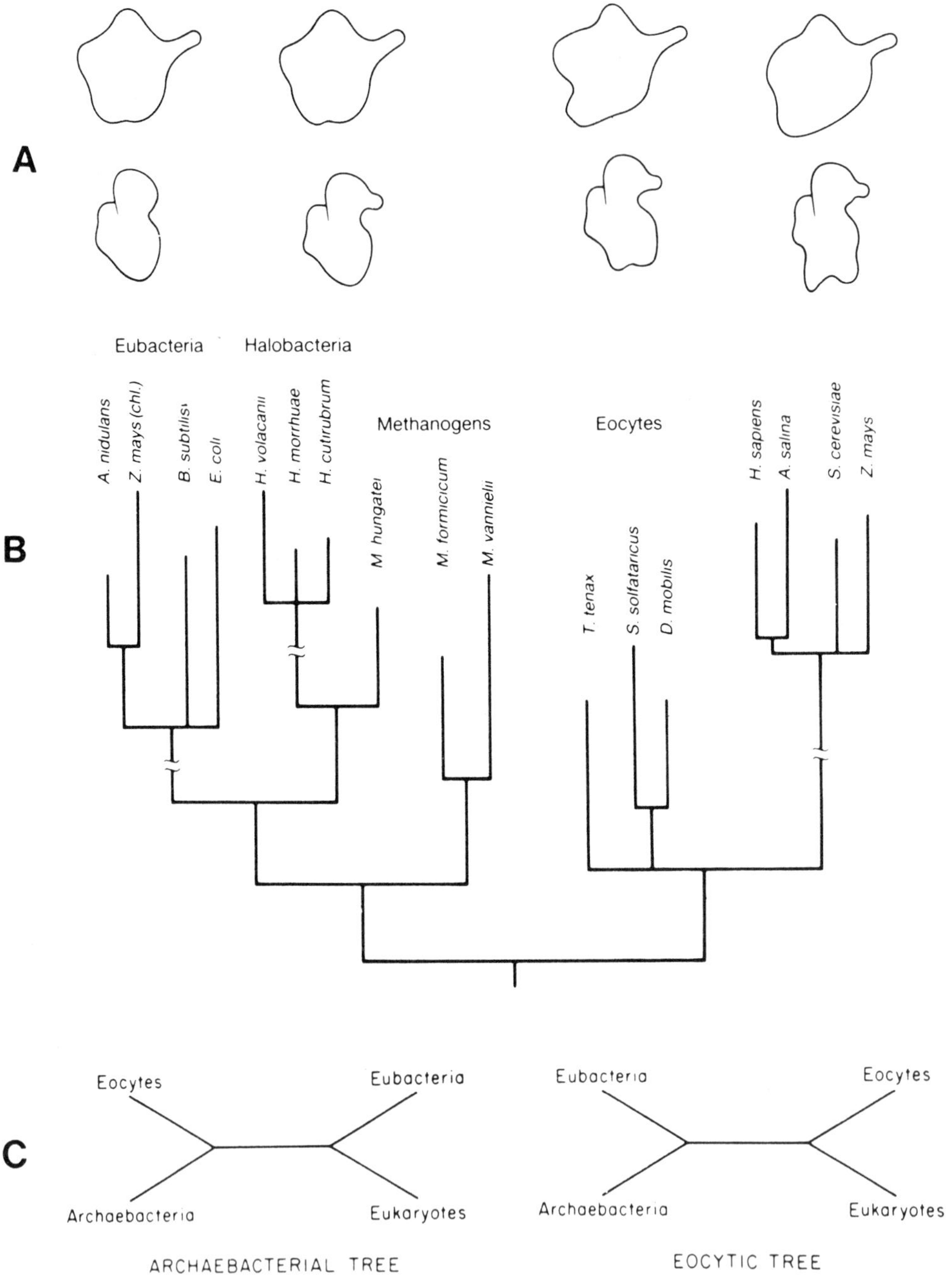

FIGURE 2. The eocyte tree. A: Shape of the large (above) and small (below) ribosomal subunits of Eubacteria, Halobacteria, Eocytes and Eukaryotes (from Lake et al. 1984). B: Rooted evolutionary tree for five groups of extant organisms: Eubacteria, Halobacteria, Methanogens, Eocytes (Sulfobacteria) and Eukaryotes (from Lake 1988). C: The archaebacterial tree and the eocyte tree, both unrooted (from Lake et al. 1984)

network, the monophyly of any group cannot be ensured (namely Archaebacteria in the Woese's network), because monophyly depends on the position of the root: if the root branches within a group, that group becomes paraphyletic. Nevertheless, whatever the position of the root, the two conflicting topologies are not congruent. Ribosomal RNA sequence data have been analyzed by others, using various methods of clustering (neighbour joining, maximum parsimony). They generally support the archaebacterial network (Cedergren et al. 1988, Gouy and Li 1989a). In fact, the Eocyte tree seems to be only favoured by the EP method applied to the SSU rRNA data.

The origin of organelles. The endosymbiotic origin of chloroplasts and mitochondria is now generally admitted. On general trees encompassing 16S-like sequences from Eubacteria, Archaebacteria, eukaryotic nucleus and organelles, the origin of organelles within defined prokaryotic groups can be observed (Fig. 3). The chloroplast 16S-like rRNA share a common ancestor with the blue-green algae *Anacystis nidulans* (Woese 1987a, Cedergren et al. 1988). The closest relation of the mitochondrial SSU rRNA sequences appears to be the alfa purple bacterium *Agrobacterium tumefasciens* (Yang et al. 1985, Gray et al. 1989). Plant mitochondrial sequences appear to be closer to *A. tumefasciens* than other mitochondrial sequences are. Mitochondrial rRNAs from other groups such as Fungi, Metazoa and Protozoa, exhibit, comparatively to plants, a "fast-clock". A striking feature is the lack of mitochondrial monophyly for the Chlorobiontes: green algae (*Chlamydomonas*) and higher plants form a single cluster for nuclear and chloroplastic rRNA but two distinct mitochondrial lineages. This could be the consequence of a very different evolutionary rate and/or of the difficulty to establish the precise relationships. However, it is suspected that the plant mitochondrial genome could have a composite origin and that their rRNA genes could originate from a more recent endosymbiotic event, subsequent to the origin of mitochondria. The propensity of plant mtDNA to integrate alien DNAs could have favored the double origin of their mitochondrial genome. However, Van de Peer et al. (1990) obtained different conclusions: only the plant mitochondria belong to the eubacteria urkingdom, whereas the other mitochondria branched off from an internal node between the three primary kingdoms.

The root of the universal tree. There is much debate about the position of the root of the universal network. It is often supposed that the eukaryotic lineage arose relatively recently, about 1.5 billion years ago, obviously from a prokaryotic ancestor. This estimate is based on the absence of earlier eukaryote-like fossils. However, in the Woese's network, the eukaryotic nuclear line seems as much ancient as the two bacterial lines. The organism ancestral to the eukaryotic nucleus could have branched off very

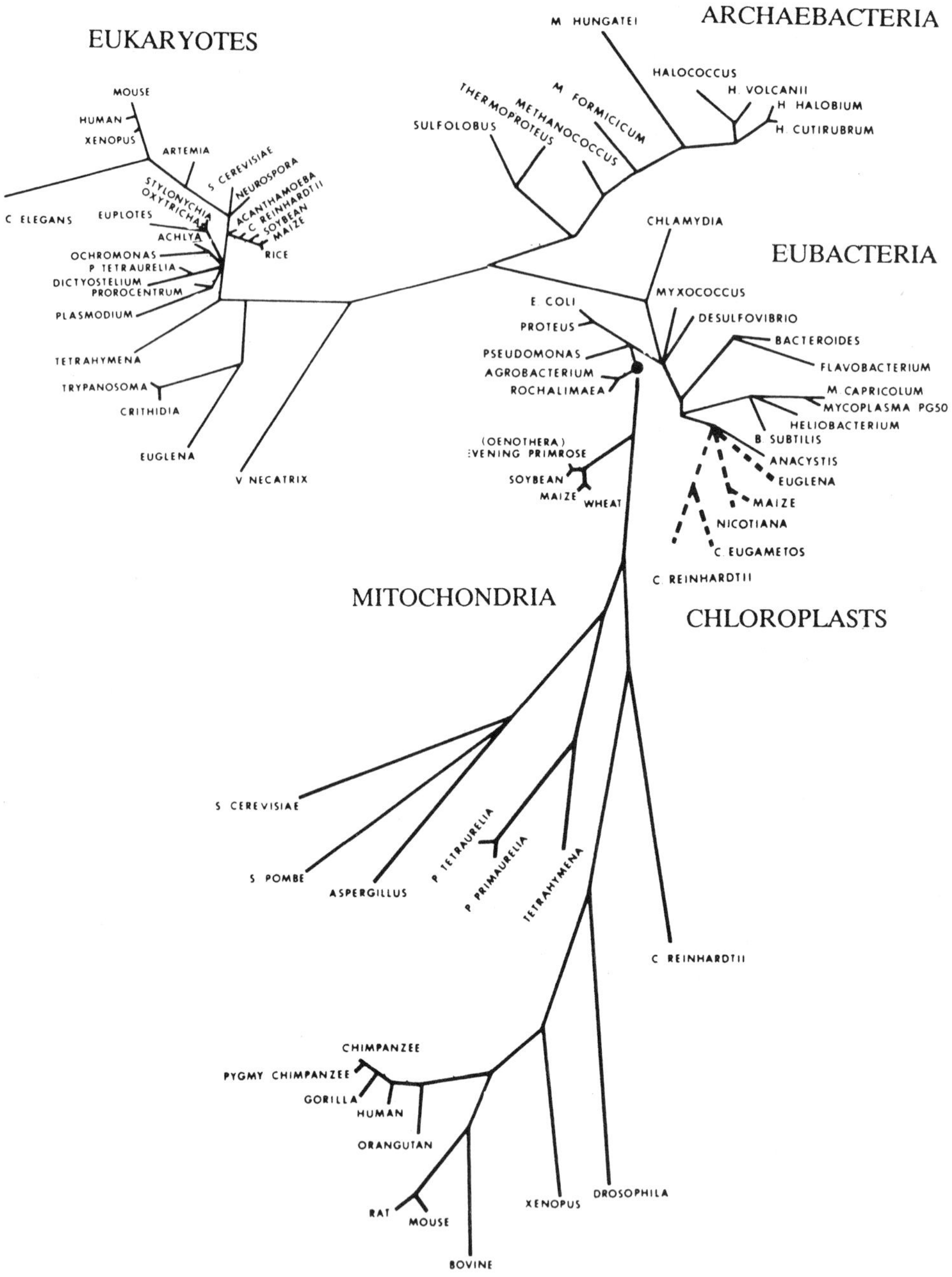

FIGURE 3. Unrooted tree from SSU rRNA genes from Archaebacteria and Eubacteria and from nucleus, mitochondria and chloroplast of Eukaryotes (from Cedergren et al. 1988).

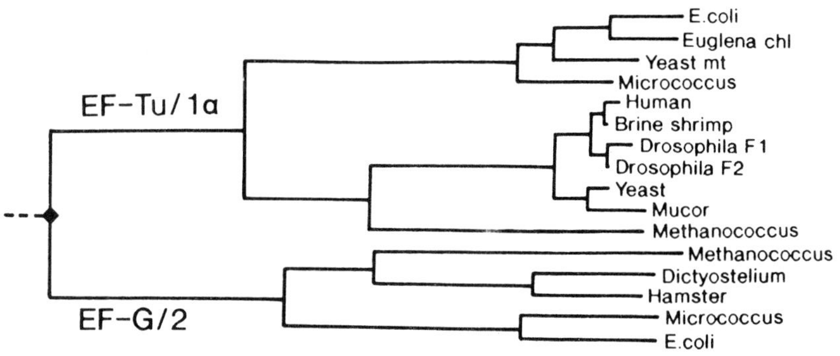

FIGURE 4. Rooting the universal phylogenetic tree by the use of a duplicated gene (elongation factor). The general tree is unrooted (in spite of the presence of the dashed line) but the event of duplication is somewhere on the branch between the two subtrees corresponding to each orthologous gene. So, this segment roots each of the subtrees; note that they have the same topology. chl: chloroplast DNA-coded gene product; mt: nuclear gene-coded mitochondrial isozyme. *Methanococcus* is an archaebacterial species. (from Iwabe et al. 1989).

early from the other lineages but would have evolved independantly as a Prokaryote for a long time. According to this view, the three lines diverged almost at the same time from a common ancestor, known as the progenote which syncretic characteristics would have been those which are common to the three urkingdoms.

Another view is that the Archaebacteria represent the urkingdom from which Eubacteria and Eukaryotes have arisen independently, the former from within the sulfur-dependant Archaebacteria (Eocytes) and the latter from the Methanogen-Halophile group. This hypothesis from Lake is not supported by other data.

The universal tree cannot be rooted, by definition, by an outgroup but recently a very clever method, based on the analysis of duplicated genes has been used. If a given pair of duplicated (paralogous) genes is present in all three primary kingdoms, that means that the duplication event has predated the first fundamental dichotomy. The duplicattion event is thus located somewhere on the branch linking the two universal trees, one for each of the orthologous genes and thus this segment roots each of them. Gogarten et al. (1989) using subunits of ATPases and Iwabe et al. (1989) using elongation factors (Fig. 4), ATPase and LDH/MDH duplications have presented the results of their analyses. In all topologies, the Archaebacteria and the Eukaryotes have a common ancestor not shared by the Eubacteria. An important taxonomical implication is that, if the results on the place of the root are confirmed, the taxon of Prokaryotae is no more valid since paraphyletic (see also Woese et al. 1990).

From the main branches to the smallest ones

The phylogenetic utility of rRNAs is not limited to the production of a universal phylogenetic tree distinguishing only the main branches of the size of primary kingdoms and or their major subdivisions. The different regions of the SSU and LSU rRNAs evolve at very different rates and it is possible to choose regions which evolutionary rate is adapted to the phylogenetic problem to be solved. The possibility of partial sequencing (using either direct sequencing or PCR amplification) is particularly useful to compare related species because most of the conserved regions (in fact the largest part of the molecule) are in this case without interest.

The sequence of SSU rRNA has revealed, within the Eukaryotes, the very high ancestry of two species of protists: the Microsporidia *Vairimorpha necatrix* (Vossbrinck, 1987) and the Diplomonade *Giardia lamblia* (Sogin et al. 1989). These two lineages had been previously distinguished from the other Eukaryotes (i.e. the Metakaryota) and classified in the subdivision Archezoa (another paraphyletic group) on the basis of some very primitive characters (Cavalier-Smith 1989). The next lineage to branch off is the taxon Euglenozoa comprising euglenoids (*Euglena gracilis*) and kinetoplastids (*Trypanosoma brucei*). Later in evolution was a nearly simultaneous splitting of animals, fungi, Chlorophytes plus plants, Chromophyte algae and ciliates plus dinoflagellates (Sogin et al. 1989, Perrasso et al. 1989). The radiation animals-plants-fungi has been carefully examined by Gouy and Li (1989b) who have presented data favouring a closer proximity between animals and plants. The study of the Metazoan phylogeny is also in progress (Field et al. 1988, Lake 1990). Several problems at lower taxonomical levels have been analysed among protist genera (Baverstock et al 1989, Preparata et al. 1989), Salamanders (Larson and Wilson 1989), Crustacea (Kim and Abele 1990) and Pentastomida (Abele et al. 1989), Fungi (Guadet et al. 1989), Diptera (Vossbrinck and Friedman 1989) and so on. In *Drosophila*, two variable domains of the 28S rRNA (about 500 nucleotides) have been sequenced in our laboratory for about 100 drosophilid species. All sequences appear to be specific and their comparison allows to propose a molecular phylogeny for the genus *Drosophila*.

It is almost incredible that the molecule which has revealed the existence of the urkingdom of Archaebacteria can also be used to distinguish very closely related species within a genus. The rRNA chronometer thus appears universal, not only in the sense it allows to build a universal tree for the most divergent organisms, but it can also yield information for a variety of situations at intermediate and lower taxonomical levels.

REFERENCES

Abele LG, Kim W, Felgenhauer BE (1989) Molecular evidence for inclusion of the phylum Pentastomida in the Crustacea. Mol Biol Evol 6:685-691

Baverstock PR, Illana S, Christy PE, Robinson BS, Johnson AM (1989) srRNA evolution and phylogenetic relationships of the genus *Naegleria* (Protista:Rhizopoda). Mol Biol Evol 6:243-257

Cavalier-Smith T (1989) Archaebacteria and Archezoa. Nature 339:100-101

Cedergren R, Gray MW, Abel Y, Sankoff D (1988) The evolutionary relationships among known life forms. J Mol Evol 28:98-112

Dayhoff MO (1976) The origin and evolution of protein superfamilies. Federation Proc 35:2132-2138

de Lanversin G, Jacq B (1989) Sequence and secondary structure of the central domain of *Drosophila* 26S rRNA: a universal model for the central domain of the large rRNA containing the region in which the central break may happen. J Mol Evol 28:403-417

Field KG, Olsen GJ, Lane DJ, Giovannoni SJ, Ghiselin MT, Raff EC, Pace NR, Raff RA (1988) Molecular phylogeny of the animal kingdom. Science 239:748-753

Fitch WM, Margoliash E (1967) Construction of phylogenetic trees. Science 155:279-284

Gerbi SA (1985) Evolution of ribosomal DNA. In McIntyre RJ (ed) Molecular Evolutionary Genetics, Plenum, New York, p 419

Gogarten JP, Kibak H, Dittrich P, Taiz L, Bowman EJ, Bowman BJ, Manolson MF, Poole RJ, Date T, Oshima T, Konishi J, Denda K, Yoshida M (1989) Evolution of the vacuolar H^+-ATPase: implications for the origin of eukaryotes. Proc Natl Acad Sci USA 86:6661-6665

Gouy M, Li W-H (1989*a*) Phylogenetic analysis based on rRNA sequences supports the archaebacterial rather than the eocyte tree. Nature 339:145-147

Gouy M, Li W-H (1989*b*) Molecular phylogeny of the kingdoms animalia, plantae and fungi. Mol Biol Evol 6:109-122

Gray MW, Cedergren R, Abel Y, Sankoff D (1989) On the evolutionary origin of plant mitochondrion and its genome. Proc Natl Acad Sci USA 86:2267-2271

Guadet J, Julien J, Lafay J-F, Brygoo Y (1989) Phylogeny of some *Fusarium* species, as determined by large-subunit rRNA sequence comparison. Mol Biol Evol 6:227-242

Gutell RR, Schnare MN, Gray MW (1990) A compilation of large subunit (23S-like) ribosomal RNA sequences presented in a secondary structure format. Nucleic Acids Res 18 [suppl]:r2319-r2330

Hendriks L, Huysmans E, Vandenberghe A, De Wachter R (1986) Primary structures of the 5S ribosomal RNAs of 11 arthropods and applicability of 5S RNA to the study of metazoan evolution. J Mol Evol 24:103-109

Hori H, Osawa S (1979) Evolutionary change in 5S RNA secondary structure and a phylogenetic tree of 54 5S RNA species. Proc Natl Acad Sci USA 76:381-385

Iwabe N, Kuma K-I, Hagesawa M, Osawa S, Miyata T (1989) Evolutionary relationship of archaebacteria, eubacteria, and aukaryotes inferred from phylogenetic trees of duplicated genes. Proc Natl Acad Sci USA 86:9355-9359

Kim W, Abele LG (1990) Molecular phylogeny of selected decapod crustaceans based on 18S rRNA nucleotide sequences. J Crust Biol 10:1-13

Lake JA (1988) Origin of the eukaryotic nucleus determined by rate-invariant analysis of rRNA sequences. Nature 331:184-186

Lake JA (1990) Origin of the Metazoa. Proc Natl Acad Sci USA 87:763-766

Lake JA, Henderson E, Oakes M, Clark MW (1984). Eocytes: a new ribosome structure indicates a kingdom with a close relationship to eukaryotes. Proc Natl Acad Sci USA 81:3786-3790

Lane DJ, Pace B, Olsen GJ, Stahl DA, Sogin ML, Pace NR (1985) Rapid determination of 16S ribosomal RNA sequences for phylogenetic analyses. Proc Natl Acad Sci USA 82:6955-6959

Larson A, Wilson A (1989) Patterns of ribosomal RNA evolution in salamanders. Mol Biol Evol 6:131-154

Neefs J-M, Van de Peer Y, Hendriks L, De Wachter R(1990) Compilation of small ribosomal subunit RNA sequences. Nucleic Acids Res 18 [suppl]:r2237-r2318

Peattie DA (1979) Direct chemical method for sequencing RNA. Proc Natl Acad Sci USA 76:1760-1764

Perrasso R, Barouin A, Qu LH, Bachellerie J-P, Adoutte A (1989) Origin of algae. Nature 339:142-144

Preparata RM, Meyer EB, Preparata FP, Simon EM, Vossbrinck CR, Nanney DL (1989) Ciliate evolution: the ribosomal phylogenies of the tetrahymenine ciliates. J Mol Evol 28:427-441

Qu LH, Michot B, Bachellerie J-P (1983) Improved methods for structure probing in large RNAs: a rapid "heterologous" sequencing approach is coupled to the direct mapping of nuclease accessible sites. Application to the 5' terminal domain of eukaryotic 28S rRNA. Nucleic Acids Res 11:5903-5919

Rothschild LJ, Ragan MA, Coleman AW, Heywood P, Gerbi SA (1986) Are rRNA sequence comparisons the Rosetta stone of phylogenetics. Cell 47:640

Raué HA, Klootwijk J, Musters W (1988) Evolutionary conservation of structure and function of high molecular weight ribosomal RNA. Prog Biophys molec Biol 51:77-129

Sogin ML, Gunderson JH, Elwood HJ, Alonso RA, Peattie DA (1989) Phylogenetic meaning of the kingdom concept: an unusual ribosomal RNA from *Giardia lamblia.* Science 243:75-77

Specht T, Wolters J, Erdman VA (1990) Compilation of 5S rRNA and 5S rRNA gene sequences. Nucleic Acids Res 18 [suppl]:r2215-r2236

Turner DH, Sugimoto N, Frier SM (1988) RNA structure prediction. Ann Rev Biophys Biophys Chem 17:167-192

Van de Peer Y, Neefs J-M, De Wachter R (1990) Small ribosomal subunit RNA sequences, evolutionary relationships among different life forms, and mitochondrial origin. J Mol Evol 30:463-476

Vossbrinck CR, Friedman S (1989) A 28S ribosomal phylogeny of certain cyclorrhaphous Diptera based upon a hypervariable region. Syst Entomol 14:417-431

Vossbrinck CR, Maddox JV, Friedman S, Debrunner-Vossbrinck BA, Woese CR (1987) Ribosomal RNA sequence suggests microsporidia are extremely ancient eukaryotes. Nature 326:411-414

Woese CR (1987a) Bacterial evolution. Microbiol Rev 51:221-271

Woese CR (1987b) Macroevolution in the microscopic world. In: Patterson C (ed) Molecules and morphology in evolution: conflict or compromise? Cambridge University Press, Cambridge, p177

Woese CR, Fox GE (1977) Phylogenetic structure of the prokaryotic domain: the primary kingdoms. Proc Natl Acad Sci USA 74:5088-5090

Woese CR, Kandler O, Wheelis ML (1990) Towards a natural system of organisms: proposal for the domains Archaea, Bacteria and Eucarya. Proc Natl Acad Sci USA 87:4576-4579

Wolters J, Erdmann VA (1986) Cladistic analysis of 5S rRNA and 16S rRNA secondary and primary structure - The evolution of eukaryotes and their relation to Archaebacteria. J Mol Evol 24:152-166

Yang D, Oyaizu Y, Oyaizu H, Olsen GJ, Woese CR (1985) Mitochondrial origins. Proc Natl Acad Sci USA 82:4443-4447

EVALUATING GENE VERSUS GENOME EVOLUTION

Robert Cedergren, Yvon Abel* and David Sankoff*
Département de biochimie
and *Centre de recherches mathématiques
Université de Montréal
Montréal, Québec
CANADA H3C 3J7

The use of DNA structural information to probe the evolutionary relationships between organisms is based on the variability of these structures with time. The explosion in the ability to extract DNA structural information from a variety of organisms has permitted inferences with respect to the type and tempo of these variations. One form of structural variation, the point mutation, is that which involves the substitution of one nucleotide for another. Point mutations, although common, are not the exclusive means by which DNA structures diverge. An important area of modern molecular biology deals exactly with these more drastic changes in DNA structure, those due to recombinatorial events including gene duplications, transpositions, deletions and insertions of genetic information. A complete description of the time-related divergence of DNA structures in various organisms must account for all forms of DNA structural variability.

The availability of gene sequence databases has been largely responsible for the development of the field of molecular evolution. An aligned homologous gene family derived from the appropriate database can be converted into a relationship diagram by any of a number of available phylogeny software packages. The topology of this sequence derived diagram is usually interpreted in terms of the phylogeny of the organisms represented by the sequences, i.e. the evolutionary history of the gene is assumed to be that of the entire genome. Strictly speaking, however, such a phylogeny represents only the

NATO ASI Series, Vol. H 57
Molecular Techniques in Taxonomy
Edited by G. M. Hewitt et al.

relationship among the sequences; this may differ from the true phylogeny of the organisms, because of the small sample size that the gene sequences represent of the genome or because of horizontal gene transfer which might have intervened since the divergence from a common ancestor. In some circumstances, then, the gene-based phylogeny is not representative of the entire genome.

Here we would like to clearly distinguish between two types of molecular evolution: gene evolution based on the analysis of micoroevolutionary events such as point mutations and short insertions/deletions and genome evolution which take into account macroevolutionary events such as transpositions, deletions, duplications and other recombination-based processes including horizontal gene transfer. We would like to be able to evaluate to what extent the identification of gene evolution and that of the organism can be verified.

Several methods could possibly be used to evaluate the relationship between gene and genome evolution. For example, one could ascertain whether phylogenetic trees obtained from a number of independent gene sequence databases have identical topologies. This strategy, however, has certain limits, since not all molecules are of equivalent value for gene evolution studies. We prefer ubiquitous, relatively long molecules which are conservative to evolutionary change in order to infer distant relationships. Ribosomal RNA has been the gene family of choice for these reasons (Gray et al., 1984; Cedergren et al., 1988), but some protein gene families can be considered as well (Iwabe et al., 1989). Another approach would be to determine phylogenies solely on the basis of genome level events, that is to say, use gene arrangement in different organisms as a probe of relatedness. These phylogenies could then be compared to gene-based phylogenies. It should be noted that the data for genome-level analysis are at a different level of abstraction then those for gene-level phylogenetic analysis. The latter is based on the local alignment of the DNA sequences, while the former involves correspondences of various types between distant regions of the genome. In fact at the present time, the complete analysis of large tracts of genome sequences is only possible with great difficulty due to the lack of appropriate techniques able to properly sort out genome level evolutionary events.

We are addressing the gene/genome evolutionary question in the following ways: 1) We are constructing gene-based phylogenies using the small subunit ribosomal RNAs, the large subunit ribosomal RNAs and tRNAs in the hopes of obtaining congruent phylogenies. and 2) We are developing quantitative criteria to evaluate the relatedness of organisms based on gene order and directionality. These latter techniques are being applied to gene mapping databases from various bacteria and eventually to mitochondria and chloroplasts.

Methods

Sequence-based phylogenies are determined using an parsimony algorithm (Cedergren et al., 1988). Most parsimonious solutions are evaluated using a bootstrapping technique and occasionally branch points are validated by the invariants method of Cavender and Felsenstein (1987) and of Lake (1987).

Genomic phylogenies are determined by a novel technique involving the pairwise comparison of the gene arrangement in different organisms. This is accomplished by an algorithm that is able to compile the number of intersections among lines drawn between homologous genes ordered on a circular genome map (Sankoff et al., 1990). The number of intersections is expressed as a fraction of the expected number of intersections between two randomly generated arrangements. In this way the relationship between organisms is represented by a number between 0 and 1; close relationships are denoted by a small fractions.

Data

One sequence database consists of an aligned compilation of known small and large subunit rRNA sequences which have been truncated to remove regions of high evolutionary divergence, the details of which have been previously published (Gray et al., 1984; Cedergren et al., 1988). The tRNA database consists of 385 sequences distributed among 27 organisms and 27 tRNA gene families (31,000 nucleotides). Where knowledge of a sequence of one of the 27 tRNA gene families in a given organism is not known, a blank sequence is entered in the database. In this way a missing sequence does not add or subtract from the mutational distance inferred among the represented organisms. tRNA sequences are aligned using the secondary structure similarity among all tRNAs, and tRNA gene families are based on the anticodon sequence of individual tRNAs (Cedergren et al., 1981). Thus there is only one $tRNA^{Phe}$ family that has a GAA anticodon, but in other cases, most notably in serine, leucine and arginine tRNAs, several families are possible. The inclusion of organisms and tRNA gene families in the database depends on the number of known tRNA sequences that could be attributed to each category. Generally, at least six sequenced tRNA sequences are required for each organism. All sequences selected from a given organism are organized end to end as if they were domains of a single gene (Cedergren and Lang, 1985).

The Normalized Gene Designation Database (NGDD) was compiled from published gene maps for five bacteria: E. coli, Salmonella typhimurium, Caulobacter crescentus, Pseudomonas aeroginosa and Bacillus subtilis (Cedergren et al., 1990). Gene designations have been normalized to a four letter code, and contradictions in nomenclature were resolved by "correcting" cases where the same gene product in one or the other bacteria have different names. In the same way, identical names corresponding to different gene products have been eliminated.

Results and Discussion

Gene-based phylogenies

The phylogenies obtained from the small and large subunit rRNA database have been previously published (Cedergren et al.,

1988). Fig. 1 shows major branch points common to both trees. Both trees fully support the monophyletic nature of the

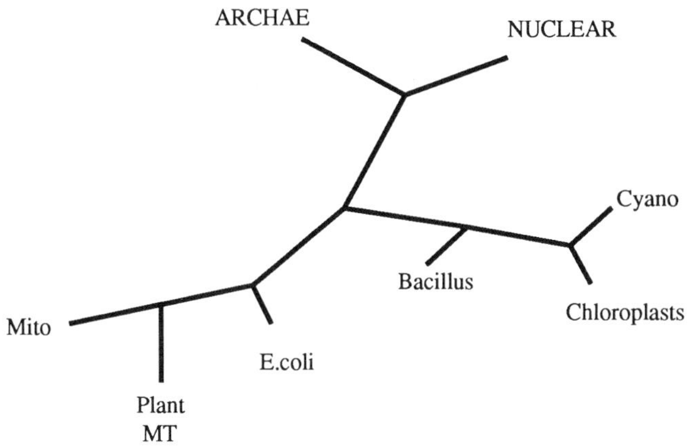

Fig. 1. Phylogenetic tree of major phyla obtained from rRNA gene sequences (Cedergren, et al., 1988).

archaebacteria, the cyanobacterial origin of chloroplasts and the -purple bacterial origin of mitochondria. The organisation of the major bacterial groups are identical in so much as data from the same organisms are available. As previously described, the plant mitochondrial rRNAs adopt a particularly curious position, especially in reference to the sequence of the Chlamydomonas, a green algal mt rRNA (Gray et al., 1989). New data not available for this analysis suggest that several other eukaryotic protists have sequences more similar to the plant mt sequence and would be expected to branch closely with these sequences at the root of the mitochondrial subtree. It is possible that the speed with which mt rRNA genes undergo change may define two mitochondrial types, those of rapid change such as those of the animal kingdom and those with more modest rates of sequence change as seen in the plant kingdom. Another interpretation would be that some protists have been improperly classified.

The use of tRNA sequence data for the determination of phylogenies is much more problematic, because they are generally considered too short to be as representative as longer genes.

Indeed, the brevity of their sequences has a variety of consequences, not the least of which is that individual tRNAs within a given family can on occasion be more closely related to a tRNA in another family than with a member of its own family (Nicoghosian et al., 1987). This is disconcerting, since if this observation has a biological base, it would mean that tRNAs could change specificity with regard to amino acid and add noise to the phylogenetic determinations that we envisage. Further credence to this hypothesis is offered by in vitro experimental results showing that on occasion as little as a 1-nucleotide change can alter the biological specificity of some tRNAs (Schulman and Abelson, 1988). On the other hand bona fide cases of an in vivo evolutionary change in specificity are difficult to find except for cases of supressor tRNAs. And in spite of many suggestions on this point, no indisputable evidence has been provided. In fact, based on the overall similarity and in particular the presence of signature sequences in all members of coherent tRNA families, we have argued that specificity change is a minor event in tRNA evolution (Nicoghosian et al., 1987). High similarity is more likely explained on the basis of the small sequence space that a tRNA can use because of its short length and the large number of constant nucleotides in the sequence.

In a study of similarity among tRNA genes of fungal mitochondria, we suggested a methodology for the possible use of tRNA sequences in phylogenetic studies (Cedergren and Lang, 1985). The methodology involved the simultaneous use of several tRNA families, a method equivalent to considering one member of all families in a given genome as a single, contiguous gene. This operation has the effect of lengthening the overall sequence and, to address the controversy referred to above, diluting out any hypothetical tRNA family switching. The strategy is thus the following: tRNAs are divided into families based on the sequence of their anticodon. In principle upwards of 40 different anticodon families are possible for a given organism based on the requirements of reading 61 codons when wobble base-pairing is invoked. In reality only some 27 families have enough members to represent a reasonable cross-section of the biological world. In like manner, many organisms are represented by so few tRNA sequences that they would add little to the precision of the phylogeny and are therefore eliminated from the compilation. The results of the selection used in our study are shown in Fig. 2. However, this selection raises another problem, that of the unrepresented tRNAS from chosen families and organisms. In these cases, we have inserted blank sequences to complete the database, and we have modified our algorithm such that missing sequences do not affect in any way the mutational cost of a given tree topology.

In all 360 real sequences (31,000 nucleotides) are present in the database which is available from the authors, and they represent 27 organisms and 27 anticodon families. This aligned database was submitted to phylogenetic analysis and the results are shown in Fig. 3. The minimal tree comprising 2174 inferred mutational events, although confirming some of the features of the rRNA derived trees fails to support the purple bacterial origin of fungal mitochondria and places Bacilli and E. coli on a common branch. Other features are more in line with other data namely the monophyletic root of both eukaryotes and archaebacteria as well as the common origin of cyanobacteria and chloroplasts (Yang et al., 1985; Pace et al., 1986; Hori and Osawa, 1987; Leffers et al., 1987; Cedergren et al., 1988).

Two novel aspects concern the origin of T4 and T5 bacteriophage tRNAs and the origin of some plant mitochondrial "chloroplast"-like sequences. The origin of the bacteriophages could not have been provided by rRNA phylogenies, since bacteriophages do not encode ribosomal RNA. Both bacteriophages are seen to share a common ancestry with mycoplasma. A group of plant mitochondrial tRNAs group very near the origin of the monocot-dicot divergence, and would confirm the chloroplast origin of these genes as has been previously postulated (Maréchal-Drouard et al., 1985; Maréchal-Drouard et al., 1990). It should be noted that we had classified this group as a special case (see Fig. 2) based on the observation that they resembled

	X PT4	X PT5	A HCU	A HVO	A MVA	P ANI	P BSU	P ECO	P MMY	P PPH	C EGR	C MAR	C PSA	C TOB	C WHT	M ALS	M SCE	M SPO	M TGL	M MAI	M MA2	M WHT	M WH2	M OEN	M OE2	E SCE	E SPO
ALA UGC	.	X	X	X	X	X	X	X	X	.	X	X	.	X	.	X	X	X	X	.	.	.	.	.	.	X	.
ARG ACG	.	.	.	.	.	.	X	X	X	.	X	X	X	X	.	.	X	.	.	.	.	.	.	.	.	X	X
ARG UCU	X	.	.	.	X	.	.	X	X	.	.	X	X	X	X	X	X	X	X	.	.	.	.	.	.	X	.
ASN GUU	.	X	X	X	X	.	X	X	X	.	X	X	X	X	.	X	X	X	X	.	.	.	X	.	.	X	.
ASP GUC	.	X	.	X	X	.	X	X	X	.	X	X	X	X	X	X	X	X	X	X	.	X	.	.	.	X	X
CYS GCA	.	.	X	X	.	.	X	X	.	.	X	X	.	X	X	X	X	X	X	.	X	.	X	.	.	X	.
GLN UUG	X	X	.	.	X	.	X	X	.	X	X	X	.	X	.	X	X	X	X	.	.	X	.	.	.	X	.
GLU UUC	.	.	.	X	X	.	X	X	X	.	X	X	X	X	X	X	X	X	X	.	.	X	.	.	.	X	X
GLY GCC	.	.	X	X	.	.	X	X	.	.	X	X	X	X	X	.	.	.	.	.	.	.	.	X	.	X	.
GLY UCC	X	X	.	X	.	.	X	X	X	.	X	X	.	X	X	X	X	X	X	.	.	.	.	.	.	.	.
HIS GUG	.	X	X	X	X	.	X	X	.	X	X	X	X	X	.	X	X	X	X	.	X	.	.	.	X	X	X
ILE GAU	.	X	.	X	.	X	X	X	X	.	X	X	.	X	.	X	X	X	X	.	.	.	.	.	.	.	.
LEU CAA	.	.	.	X	.	X	X	X	.	.	X	X	X	X	.	.	.	.	.	.	.	.	.	.	.	X	.
LEU UAA	X	.	.	X	.	.	X	.	.	.	X	X	.	X	.	X	X	X	X	.	.	.	.	.	.	.	X
LEU UAG	.	X	.	.	X	.	.	X	.	X	X	X	.	X	.	X	.	X	.	.	.	.	.	.	.	X	.
LYS UUU	.	X	.	X	X	.	X	X	.	.	X	X	X	X	.	X	X	X	X	.	.	X	.	X	.	X	.
MET CAUF	.	.	X	X	.	X	X	X	X	.	X	X	X	X	X	X	X	X	X	X	.	X	.	.	.	X	X
MET CAUM	.	X	.	X	.	.	X	X	X	X	.	X	.	X	X	X	X	X	X	X	.	.	.	.	.	X	.
PHE GAA	.	.	.	X	X	X	X	X	X	.	X	X	.	X	.	X	X	X	X	.	.	.	X	.	X	X	X
PRO UGG	X	X	.	X	X	.	X	X	X	X	X	X	X	X	X	X	X	X	X	.	.	X	.	.	.	X	.
SER GCU	.	.	.	X	.	.	X	X	.	.	X	X	.	X	X	X	X	X	X	X	.	X	.	X	.	.	.
SER GGA	.	X	.	X	.	.	X	X	.	.	.	X	.	X	.	.	.	.	.	.	.	.	X	.	X	.	.
SER UGA	X	X	.	.	.	.	X	X	X	.	X	X	X	X	.	X	X	X	X	.	.	X	.	.	.	X	X
THR UGU	X	.	.	.	X	.	X	X	X	.	X	X	.	X	.	X	X	X	X	.	.	.	.	.	.	.	.
TRP CCA	.	.	.	X	.	.	X	X	.	.	X	X	X	X	X	.	.	X	.	.	X	.	X	.	X	X	.
TYR GUA	.	.	.	X	X	.	X	X	.	.	X	X	X	X	X	X	X	X	X	.	.	X	.	X	.	X	X
VAL UAC	.	.	.	.	X	.	X	X	X	.	X	X	.	X	.	X	X	X	X	.	.	.	.	.	.	X	.

Fig. 2. tRNAs used for the phylogeny study. Symbols are the following: X : Virus, A : Archaebacteria, P : Eubacteria, C : Chloroplast, M : Mitochondria, and E : Nuclear (eukaryote). Abbreviations for organisms are the following: X PT4 : Phage T4, PT5 : Phage T5 -- A HCU : Halobacterium cutirubrum, HVO : Halobacterium volcanii, MVA : Methanococcus vannielii -- P ANI : Anacystis nidulans, BSU : Bacillus subtilis, ECO : Escherichia coli, MMY : Mycoplasma mycoid, PPH : Photobacterium phosphoreum -- C EGR : Euglena gracilis, MAR : Marchantia polymorpha (liverwort), PSA : Pisum sativum (pea), TOB : Nicotiana tabacum, WHT : Triticum aestivum -- M ALS : Aspergillus nidulans, SCE : Saccharomyces cerevisiae, SPO : Schizosaccharomyces pombe, TGL : Torulopsis glab., MAI : Zea mays, MA2 : Zea mays (chloroplast-like), WHT : Triticum aestivum, WH2 : Triticum aestivum (chloroplast-like), OEN : Oenothera berteriana (evening primrose), OE2 : Oenothera berteriana (chloroplast-like);
E SCE : Saccharomyces cerevisiae, SPO : Schizosaccharomyces pombe.

chloroplast tRNAs. We did however allow them to migrate to the part of the tree that produced the most parsimonious solution; they could have betrayed a mitochondrial origin, if indeed this were the case.

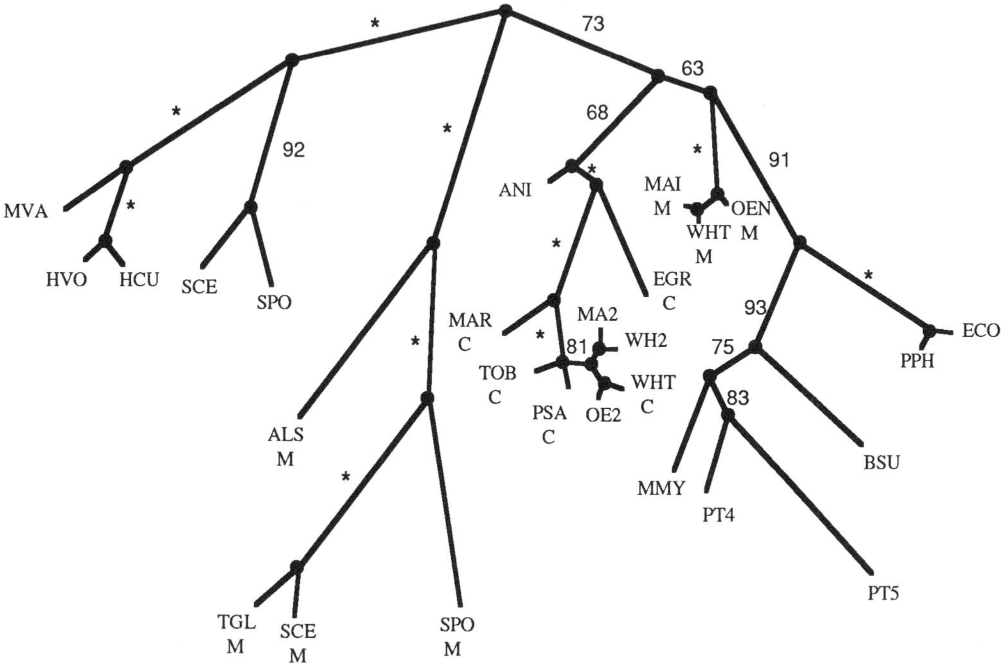

Fig. 3. Minimum phylogenetic tree from tRNA sequences. The tRNA sequences from Fig. 2 were analyzed by a maximum parsimony algorithm. Organisms are abbreviated as in Fig. 2. The branch lengths are proportional to mutational distance and numbers above branches refer to the percentage of bootstrap trees which confirm that branch. *, is used to represent over 95%. Branches which are not confirmed to more than 50% are collapsed to the next nearest node.

Since our phylogenetic analysis system permits the predetermination of all or part of the tree structure, we are able to evaluate alternate, sub-optimal topologies. In such an experiment with the tRNA data, we have placed the Bacillus branch (with Mycoplasma) on the principal bacterial root and put the mitochondria under the constraint that they are monophyletic. With these two constraints, the minimized topology of Fig. 4 was obtained. The resulting tree is now congruent with the rRNA tree. Although this arrangement is not the global minimum for the data, it is a local minimum having the mutational cost only 1% higher than the minimal tree in Fig. 3. It is of considerable interest that this new tree places the bacteriophages at the origin of the bacterial root. This arrangement is consistent with an ancient origin of bacteriophages and suggests that bacteria and bacteriophages have co-evolved. The characterization of bacteriophages using different bacterial hosts will be an essential step in the verification of this hypothesis.

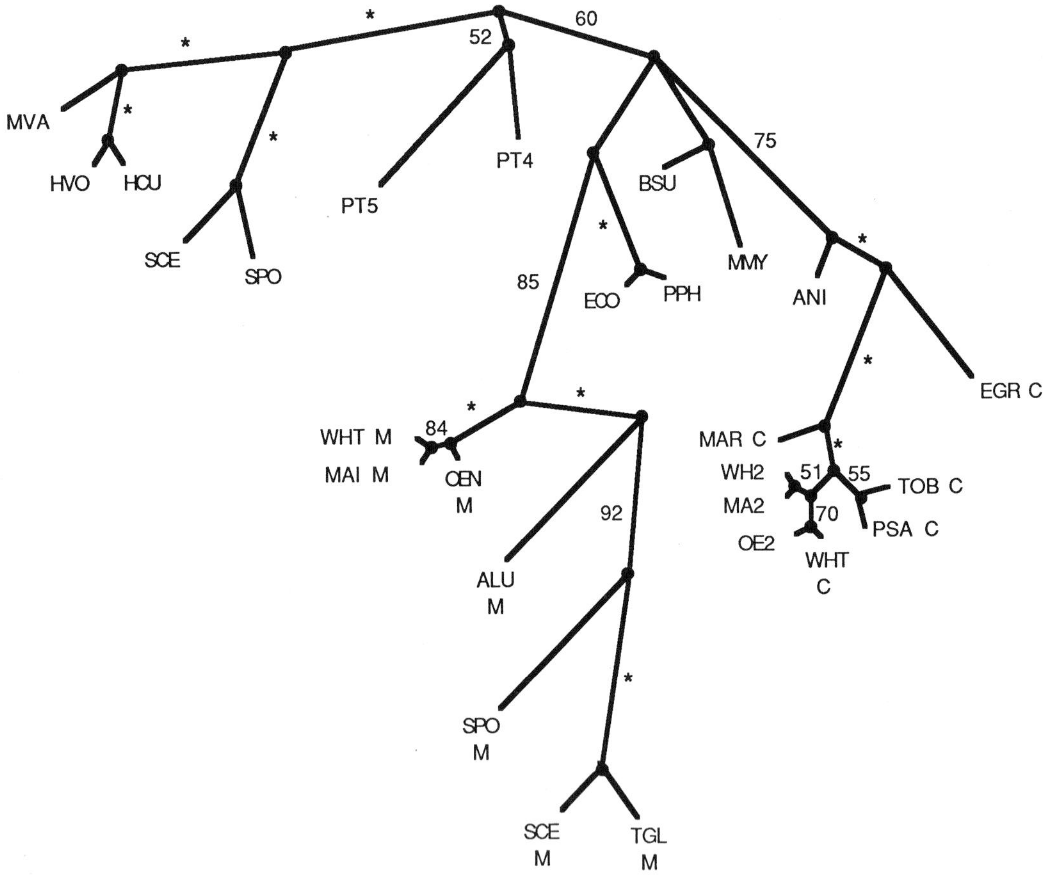

Fig. 4. Modified phylogenetic tree of tRNA sequences. This tree was obtained under the constraints described in the paper. Branch lengths are proportional to mutational distance and numbers above branches refer to the percentage of bootstrap trees which confirm that branch.

Genome-based phylogenies

Evaluation of the role of macroevolutionary events involving the rearrangement of genomes was made possible by the implementation of an algorithm for assessing the overall difference in gene order between two genomes. In order to accomplish this task for the bacterial genomes, a normalized databank of gene markers was necessary so that all genes would be represented in a consistent manner. The use of the Normalized Gene Designation Database (NGDD, Cedergren et al., 1990) allowed the determination of the relationship between the five bacteria shown in Fig. 5. This arrangement is consistent with that expected from both gene sequence data and more classical determinants. The degree of relationship calculated as the fraction of improper gene alignments between two species divided by that expected between randomly ordered genomes show that all comparisons except those involving Bacillus is significant.

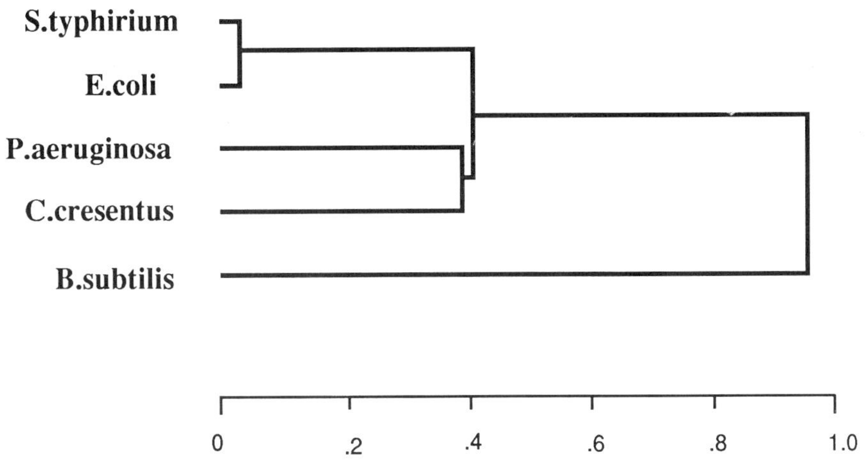

Fig. 5. Tree diagram of relationships among five bacteria as determined by gene order.

The expected availability of genome length sequences in the near future requires the implementation of new techniques for the analysis of genome level evolutionary events. Based on the fragmentary data presented here, phylogenetic trees of gene sequences correspond with that based on genome order. It remains to be seen whether this pattern will be reproduced on more extensive data sets. Future work will be directed to the use of other gene order databases such as those for mitochondria and chloroplasts which we are currently establishing to further evaluate the importance of gene order in evolution.

Acknowledgements

This work was supported by a grant from NSERC of Canada. Robert Cedergren and David Sankoff are Fellows of the Canadian Institute for Advanced Research. The authors thank Jing-hua Yang for the preparation of the figures and Professor Mathias Sprinzl and Michael Gray for providing tRNA sequence databases from which ours was constructed.

References

Cavender JA, Felsenstein J (1987) Invariants of phylogenies: simple case with discrete states. J Classif 4:57-71

Cedergren R, LaRue B, Sankoff D, Grosjean H (1981) The evolving tRNA Molecule. CRC Crit Rev Biochem 11:35-104

Cedergren R, Lang BF (1985) Probing fungal mitochondrial evolution with tRNA. BioSyst 18:263-267

Cedergren R, Gray MW, Abel Y, Sankoff D (1988) The evolutionary relationships among known life forms. J Mol Evol 28:98-112

Cedergren R, Abel Y, Sankoff D (1990) Normalized Gene Designation Database. Nature 345:484

Gray MW, Sankoff D, Cedergren R (1984) On the evolutionary descent of organisms and organelles. Nucl Acids Res 12:5837-5852

Gray MW, Cedergren R, Abel Y, Sankoff D (1989) On the evolutionary origin of the plant mitochondrion and its genome. Proc Natl Acad Sci (USA) 86:2267-2271

Hori H, Osawa S (1987) Origin and evolution of organisms as deduced from 5s ribosomal RNA sequences. Mol Biol Evol 3:445-472

Iwabe N, Kuma K, Hasegawa M, Osawa S, Miyata T (1989) Evolutionary relationship of archaebacteria, eubacteria and eukaryotes inferred from phylogenetic trees of duplicated genes. Proc Natl Acad Sci (USA) 86:9355-9359

Lake JA (1987) A rate-independent technique for analysis of nucleic acid sequences: evolutionary parsimony. Mol Biol Evol 4:167-191

Leffers H, Kjems J, Ostergaard L, Larsen N, Garrett RA (1987) Evolutionary relationships amongst archaebacteria. A comparative study of a sulphur-dependent extreme thermophile, an extreme halophile and a thermophilic methanogen. J Mol Biol 195:43-61
Maréchal-Drouard L, Guillemaut P, Grienenberger J-M, Jeannin G, Weil J-H (1985) Sequence and codon recognition of bean mitochondria and chloroplast tRNAsTrp; evidence for a high degree of homology. Nucl Acids Res 13:4411-4416
Maréchal-Drouard L, Guillemaut P, Cosset A, Arbogast M, Weber F, Weil J-H, Dietrich A (1990) Transfer RNAs of potato mitochondria have different genetic origins. Nucl Acids Res 18:3689-3696
Nicoghosian K, Bigras M, Sankoff D, Cedergren R (1987) Archetypical features of tRNA families. J Mol Evol 26: 341-346
Pace NR, Olsen GJ, Woese CR (1986) Ribosomal RNA phylogeny and the primary lines of evolutionary descent. Cell 45:325-326
Sankoff D, Cedergren R, Abel D (1990) Genomic Divergence through gene rearrangement. Methods in Enzymol 183:428-438
Schulman LH and Abelson J (1988) Recent excitement in understanding transfer RNA identity. Science 240:1591-1592
Yang D, Oyaizu Y, Olsen GJ, Woese CR (1985) Mitochondrial origins. Proc Natl Acad Sci (USA) 82:4443-4447

DNA AND HIGHER PLANT SYSTEMATICS: SOME EXAMPLES FROM THE LEGUMES

Jeff J. Doyle & Jane L. Doyle
L. H. Bailey Hortorium
Cornell University
Ithaca, NY 14850
USA

Introduction

In the last few years, the DNA revolution has begun to have a dramatic impact on the field of plant systematics. Access to the various plant genomes--chloroplast, mitochondrial, and nuclear--has provided the systematist with a virtually inexhaustible source of characters for phylogenetic analysis. The interaction of DNA technology and cladistic analysis has been particularly powerful: the tools for producing empirical data relevant to phylogenetic relationship have been complemented by a rigorous theoretical framework on which to build explicit hypotheses of homology and phylogeny. This potent combination has led to a rapid acceptance of molecular approaches in "mainstream" plant systematics, which has been manifested in a large number of publications and papers at plant systematics meetings (at least in the USA) involving DNA phylogenies. Not all of the three genomes have received equal attention, however, nor have the tools that are currently most widely used been shown to be useful at all taxonomic levels. Furthermore, the role of polymorphism, of confidence in phylogenies, and in general of the difference between gene trees and species trees is only slowly becoming appreciated among plant systematists and clearly should have some impact on the development of the field.

In this paper, we will draw from the work in our laboratory on the flowering plant family Leguminosae to illustrate the utility of a variety of DNA approaches for addressing systematic questions in plants. The legumes are considered the third largest family of flowering plants, behind only the Asteraceae (daisy family) and the Orchidaceae (orchids) in number of species (Polhill, 1981). It is a diverse family, whose members range from annual alpine herbs to massive trees of the rainforest canopy. Its largest subfamily, the Papilionoideae, includes numerous economically important members, such as pea (*Pisum*), broad bean (*Vicia*), alfalfa (*Medicago*), clovers (*Trifolium*), soybean (*Glycine*), common beans (*Phaseolus*), mung bean and cowpea (*Vigna*), and lupine (*Lupinus*). Many of these plants have also figured prominently in studies of gene structure and regulation. The relationships among the over 400 genera and larger groupings (subtribes and tribes) of the subfamily Papilionoideae have received much attention in recent years, and hypotheses of higher level phylogenetic relationship have been propounded based on a diversity of taxonomic characters, from morphology and anatomy to the chemistry of secondary metabolites (Polhill, 1981). For all of these reasons the family is an attractive group on which to conduct molecular phylogenetic studies.

NATO ASI Series, Vol. H 57
Molecular Techniques in Taxonomy
Edited by G. M. Hewitt et al.

The mitochondrial genome

The plant mitochondrial genome bears little structural resemblance to the compact mitochondrial genome of most animals, despite sharing most of the same small complement of protein encoding genes. Flowering plant mitochondrial genomes range in size from around 200 kilobase pairs (kb) to over 2000 kb, one to two orders of magnitude larger than those of vertebrate animals (reviewed in Palmer 1985). Most of this genome does not appear to encode protein or RNA products; repeated sequences occur but not in high copy number, and a large percentage of the genome appears to be composed of promiscuous DNA, presumably non-functional sequences with functional counterparts in the nucleus or chloroplast. The mitochondrial genome is highly recombinagenic, and a single genome such as that of *Zea* can be composed of a master circle and several subgenomic circular molecules, as well as linear or circular plasmids. Despite this apparently rapid structural evolution, the sequences of mitochondrial genes appear to evolve slowly (Wolfe et al., 1987), in contrast to the mitochondrial genes of vertebrate animals. This slow evolutionary rate is also reflected in conservation of most restriction sites between species within the genus *Brassica*, despite large differences in their maps caused by recombination mediated by repeated sequences (Palmer and Herbon 1988). Because the mitochondrial genomes of related species can be compared only through laborious and precise restriction mapping or by sequencing of highly conserved genes, there has been little use of this compartment as a source of phylogenetic information at any level in plants, including legumes, though some use has been made of the highly variable restriction fragment patterns as a means of assessing germplasm diversity in such legume genera as soybean (Grabau et al. 1989).

The nuclear genome

Though not as poorly represented as the mitochondrial genome, relatively few studies have been conducted in which nuclear DNA variation has played a predominant role. Furthermore, despite the fact that the nucleus has by far the largest genome, few classes of sequences have even been assayed as a source of phylogenetic markers. Like other plant genomes, those of legumes contain large amounts of repeated sequences, within which are interspersed low copy number genes encoding protein products (reviewed by Walbot and Goldberg 1979).

The nuclear multigene families encoding ribosomal RNAs are representatives of the highly-repeated to intermediately-repeated classes of nuclear sequences, and have provided characters at several taxonomic levels. This is particularly true of the family encoding the 18S, 5.8S, and 25-28S ribosomal RNAs (rDNA; reviewed for legumes by Jorgensen and Cluster 1988), which is formed of up to several thousand nearly identical units composed of highly conserved coding regions and more labile intergenic spacers. Within these units, the RNA coding regions have proven useful for studies involving long evolutionary divergences. For example, Zimmer and colleagues have attempted to use sequence variation in these RNAs to elucidate the relationships of flowering plants to "gymnosperms", and to investigate relationships within angiosperm families (Zimmer et al. 1989).

While the RNAs or the genes encoding them provide characters for broad phylogenetic comparisons, the spacers that separate the gene-bearing units are far more variable over shorter evolutionary time scales, down to the level of individuals within populations (reviewed for legumes by Doyle 1987 and Jorgensen and Cluster 1988). Restriction fragment length polymorphism (RFLP) studies, mainly involving changes in the large intergenic spacer, have found use in studies within genera and species of legumes, such as, for example, *Pisum* (Ellis et al. 1984) and *Glycine* (Doyle and Beachy 1985). In *Glycine*, rDNA variation has been particularly useful as a source of markers for studying allopolyploid evolution (see below); in this capacity, a second nuclear ribosomal gene family, that encoding the much smaller 5S rRNAs, also has been used (Doyle and Brown 1989).

Although in at least some plants the low copy fraction of the plant genome is comprised of truly "single copy" sequences (Bernatzky and Tanksley 1986), most of the protein coding genes thus far studied belong to small or medium sized multigene families. Examples of these that have received study and include legume sequences are *rbc*S (small subunit of ribulose bisphosphate carboxylase; Meagher et al. 1989), *cab* (chlorophyll a/b binding protein; Demmin et al. 1989), and actins (Hightower and Meagher 1985). That these and other low copy genes have yet to find wide use in addressing phylogenetic questions in higher plants is in part due to the difficulty of determining which genes among the various members of multigene families across taxa are appropriate for organismal phylogenetic comparison. Only orthologous genes (Fitch 1971) provide the opportunity for such legitimate comparison, as these are the genes in which divergence is traced back to a speciation event and not to a more ancient gene duplication event, as is true for paralogous genes. Use of a mixture of paralogous and orthologous genes will result in incorrect organismal phylogenies.

Orthology is best determined by sequencing all of the members of a particular multigene family from each of the various taxa under consideration and constructing a full gene tree. At this point the organismal phylogeny should emerge from each of the sets of orthologues. Needless to say, this laborious method is not likely to be attractive to those whose primary interest in sequence data is the construction of organismal phylogenies. Orthology may be hypothesized *a priori* using such evidence as structural features (e.g., intron number and position), by chromosome map location, or by functional and/or regulatory specificity. However, any type of evidence may be misleading, or may only apply to some members of a given family.

Of the possible clues, regulatory information is the most readily available, as the construction of cDNA libraries from mRNA in individual tissues is common practice. The small multigene family encoding flowering plant glutamine synthetase (GS) is a case in point. Different isoforms of this enzyme, which plays a major role in plant nitrogen metabolism, are most prevalent in chloroplasts, the cytosol of various organs, or the nodules in which symbiotic nitrogen fixation takes place. cDNA sequences from these various compartments are available from a number of legumes and other flowering plants, and cladistic analyses of these sequences have been conducted in an effort to test whether the site of expression (perhaps synonymous with functional differentiation) is an accurate predictor of orthology (Doyle submitted). The most robust results of various analyses on the full data set or subsets, summarized in Fig. 1, separate the chloroplast sequences from all of the cytosolic sequences, indicating that there are, broadly, two groups of orthologous genes. This is not surprising, particularly for the chloroplast genes, since not only is their pattern of regulation quite distinct from that of the cytosolic sequences, but they have distinctive structural features (3' extensions, 5' sequences encoding putative transit peptides) as well.

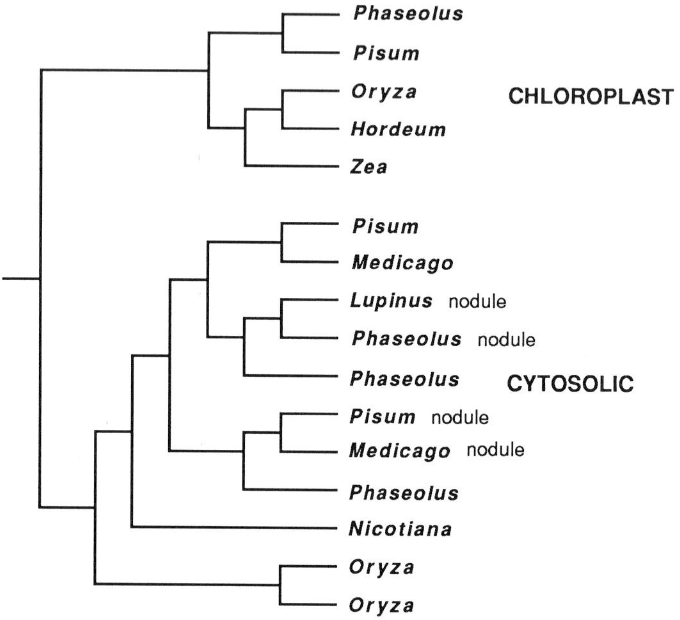

Fig. 1. Summary of cladistic relationships among flowering plant glutamine synthetase genes, based on analyses of all codon positions and substitution classes or data subsets in which third codon positions, first and second codon positions, and/or transitions were deleted. Within the cytosolic clade, genes encoding GS isoforms expressed predominantly in the nodule are indicated. Characters were polarized and the tree rooted using a rat GS sequence.

Within the cytosolic sequences, however, orthologous relationships, though apparent at the sequence level as revealed by good support for several clades, do not agree with regulatory pattern. The four sequences from mRNAs expressed predominantly in nodules do not form a clade but rather belong to two different groups of genes (Fig. 1). Biologically, and indeed evolutionarily, this is an interesting result, as it bears on the evolution of gene regulation; however, it suggests that caution must be exercised in using functional or regulatory properties of genes within multigene families as predictors of orthology.

In other multigene families, concerted evolution produces gene families in which all of the members are more similar in sequence to one another than to corresponding genes in other taxa, resulting in a loss of the distinction between orthologous and paralogous copies. Though perhaps the classic examples of this phenomenon are the nuclear ribosomal genes (e.g., Jorgensen and Cluster 1988), concerted evolution is observed in genes of the lower copy fraction of the genome as well. Analyses of *rbc*S genes using both distance measures (Meagher et al. 1989) and cladistic methods (Doyle and Meagher unpublished data) show a clustering of genes within genera as would be expected for concerted evolution of this family. The glycosylated seed storage proteins (vicilins) of legumes also appear to evolve in a concerted fashion; for example, the vicilin type genes of *Glycine* and *Phaseolus* are all more similar within these two taxa (Doyle et al. 1986), as is also shown by cladistic analysis of vicilin proteins (Gibbs et al. 1989). However, when gene sequences for a larger number of taxa are compared cladistically, it is apparent that although concerted evolution may act effectively enough to homogenize the vicilin families of these two genera, it appears to have had less effect on the genes of two other legumes, *Vicia* and *Pisum* (Fig. 2; Doyle unpublished data). The single gene thus far available from *Vicia* is nested within the several *Pisum* genes, and appears to be orthologous to one of these. Furthermore, as the cladogram of Gibbs et al. (1989) indicates, concerted evolution is much less prevalent in the other major seed storage protein multigene family of legumes, the non-glycosylated glycinin family.

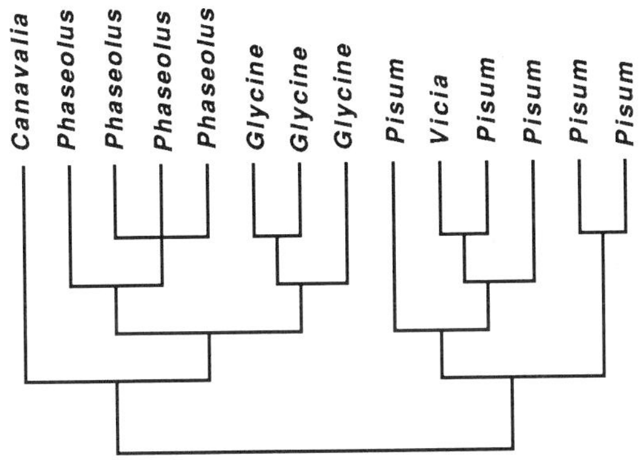

Fig. 2. Cladistic relationships among genes encoding glycosylated seed storage proteins of legumes. A storage protein gene from cotton was used as outgroup for these analyses. The *Pisum* genes that form a clade with the single *Vicia* gene belong to the vicilin class, while the remaining two *Pisum* sequences encode convicilin proteins.

This apparent variation in the efficiency of concerted evolution within gene families and between families whose members presumably experience the same types of evolutionary constraints suggests that it is dangerous to assume that the question of orthology can be ignored in families in which concerted evolution is thought to be the rule. In addition to this problem, the process of concerted evolution may itself result in a loss of cladistic information. The primary mechanisms of concerted evolution, such as unequal crossover and gene conversion, involve recombinational mechanisms; recombination in gene trees is analogous to hybridization in organismal phylogeny, and both result in increased homoplasy.

These considerations may help explain why some genes do not appear to contain much phylogenetic information (Bremer 1988; Archie 1989), and why phylogenetic analyses of some gene families produce such unexpected groupings. In studies of *rbc*S, for example, the two legumes, *Pisum* and *Glycine*, do not group together in either distance (Meagher et al. 1989) or cladistic analyses (Doyle and Meagher unpublished data). In any case, there are too few nuclear protein-coding gene sequences available, representing only a small number of taxa, in order to be able to address phylogenetic questions in any detail. This is particularly true given that it is clear from both practical and theoretical considerations that both character-based parsimony approaches and tree building methods using distances produce topologies that are dependent on the taxa included in the analysis (discussed for cladistic analysis by Donoghue et al. 1989; example from distance data in Wolfe et al. 1989). Furthermore, there is an increasing emphasis on assessing the confidence in phylogenies estimated from molecular data (Felsenstein 1988), and it simply is not clear at this stage that the majority of nuclear gene sequences will yield data of sufficient quality for phylogenetic analysis.

Apart from gene sequences, other types of information are likely to be useful for studies of phylogeny in the legumes and other plants. Gene duplications have provided evolutionary markers in groups such as *Clarkia* of the Onagraceae (reviewed by Gottlieb 1988); a duplication of the gene encoding cytosolic glucosephosphate isomerase (GPI) appears to have occurred early in the history of the legumes, with subsequent silencing of one of the resulting pair in a variety of distinct lineages (Weeden et al. 1989). Other low copy sequences, many of unknown function, are the focus of numerous chromosome mapping studies in legume genera that should eventually provide information on evolution at the level of the entire nuclear genome. Comparisons of the linkage maps of *Pisum*, *Lens* (lentil), and *Cicer* (chickpea), for example, show varying levels of linkage conservation among these related genera (Weeden and Wolko 1988; Weeden personal communication). RFLP variation, which is the source of the data for construction of linkage maps, is also of potential use as a source of phylogenetic characters, though there are significant problems in assessing the homology of fragments detected with the small (generally less than 1 kb) probes used in such studies. Early studies of this type have suggested relationships among perennial *Glycine* species (Menacio et al. 1990).

The chloroplast genome

The chloroplast genome has thus far been the greatest source of phylogenetic information for flowering plant molecular systematists. This genome has many properties that make it particularly useful in this regard (reviewed in Palmer et al. 1988a), including a nearly constant size (around 150 kb), gene content, and gene arrangement among flowering plants, conservative rate of evolution, and high copy number of chloroplast genomes per cell. Nearly all land plant chloroplast genomes contain two copies of an approximately 20 kb segment bearing, among other things, the 16S and 23S ribosomal RNA genes. These two copies are separated from one another by two single copy regions of unequal size (large single copy [LSC] and small single copy [SSC] regions) and are inverted in orientation relative to one another. The two copies of this inverted repeat (IR) region evolve in complete concert with one another, and at a slower rate than the single copy regions. Apart from sequences making up the IR, which occur in two copies, repeated sequences are relatively rare in flowering plant chloroplast genomes, though in some cases such as sub-clover (*Trifolium subterraneum*) dispersed repeats are present and appear to mediate rearrangement (Milligan et al. 1989). Because all known chloroplast genes occur as either single copies or as two tightly concerted copies, the problem of establishing orthology is minimized.

Because the gene content and order of flowering plant chloroplast genomes are highly conserved, rare gene losses or structural rearrangements provide characters whose homology is presumably easy to determine. The legumes are particularly rich in structural rearrangements. The gene *rpl*22 has been lost from all legume chloroplast genomes thus far studied (Palmer and Doyle, unpublished data, cited in Palmer et al. 1988a), and may eventually serve as a character for elucidating the relationships of the family to other Rosidae. Similarly, a 50 kb inversion in the LSC region appears to occur in legume genera representing all three subfamilies (Palmer et al. 1988b). A second inversion, of 78 kb, has been used as a marker at the subtribal level in the tribe Phaseoleae (Palmer et al. 1988b; Bruneau et al. 1990).

The absence of the IR from the chloroplast DNA (cpDNA) of *Pisum* was one of the earliest observed cpDNA structural mutations (Kolodner and Tewari 1979; Palmer and Thompson 1981). Mapping of additional legume cpDNAs indicated that this character was largely confined to genera belonging to tribes that had long been considered related (Palmer et al. 1987). A simple method of testing for RFLP phenotypes consistent with the presence or absence of the IR was developed as a means of rapidly screening large numbers of plants for this character (Fig. 3).

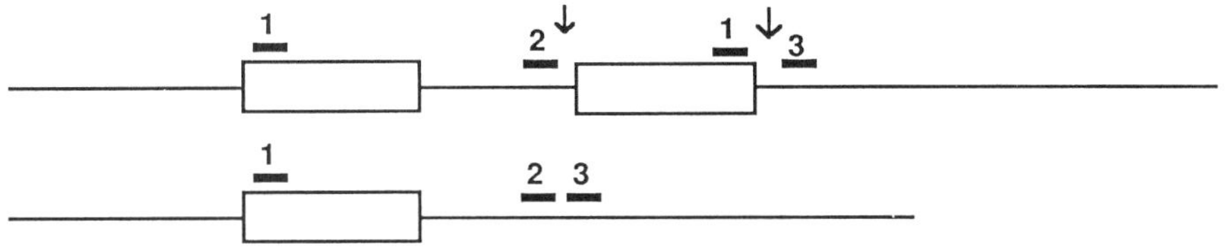

Fig. 3. Screening strategy for detecting presence/absence of the chloroplast inverted repeat. In this diagram, the circular chloroplast genome is shown as a linear molecule. In the top diagram both copies of the inverted repeat are present (shown as open boxes), while in the lower diagram one copy has been deleted at the sites indicated by arrows. The presence of the IR is determined by use of probe 1, representing a small (ca. 1 kb) region of the inverted repeat near the boundary with the large single copy region. In plants that possess the inverted repeat, probe 1 occurs twice, and will generally detect two different restriction fragments; only one fragment will be detected in plants that have suffered the IR deletion. The loss of the IR is confirmed by probing digests sequentially with probes 2 and 3, which flank the breakpoints of the deletion. In plants possessing both copies of the IR, these probes are ca. 20 kb apart, and generally detect two different restriction fragments; when one copy is deleted, the two probes are adjacent and will in most cases detect the same fragment.

A survey of 95 species from 77 genera, and representing 25 of the 31 tribes of papilionoid legumes indicated that the IR loss can be considered to be a character of unique origin uniting six temperate, largely herbaceous tribes (Lavin et al. 1990). The presence of the IR in two tribes (Loteae and Coronilleae) that have been considered related to this temperate herbaceous group has led to a reconsideration of morphological and chemical characters of these two tribes, which suggests that they may indeed have their closest phylogenetic affinities outside this group. *Wisteria* also lacks the IR (Palmer et al. 1987), and was the sole example of IR loss outside the temperate herbaceous group; a member of the tropical woody tribe Millettieae, certain morphological and chemical features of *Wisteria* suggest that it may be transitional between this basal tribe and the temperate herbaceous group (Lavin et al. 1990).

While structural mutations have been shown to be useful in the legumes at the subtribal level and above, each represents only a single character per chloroplast genome. Sequencing or restriction mapping potentially can provide numerous characters, and thus more detailed resolution of taxonomic questions at a diversity of levels. Of these two approaches, restriction mapping has been by far the more widely used, and is likely to retain a prominent place in the molecular systematist's toolkit, even as polymerase chain reaction and sequencing methods become more refined and accessible. This is because RFLP mapping studies on cpDNA are simple and straightforward, relatively inexpensive, and yield data which are readily amenable to phylogenetic analysis. Cladistic analyses of restriction site data from moderate numbers of species within genera have in many cases produced one or a few competing trees with low levels of homoplasy (reviewed by Palmer et al. 1988a). Moreover, due to the variation in evolutionary rates among regions of the chloroplast genome, it is possible to obtain useful characters at a diversity of taxonomic levels from the same genomes, simply by using probes for these different regions.

Legumes were among the first plants for which cpDNA restriction maps were produced, and early comparisons were made among the chloroplast genomes of *Vigna*, *Phaseolus*, and *Glycine* by Palmer et al. (1983). A legume, *Pisum*, was also among the first genera in which phylogenetic relationships were studied at and below the species level using cpDNA RFLP methods (Palmer et al. 1985). Our studies on cpDNA variation in several legume groups illustrate both the advantages and potential pitfalls of this approach.

To date it has been found generally that RFLP variation is most useful at the level of species within a genus. In *Glycine* subgenus *Glycine* (the wild perennial relatives of soybean), over 150 restriction site gains or losses were observed among the dozen species when surveyed with 29 different restriction enzymes and probes spanning the entire chloroplast genome (Doyle et al. 1990a). Cladistic analysis produced two equally parsimonious trees with only a very low level of homoplasy (< 10%). The strict consensus tree from these two trees (Fig. 4) showed three major clades (plastome groups A, B, and C) within the subgenus, each of which was supported by a large number of restriction site characters on the shortest trees.

Moreover, these three clades are largely congruent with groupings hypothesized on the basis of morphology and crossing relationships (reviewed in Doyle et al. 1990a). A subset of these characters was used recently to determine the cpDNA affinities of newly discovered species in the subgenus (Doyle et al. 1990b), establishing cpDNA variation as a useful resource for classifying germplasm in this genus.

Resolution within clades is limited, however, and the original study utilized a relatively small number of accessions per species. Although it has been suggested that cpDNA is not polymorphic in plant populations, a more careful reading of the literature shows that few workers have looked for it, and that where adequate sampling has been carried out polymorphism is the rule, rather than the exception (Birky, 1988); this is certainly true in *Glycine* (Fig. 4). The effect on phylogeny reconstruction of polymorphisms shared across species boundaries has received theoretical study and is known to be a major potential source of phylogenetic error (Neigel and Avise 1986; Pamilo and Nei 1987).

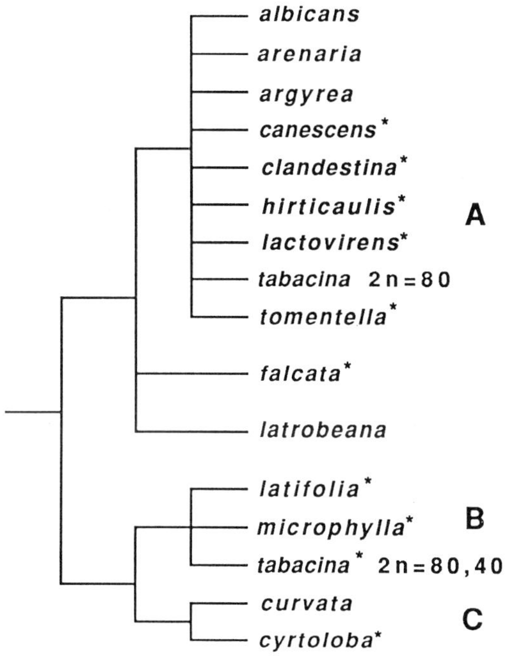

Fig. 4. Strict consensus tree for chloroplast genomes of *Glycine* subgenus *Glycine*, based on cladistic analysis of 159 restriction site characters. Plastome groups are designated by capital letters (A-C). The two shortest trees summarized here differed only in the placement of *G. falcata* and *G. latrobeana* relative to the remaining A plastome species. Asterisks indicate species in which chloroplast DNA polymorphisms have been encountered. *Glycine tomentella* includes both diploid and polyploid cytotypes; *G. tabacina* polyploids occur in both the A and B plastome groups, while diploids of this species possess only B type plastomes. All other species are exclusively diploid ($2n = 40$), except *G. hirticaulis*, which is polyploid. Cladograms were polarized using *G. soja*, of the annual subgenus, as outgroup.

For these two reasons we studied over 70 accessions (populations) representing the diploid species comprising the B genome, which also correspond to the B plastome group (Doyle et al. 1990c). Increasing the number of accessions, combined with using additional frequent cutting restriction enzymes not used in the initial survey, resulted in finding numerous additional restriction site characters. The survey also showed that all of the taxa studied were polymorphic for cpDNA. Cladograms constructed from these data showed an increase in homoplasy (to around 15%) relative to the original survey; more importantly, several instances were found in which cpDNA groupings were not consistent with taxonomic placement, which was based largely on morphology. In some cases morphologically and isozymically similar accessions were grouped in completely different regions of the cpDNA tree, while plastome clades included morphologically divergent accessions (Fig. 5).

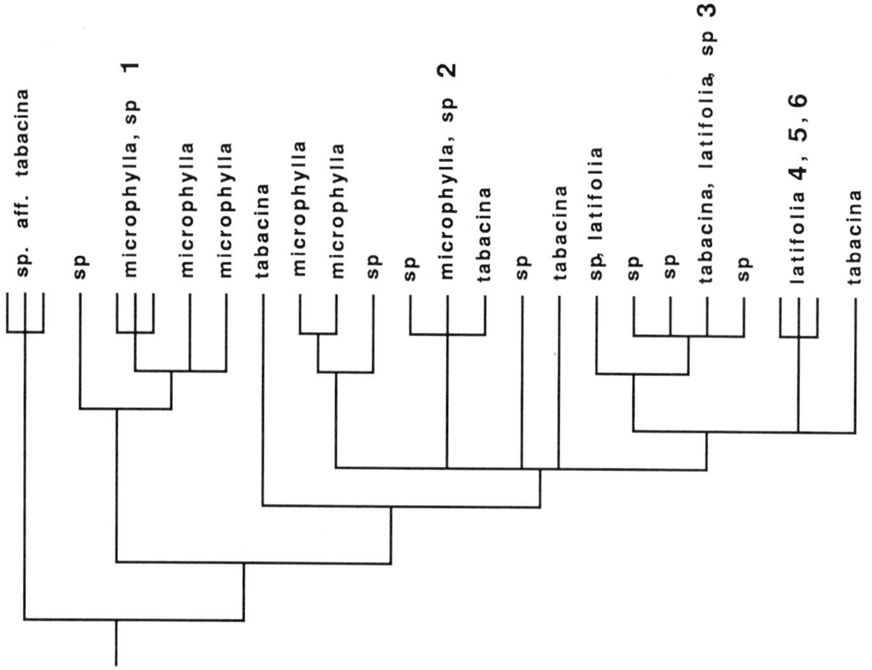

Fig. 5. Cladistic relationships among 74 B plastome accessions representing the currently recognized diploid species *G. latifolia*, *G. microphylla*, and *G. tabacina*, a distinctive but presently not formally recognized taxon, *G.* sp. aff. *tabacina*, and a heterogeneous group of morphologically distinct B nuclear genome diploids indicated here by "sp". The tree shown is a strict consensus tree from the large number of equal length trees generated when the entire data set, with numerous missing characters due to sampling strategy, was used. Numbers next to six plastomes indicate that these types were also found in polyploid accessions of *G. tabacina*.

It is likely that the disagreement between the cpDNA tree and other types of data is due to the recency and incompleteness of divergence of some of these taxa, resulting both in the maintenance of ancestral cpDNA polymorphisms and incomplete barriers to gene flow (Doyle et al. 1990c). These results indicate that polymorphism and its consequences cannot safely be ignored if cpDNA variation is to be used at lower taxonomic levels, just as is the case with animal mtDNA.

The cpDNA characters differentiating the major plastome groups have been used in concert with nuclear rDNA variation to study one of the two large polyploid complexes in subgenus *Glycine*, the *G. tabacina* complex. Two major classes of *G. tabacina* tetraploids ($2n = 80$) are intersterile and morphologically distinct, and are thought to share only one of their diploid progenitor genomes (Singh et al. 1987). Representatives of both types were included in the original cpDNA survey, which showed that one belonged to the A plastome group while the second belonged to the B group (Doyle et al. 1990a; Fig. 4). Subsequently, a screening of 91 accessions, many of which had been used in artificial hybridization studies, was conducted using a small number of cpDNA characters diagnostic for the A or B plastome types. Results showed that all accessions of one polyploid race, which had been hypothesized based on crossing studies to have a nuclear genome constitution AAB_2B_2, possessed the A-type plastome, while the second, a BBB_2B_2 genome, had a B-type plastome (Doyle et al. 1990e). Studies of the nuclear ribosomal genes showed that both types were fixed hybrids for rDNA repeat classes found in diploid *Glycine* species, and shared one repeat type. These results supported the hypotheses of nuclear genome origins, and were consistent with cpDNA results in predicting diploid progenitors (Doyle et al. 1990e, 1990f).

The BBB_2B_2 polyploid was studied further, using the cpDNA characters discovered among the polymorphic B plastome diploids mentioned above. Although the initial expectation of these studies was that a single diploid plastome donor would be identified, it was found instead that several different plastomes of diploid accessions occurred in the polyploids (Doyle et al. 1990c, 1990f; Fig. 5). The most likely explanation is that the polyploids originated several times through hybridizations involving different diploid accessions. Similar results have been reported recently for other polyploid flowering plants, also using cpDNA variation (Soltis and Soltis 1989; Soltis et al. 1989)--further evidence of the potential for polymorphism in this genome.

At higher taxonomic levels, the utility of the restriction site approach is expected to be diminished, in large part because multiple site mutations in restriction fragments makes establishing homology of restriction sites virtually impossible without detailed mapping. However, this expectation is based on studies using herbaceous plant taxa, and our studies and those of others have shown that many woody plants exhibit much lower levels of cpDNA divergence within and between genera. This makes the RFLP approach potentially feasible at higher taxonomic levels for groups such as the many woody legume tribes, particularly if only the more slowly evolving inverted repeat region is used and detailed double digest restriction maps are prepared. This strategy has been used successfully in a study that included a number of genera from two woody papilionoid tribes, Robinieae and Millettieae (Lavin and Doyle 1990). Many restriction sites in the IR region were found to be shared among all of these woody taxa (Fig. 6), and two major map types were found that were congruent with tribal classification.

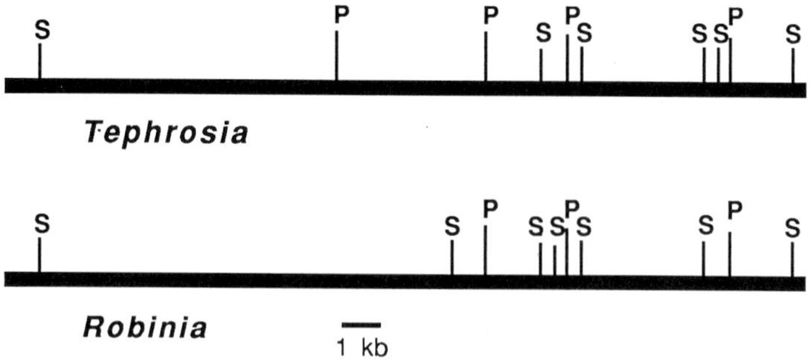

Fig. 6. Restriction maps of *Tephrosia* and *Robinia* for the enzymes *Pvu*II (P) and *Sac*I (S) in the chloroplast inverted repeat region. Note the similarity of the two maps in these representatives of two different woody tribes.

The IR regions of Millettieae were all quite similar to one another; greater divergence was found among Robinieae, with two subgroups supported in cladistic analyses that corresponded to subtribes identified on the basis of morphological characters. These results also provided evidence bearing on the tribal affinities of certain problematic genera. For example, *Sphinctospermum* possesses a chloroplast genome that has its closest relatives among Robninieae subtribe Robiniinae, and is quite divergent from members of Millettieae, where it has also been placed.

Chloroplast DNA sequences also provide a potentially important source of phylogenetic data. As mentioned above, most chloroplast genes occur as single copies, and thus using these genes in phylogeny reconstruction avoids the obvious problems of identifying orthologous genes. The chloroplast gene encoding the large subunit of ribulose bisphosphate carboxylase (*rbc*L) is being sequenced in many laboratories, though as yet comparatively few of these studies have been published (e.g. references in Palmer et al. 1988a) and it is thus difficult to assess the quality of the data. An extensive data base should become available for this gene within the next few years, one which will also include data from extinct plants, given the recent publication of an *rbc*L sequence from fossil *Magnolia* (Golenberg et al. 1990).

Conclusions

Analysis of variation at the genome level avoids many of the long-recognized problems of using phenotypic variation to assess "relationship". In this sense DNA systematics would appear to represent the final stage in a quest that has led from morphological characters controlled by many genes, to secondary metabolites that seemed less distant from the genotype, to proteins encoded by single genes, and now finally to the gene itself. Having arrived at this ultimate stage, however, systematists must now consider the forces that shape the patterns of variation at this level, and must recognize that DNA sequences are by no means free of homoplasy. The alignment of sequences for analysis represents, in a single step, a global homology decision whose impact is to correlate characters across the entire data set. Molecular polymorphism and its consequences on species phylogeny confront the systematist with situations in which high confidence in gene tree topologies is totally misleading as a measure of organismal phylogeny. Similarly, orthology and paralogy represent special cases of homology that must be considered for phylogeny reconstruction. Although choice of characters would appear to be more straightforward for DNA sequences than for morphological characters, decisions about whether to use or exclude third codons or transitions may have a dramatic effect on the topologies obtained from analyses.

Expectations that DNA sequence variation will provide a phylogenetic panacea are at best naively harmless, but at worst represent a threat to the field of systematics, by siphoning interest and attention away from important issues concerning morphological and developmental characters. Perhaps the greatest impact of the molecular revolution on plant systematics will be the integration of molecular developmental studies with phylogenetic theory, making possible a far more precise definition of homology than has been available to date for phenotypic characters. Continued development of the relationship between molecular biology and phylogenetic theory promises not only to furnish systematists with hypotheses of evolutionary pattern but could provide a character-based method for studying the evolution of gene sequences that could potentially be far superior to the current distance methods. This type of synergism will only be possible if systematics can maintain its traditionally broad multidisciplinary perspective on evolution and phylogeny.

Meanwhile, molecular data are indeed proving their utility for phylogeny reconstruction, and in the process often are confirming the superb taxonomic insights of earlier, non-molecular workers. In the legumes, the major structural mutations of the chloroplast genome described above have supported the monophyly of major groups of genera that in many cases have been considered related on the basis of morphological characters since Bentham's work in the 1860's. Relationships among species of subgenus *Glycine* are largely congruent, whether assessed using artificial hybridization, morphology, isozymes, or various classes of DNA variation. These results, taken with similar findings in other plant groups, show how absurd a "molecules vs. morphology" attitude truly is.

Acknowledgements

We wish to acknowledge the many people who have made contributions to the work mentioned here, including Tony Brown, Jeff Palmer, Norm Weeden, Anne Bruneau, Dan Potter, Matt Lavin, and Michael Sanderson. Research was supported by grants NSF BSR-8516630 and BSR-8805630 to JJD.

Literature Cited

Archie JW (1989) Phylogenies of plant families: a demonstration of phylogenetic randomness in DNA sequence data derived from proteins. Evolution 43:1796-1800

Banks JA, Birky CW Jr. (1985) Chloroplast DNA diversity is low in a wild plant, *Lupinus texensis*. Proc Natl Acad Sci USA 82:6950-6954

Bernatzky R, Tanksley SD (1986) Majority of random cDNA clones correspond to single loci in the tomato genome. Mol Gen Genet 203:8-14

Birky CW Jr. (1988) Evolution and variation in plant chloroplast and mitochondrial genomes. In: Gottlieb LD, Jain SK (eds) Plant Evolutionary Biology. Chapman and Hall, New York, pp 23-53

Bremer K (1988) The limits of amino acid sequence data in Angiosperm phylogenetic reconstruction. Evolution 42:795-803

Bruneau A, Doyle JJ Palmer JD (1990) A chloroplast DNA inversion as a subtribal character in the Phaseoleae (Leguminosae). Syst Bot 15:378-386

Demmin DS, Stockinger EJ, Chang YC, Walling LL (1989) Phylogenetic relationships between the chlorophyll a/b binding protein (*CAB*) multigene family: an intra- and interspecies study. J Mol Evol 29:266-279.

Donoghue MJ, Doyle JA, Gauthier JA, Kluge AG, Rowe T (1989) The importance of fossils in phylogeny reconstruction. Annu Rev Ecol Syst 20:431-460.

Doyle JJ (1987) Variation at the DNA level: uses and potential in legume systematics. In: Stirton CH (ed) Advances in Legume Systematics, Part 3. Royal Botanic Gardens, Kew, pp 1-30

Doyle JJ, Beachy RN (1985) Ribosomal gene variation in soybean (*Glycine*) and its relatives. Theor Appl Genet 70:369-376.

Doyle JJ, Brown AHD (1989) 5S nuclear ribosomal gene variation in the *Glycine tomentella* polyploid complex (Leguminosae). Syst Bot 14:398-407

Doyle JJ, Doyle JL, Brown AHD (1990a) A chloroplast DNA phylogeny of the wild perennial relatives of the soybean (*Glycine* subgenus *Glycine*): congruence with morphological and crossing groups. Evolution 44:371-389

Doyle JJ, Doyle JL, Brown AHD (1990b) Chloroplast DNA phylogenetic affinities of newly discovered species in *Glycine* (Leguminosae: Phaseoleae). Syst Bot 15:466-471

Doyle JJ, Doyle JL, Brown AHD (1990c) Chloroplast DNA polymorphism and phylogeny in the B genome of *Glycine* subgenus *Glycine* (Leguminosae). Amer J Bot (in press)

Doyle JJ, Doyle JL, Brown AHD, Grace JP (1990d) Multiple origins of polyploids in the *Glycine tabacina* complex inferred from chloroplast DNA polymorphism. Proc Natl Acad Sci USA 87:714-717

Doyle JJ, Doyle JL, Grace JP , Brown AHD (1990e) Reproductively isolated polyploid races of *Glycine tabacina* (Leguminosae) had different chloroplast genome donors. Syst Bot 15:173-181

Doyle JJ, Doyle JL (1990) Analysis of a polyploid complex in *Glycine* with chloroplast and nuclear DNA. Austral J. Syst Bot (in press)

Doyle JJ, Schuler MA, Godette WD, Zenger V, Beachy RN, Slightom J (1986) The glycosylated seed storage proteins of *Glycine max* and *Phaseolus vulgaris*. J Biol Chem 261:9228-9238

Ellis THN, Davies DR, Castleton JA, Bedford ID (1984) The organization and genetics of rDNA length variants in peas. Chromosoma 91:74-81.

Felsenstein J (1988) Phylogenies from molecular sequences: inference and reliability. Annu Rev Genet 22:521-565.

Fitch WM (1970) Distinguishing homologous from analogous proteins. Syst Zool 19:99-113

Gibbs PE, Strongin KB , McPherson A (1989) Evolution of legume seed storage proteins--a domain common to legumins and vicilins is duplicated in vicilins. Mol Biol Evol 6:614-623

Golenberg EM, Giannasi DE, Clegg MT, Smiley CJ, Durbin M, Henderson D, Zurawski G (1990) Chloroplast DNA sequence from a Miocene *Magnolia* species. Nature 344:656-658.

Gottlieb LD (1988) Towards molecular genetics in *Clarkia*: gene duplications and molecular characterization of PGI genes. Ann Missouri Bot Gard 75:1169-1179

Grabau EA, Davis WH , Gengenbach BG (1989) Restriction fragment length polymorphism in a subclass of the 'Mandarin' soybean cytoplasm. Crop Sci 29:1554-1559

Hightower RC, Meagher RB (1985) Divergence and differential expression of soybean actin genes. EMBO J 4:1-8.

Jorgensen RA, Cluster PD (1988) Modes and tempos in the evolution of nuclear ribosomal DNA: new characters for evolutionary studies and new markers for genetic and population studies. Ann Missouri Bot Gard 75:1238-1247

Kolodner R, Tewari KK (1979) Inverted repeats in chloroplast DNA from higher plants. Proc Natl Acad Sci USA 76:41-45

Lavin M, Doyle JJ, Palmer JD (1990) Evolutionary significance of the loss of the chloroplast DNA inverted repeat in the Leguminosae subfamily Papilionoideae. Evolution 44:390-402

Lavin M, Doyle JJ (1990) Tribal relationships of *Sphinctospermum* (Leguminosae): integration of traditional and chloroplast DNA data. Syst Bot (in press)

Meagher RB, Berry-Lowe S, Rice K (1989) Molecular evolution of the small subunit of ribulose bisphosphate carboxylase: nucleotide substitution and gene conversion. Genetics 123:845-863

Menacio DI, Hepburn AG, Hymowitz T (1989) Restriction fragment length polymorphism (RFLP) of wild perennial relatives of soybean. Theor Appl Genet 79:235-240

Milligan BG, Hampton JN , Palmer JD (1989) Dispersed repeats and structural reorganization in subclover chloroplast DNA. Mol Biol Evol 6:355-368

Neigel JE, Avise JC (1986) Phylogenetic relationships of mitochondrial DNA under various demographic models of speciation. In: Karlin S, Nevo E (eds) Evolutionary Processes and Theory. Academic Press, New York, pp 515-534

Palmer JD (1985) Evolution of chloroplast and mitochondrial DNA in plants and algae. In: MacIntyre RJ (ed) Molecular Evolutionary Genetics. Plenum, New York, pp. 131-240

Palmer JD, Herbon LA (1988) Plant mitochondrial DNA evolves rapidly in structure but slowly in sequence. J Mol Evol 28:87-97

Palmer JD, Jansen RK, Michaels HJ, Chase MW, Manhart JR (1988a) Chloroplast DNA variation and plant phylogeny. Ann Missouri Bot Gard 75:1180-1206

Palmer JD, Jorgensen RA, Thompson WF (1985) Chloroplast DNA variation and evolution in *Pisum*: patterns of change and phylogenetic analysis. Genetics 109:195-213

Palmer JD, Osorio B, Thompson WF (1988b). Evolutionary significance of inversions in legume chloroplast DNAs Curr Genet 14:65-74

Palmer JD, Osorio B, Aldrich J, Thompson WF (1987) Chloroplast DNA evolution among legumes: loss of a large inverted repeat occurred prior to other sequence rearrangements. Curr Genet 11:275-286

Palmer JD, Singh GP, Pillay DTN (1983) Structure and sequence evolution of three legume chloroplast DNAs. Mol Gen Genet 190:13-19

Palmer JD, Thompson WF (1981) Rearrangements in the chloroplast genomes of mung bean and pea. Proc Natl Acad Sci USA 78:5533-5537

Pamilo P, Nei M (1988) Relationships between gene trees and species trees. Mol Biol Evol 5:568-583

Polhill RM (1981) Papilionoideae. In: Polhill RM, Raven PH (eds) Advances in Legume Systematics, Part 1. Royal Botanic Gardens, Kew, pp 191-208

Singh RJ, Kollipara KP, Hymowitz T (1987) Polyploid complexes of *Glycine tabacina* (Labill.) Benth. and *G. tomentella* Hayata revealed by cytogenetic analysis. Genome 29:490-497

Soltis DE, Soltis PS (1989) Allopolyploid speciation in *Tragopogon*: insights from chloroplast DNA. Amer J Bot 76:1114-1118

Soltis DE, Soltis PS, Ness B (1989) Chloroplast DNA variation and multiple origins of autopolyploidy in *Heuchera micrantha* (Saxifragaceae). Evolution 43:650-656

Walbot V, Goldberg RB (1979) Plant genome organization and its relationship to classical plant genetics. In: Hall TC, Davies JW (eds) Nucleic Acids in Plants. CRC Press, Boca Raton, pp 3-40.

Weeden NF, Doyle JJ , Lavin M (1989) Distribution and evolution of a glucosephosphate isomerase duplication in the Leguminosae *Evolution* 43:1637-1651

Weeden NF , Wolko B (1990) Linkage map for the garden pea. In: O'Brien S (ed) Genetic Maps. Cold Spring Harbor, New York, pp 6106-6112

Wolfe KH, Gouy M, Yang Y-W, Sharp PM, Li W-H, (1989) Dates of the monocot-dicot divergence estimated from chloroplast DNA sequence data. Proc Natl Acad Sci USA 86:6201-6205.

Wolfe KH, Li W-H, Sharp PM (1987) Rates of nucleotide substitution vary greatly among plant mitochondrial, chloroplast and nuclear DNAs. Proc. Natl. Acad. Sci. USA 84, 9054-9058.

Zimmer EA, Hamby RK, Arnold ML, LeBlanc DA, Theriot EC (1989) Ribosomal RNA phylogenies and flowering plant evolution. In: Fernholm B, Bremer K, Jornvall H (eds) The Hierarchy of Life. Elsevier, Amsterdam, pp 205-214.

DNA-DNA HYBRIDIZATION: PRINCIPLES AND RESULTS

Jeffrey R. Powell
Department of Biology
Yale University
P.O. Box 6666
New Haven, CT 06511 USA

Adalgisa Caccone
Dipartimento di Biologia
II Universita di Roma 'Tor Vergata'
00173 Roma, ITALY

The purpose of this chapter is to present some principles of DNA-DNA hybridization especially as used in phylogenetic reconstruction. We will also present some results of general interest for workers using or intending to use this technique. In particular, we will discuss primarily the so-called TEACL method of performing hybridization experiments. Elsewhere in this volume we give in some detail the protocols we have developed in our laboratory (Caccone and Powell, this volume). We will not present details of our own work as this can be found in Caccone et al. (1987, 1988a), Caccone and Powell (1987, 1989, 1990), Powell and Caccone (1989, 1990) and Goddard et al. (1990). Another good source of information on principles and practical application of DNA-DNA hybridization is Werman et al. (1990).

Gene trees *versus* species trees

A critical issue in correctly reconstructing phylogeny from molecular data is the "gene tree *versus* species tree" problem. The problem is at least two-fold. One is the issue of paralogous and orthologous gene comparisons. It is quite clear that a large number of coding sequences in eukaryotic genomes are represented more than once, i.e. multi-gene families are quite common. J. Doyle, in these proceedings, has addressed the possible problems in deducing phylogenetic relationships when more than one copy of a gene (or closely related sequence) is present in a genome.

A related problem occurs due to polymorphism; the paper of Pamilo and Nei (1988) is particularly seminal in this regard. Figure 1, modified from Pamilo and Nei, illustrates the issue.

NATO ASI Series, Vol. H 57
Molecular Techniques in Taxonomy
Edited by G. M. Hewitt et al.

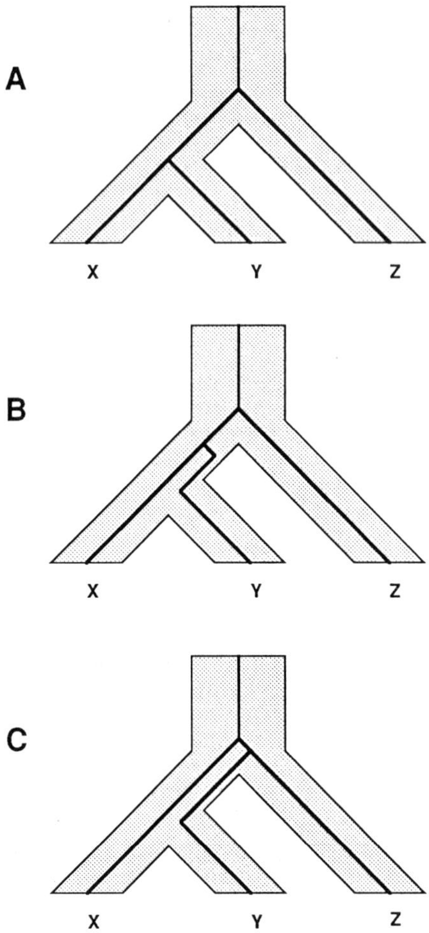

Figure 1. Illustration of "gene trees" *versus* "species trees". The lines represent alleles within lineages while the inclusive shaded areas represent the species or populations. In case A the species tree and gene tree coincide both in terms of topology and times of divergence. In case B they coincide with respect to topology but not in time of divergence. In case C, the gene tree and species tree have different topologies. Modified from Pamillo and Nei (1988).

Sometimes the study of a single allele from a single gene from each taxon being compared may correctly reflect the phylogenetic history of the species (Figure 1a). Because of the ubiquity of polymorphisms within species (especially on the nucleotide level), a single allele from each taxon may not reflect the phylogeny of the species. Figure 1b indicates a situation where the branching pattern of the gene and species is the same, but the timing of the splits are very different. Figure 1c indicates a case where even the branching patterns of the gene and species trees do not coincide. Estimates of the probability of gene trees not accurately reflecting species trees depend upon several variables such as times between splits, levels of polymorphism, etc.

To quote from Pamilo and Nei (1988) the probability of obtaining the correct species tree "cannot be substantially increased by increasing the number of alleles sampled from a locus. To increase the probability, one has to use DNA sequences from many different loci that have evolved independently of each other."

DNA-DNA Hybridization

This last point in the above paragraph describes precisely the power of DNA-DNA hybridization in deducing correct phylogenetic relationships. The major advantage of the technique is that it measures in a single experiment the degree of nucleotide divergence over thousands of gene loci and millions of nucleotide sites. It is highly unlikely that an average divergence over so much of the genome would not correctly reflect the genetic divergence between taxa which would, in turn, reflect the relative time of divergence.

This is not to imply that DNA-DNA hybridization is the ultimate tool for phylogenetics. As with all techniques there are advantages and disadvantages. Of most interest is a comparison of results from DNA-DNA hybridization with DNA sequence data. Table 1 summarizes our view of the relative merits of each source of data.

In this table, we use the term "biological error" to indicate that the genomes of the taxa may contain misleading evidence due to the problems just mentioned: the problem of gene trees not reflecting species trees due to polymorphisms and multi-gene families (orthologous *versus* paralogous comparisons). Because of the amount of labor involved, for most taxa there will only be a very limited number of genes (often only one) for which the DNA sequence has been determined and thus these problems may be serious. DNA-DNA hybridization overcomes these problems to a large extent. On the other hand, there is inevitably an error associated with measurements of the thermal stability of DNA duplexes (the "experimental error" in Table 1). When done correctly (see Kreitman, this volume) DNA sequencing should be virtually free of experimental error, i.e. the exact sequence of a gene (or single allele) is known.

Table 1. A comparison of DNA-DNA hybridization and DNA sequencing for use in phylogenetic studies. See text for details.

	DNA-DNA Hybridization	Sequencing
Type of Error		
Biological	virtually zero	possible
Experimental	present	virtually zero
Type of Data		
Distances	yes	yes
Character state changes	no	yes

Analytical procedures also vary. Obviously only techniques applicable to distance data may be used for results from DNA-DNA hybridization. Both distance and character state methods can be applied to sequence data (Table 1). It is not our intention to discuss the relative merits of these analytical techniques; they have been amply discussed most recently by Felsenstein (1988) and Swofford and Olson (1990; see also their contribution to this volume). The commonly used algorithms for using distance to produce a tree of relationship are UPGMA (Sokal and Michener, 1858), Fitch and Margoliash (1967) and Neighbor Joining (Saitou and Nei, 1967). The Fitch and Margoliash procedures can be used either assuming equal rates in all branches (i.e. a clock) or without this assumption (these are the programs KITSCH and FITCH on Felsenstein's PHYLIP package).

In all of the DNA-DNA hybridization data sets we have collected, all four tree-building algorithms yield the same topology with nearly identical branch lengths. This adds to our confidence of the robustness of the data.

Assuming there is a single correct branching pattern for any given set of taxa, different techniques of molecular analysis should converge on the correct pattern. Thus, the disadvantages of any single technique should cancel each other while the advantages should be mutually supportive. To our knowledge, in cases where sufficient data are available, this has been the case with DNA-DNA hybridization and DNA sequence data, as well as with restriction endonuclease analysis. Thus these techniques should not be viewed as "competitors" but rather as complementary, keeping in mind the practical and theoretical advantages of each.

Repeats *versus* uniques

A crucial aspect of virtually all DNA-DNA hybridization studies is that they involve only so-called single-copy DNA (scDNA) or "unique" DNA. Total genomic DNA is made single-stranded by heating and then allowed to reassociate into duplex DNA after cooling. Repeated copies of DNA sequences re-anneal faster than single-copy sequences due to their higher concentration in the solution. Thus, under controlled conditions and time, it is possible to remove excess copies of repeated sequences. Figure 2 is a schematic representation of a "C_ot curve" for a typical eukaryote. In almost all DNA-DNA hybridization experiments, only the final fraction of such curves is used, i.e. at least 50% of the genome is allowed to re-associate and the remaining single-stranded fraction is used in the actual experiments.

An important point in understanding such experiments is that repeat sequences of DNA are not totally removed. All sequences are in molar ratio. This is because after excess copies of repeat sequences have reassociated, their dynamics of reassociation are similar to truly unique sequences. Thus no information is lost by discarding repetitive sequences, just redundant information due to repetition is discarded. This is why most authors using the technique state that "single-copy" DNA is used, not "unique" DNA. For systematic purposes, this is equivalent to equally weighting repetitive and unique sequences.

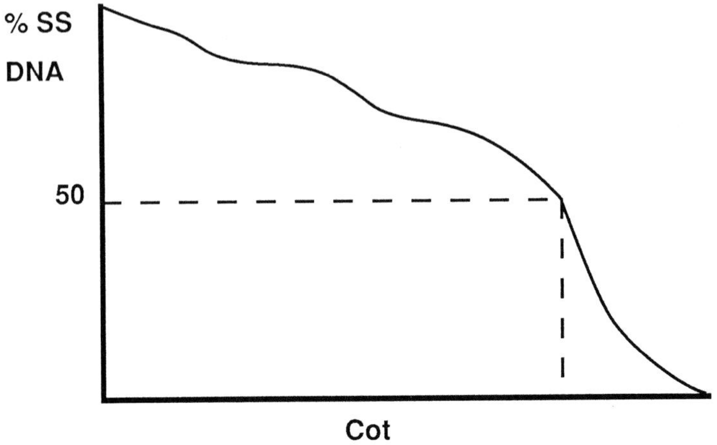

Figure 2. Schematic illustration of a C_ot curve. After dissociating a eukaryotic genome, the highly repetitive sequences reanneal very quickly and are represented in the left-most hump. Sequences of moderate repetition are represented in the next hump, while "unique" sequences reanneal slowest and are in the final fraction on the right. The actual proportion of the genome in each fraction varies considerably, but, with some exceptions, the repetitive fraction is seldom greater than 50%.

A few DNA-DNA hybridization experiments have been performed with the repetitive fraction as defined by C_ot curves (Caccone, unpublished; Schulze and Lee, 1986). The results have been reasonably consistent with those derived from the single-copy fraction, although with some discrepancies. Especially if used for systematic (phylogenetic) purposes, we would strongly suggest using only the single-copy fraction, scDNA. There are at least two reasons. One is the problem of concerted evolution of repeat sequences which may complicate phylogenetic interpretation. A second potentially confusing phenomenon is the fact that many, perhaps most, repetitive sequences are (or were) mobile pieces of DNA that have the ability to move within genomes and likely between genomes. If such sequences can move across species boundaries, i.e. be horizontally transfered, they would be misleading in analyses of phylogenies.

METHODOLOGIES

A major technical consideration in DNA-DNA hybridization is choice of a method for determining the amount of DNA which is single-stranded and double-stranded at various temperatures. Two methods have been most often employed: the binding of double-stranded DNA to HAP (hydroxylapatite) and digestion with the enzyme S1 DNA nuclease which cleaves single-stranded DNA into mono-nucleotides. We detail our procedures for the latter method in another chapter in this volume (Caccone and Powell).

To understand the relative merits of the two methods, it is important to recall that the thermal stability of DNA duplexes is determined by factors other than simply base pair match (i.e. A with T and C with G). One factor is the base composition. G-C pairing forms three hydrogen bonds while A-T pairs form two. Thus sequences rich in C-G will have a greater thermal stability than those rich in A-T. TEACL (tetraethylammonium chloride) is a coumpound which destabalizes the hydrogen bonds in DNA, but does so more for G-C pairs than for A-T pairs. At a certain concentration (2.4 *M*) both types of pairs are equally stable and thus the melting is independent of base composition. Another variable which enters in is the size of the duplex. Longer molecules have greater thermal stabiity than shorter ones (Crothers et al., 1965). With the TEACL method, one can measure the sizes of the molecules being melted and make appropriate corrections (Caccone and Powell, this volume). By negating these variables in the TEACL procedure we are able to more accurately measure variation in thermal stability due to base-pair mismatch.

The HAP method cannot (at present) control for the two variables just mentioned. However, the HAP method has one advantage over the TEACL method: it can and has been automated. Thus more replicates can be performed for a given amount of effort. To date, no one has been able to automate the TEACL procedure. However, there are attempts to try to combine the two procedures such that the advantages of TEACL can be use with the automated machines using HAP (F. Sheldon, C. Sibley, and J. Comstock, personal communications; Werman et al. 1990).

To summarize, the TEACL method allows for more accurate measurements of DNA duplex thermal stability while the HAP method, if automated, allows more replicates per unit effort. Finally, in the only case of which we are aware for which both the TEACL and HAP methods have been applied to the same taxa, the results are very similar [compare Sibley and Ahlquist (1987) with Caccone and Powell (1989)].

Measurements

Various parameters derived from DNA-DNA hybridization studies can be used as measures of the genetic divergence between the taxa being compared. The most commonly used is the change in the median melting temperature, T_m. This is ususally given as ΔT_m which is the T_m of the homoduplex minus the T_m of the heteroduplex. While this type of measure is very useful, especially with complex eukaryotic genomes, matters can become more complex.

When reassociating single-stranded DNA of two different sources, it is necessary to carefully control the conditions of the reassociation reaction. Generally conditions are used (i.e. salt concentration and temperaturue) which requires about 70-75% or greater base-pair match to form a stable duplex. Thus if there are some sequences more than 30% divergent, they will not reassociate. In order to describe the degree of cross-hybridzation between two taxa, a measure called the NPH (normalized percent hybridization) or NPR (normalized percent reassociation) is used. This is simply the percent of reassociation in the heteroduplex reaction divided by the percent reassociation in the homoduplex reaction. Thus another measure of the degree of divergence might be the ΔNPR, that is, the change in the percent of the genome which will cross-hybridize.

Thus ΔT_m measures only the degree of divergence between those sequences which actually hybridized under the conditions used. Some authors have combined the ΔT_m with NPR in an attempt to give one measure which reflects the total genomic divergence, that part which did hybridize and that part which did not. This has been variously termed the ΔT_{median} (Britten, 1986), $\Delta T_{50}H$ (Sibley and Ahlquist, 1987), or ΔT_mR (Benveniste,

1985). Basically what is done is to asssume the decrease in reassociation represents DNA already "melted" and the ΔT_m calculated accordingly. As we (Caccone et al. 1988a; Caccone and Powell, 1990) and others (e.g. Benveniste, 1985) have pointed out, there is generally a linear correlation between ΔT_m and NPR. Thus there is a linear relationship between ΔT_m and $\Delta T_{50}H$. Phylogenetic trees based on each should be congruent with respect to branching pattern and *relative* branch lengths. Because there is often considerable error in the measurement of NPR, we prefer to use ΔT_m.

Another proposed measure is the change in the modal melting temperature, i.e. the Δmode (Brownell, 1983; Sarich et al. 1989). This may be useful under certain circumstances, however as a general measure it is probably not as accurate as ΔT_m. One problem with this measure is it requires extrapolation based on certain assumptions. Also, unlike the HAP method, the TEACL method produces a cumulative melting curve which would require statistical manipulation in order to extract a mode.

Further detailed discussion of the merits and problems with measures can be found in Sheldon and Bledsoe (1989), Bledsoe and Sheldon (1989), Werman et al. (1990), and Kirsch et al. (1990).

Accuracy of ΔT_m

An obvious question about DNA-DNA hybridization is: How accurately does the thermal stability of DNA duplex measure the degree off base-pair mismatch? Several authors have addressed this problem using various techniques such as employing synthetic oligonucleotides or *in vitro* chemical modification. Given the large amount of data now available on the nucleotide sequence of naturally occuring DNA, we felt it would be useful to measure the ΔT_m between sequences with a known amount of base-pair mismatch (Caccone et al. 1988b). Figure 3 presents the results. When the size correction is taken into account, the correlation between base-pair mismatch and ΔT_m is 0.98 or greater (see Caccone et al. for details). Thus we are quite confident that ΔT_m does accurately reflect base-pair mismatch and thus measures genetic divergence.

Two other points are worthy of mention. One is the conversion of ΔT_m to base-pair mismatch, i.e. What percent of base-pair mismatch is necessary to cause a one degree change in the T_m? The usual conversion had been 1:1; a ΔT_m of one degree represents 1% base-pair mismatch. The results in Figure 2 indicate a ΔT_m of one degree corresponds to 1.7% mismatch.

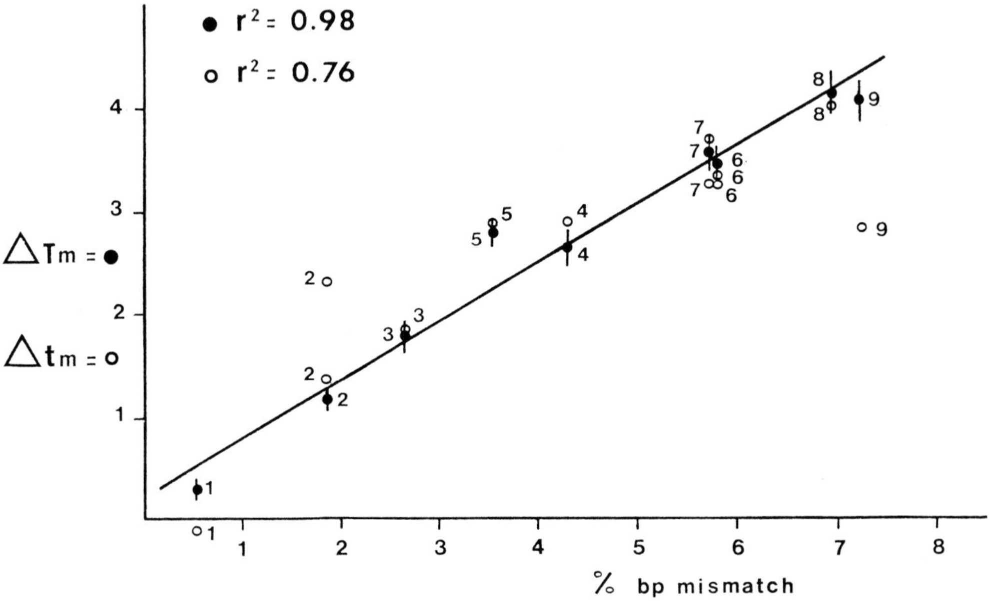

Figure 3. ΔTm *versus* percent base pair mismatch. This is based on cloned sequences of known base sequence. The solid dots are for size-corrected ΔTm's, while the open circles are for uncorrected Δtm's. Details are in Caccone et al. (1987).

The second point of interest concerns the importance and accuracy of size correction. In Figure 3 we also plotted the uncorrected (for size) Δt_m's (we use uppercase T to denote size corrected values and lowercase t to denote uncorrected values). The correlation between Δt_m and percent base-pair mismatch is 0.76. Thus using duplex size correction significantly increases the accuracy of measuring base-pair mismatch, which is, afterall, what we really want to measure.

Intraspecific Variation

In interpreting the significance of interspecific genetic divergence it is important to consider intraspecific variability. Using homozygous "haploid" strains of *Drosophila* we have shown that two genomes within a species may differ at a little over 2% of their nuucleotide sites (Caccone et al., 1987). Britten et al. (1978) present evidence for ΔT_m's as high as five between individuals from the same population. Thus intraspecific DNA variation may not be trivial.

There are two ways to avoid confusing intraspecific variation with interspecific divergence. In our studies of insects (*Drosophila* and cave crickets) of necessity we use many individuals in a DNA preparation. In the case of *Drosophila* we usually use a so-called isofemale line begun by a single inseminated female from nature; thus these lines carry a minimum of four genomes. In the case of the cricket study, several animals captured in the wild were combined. Thus when performaing the homoduplex studies, polymorphism within the DNA preparation will lower both the NPR and T_m proportional to the level of polymorphism. Because in obtaining any of the measures of divergence one subtracts the heteroduplex reaction parameters from the homoduplex parameters at least some of the intraspecific polymorphism is already corrected for. While between haploid genomes of parthenogenetic *Drosophila* we have measured ΔT_m's of about 1.7°C, ΔT_m's between diploid sexually reproducing strains of *Drosophila* are less than 0.5.

A second manner in which to control for intraspecific variation is to use more than one individual or strain in the studies. If one consistently obtains the same ΔT_m between taxa despite the indivivdual or strain used, one can be quite confident significant interspecific divergence is measured independently of intraspecifc variation.

Resolution

How closey related can two taxa be and still have their genetic relationship resolved by DNA-DNA hybridization? Another way of stating this is to ask How far apart do nodes on a dendrogram need to be in order to be statistically distinct? We have twice faced this problem in our *Drosophila* work. The first case was with the *melanogaster* subgroup and concerned the relationship among three closely related sibling species, *simulans, mauritiana*, and *sechellia* (Figure 4).

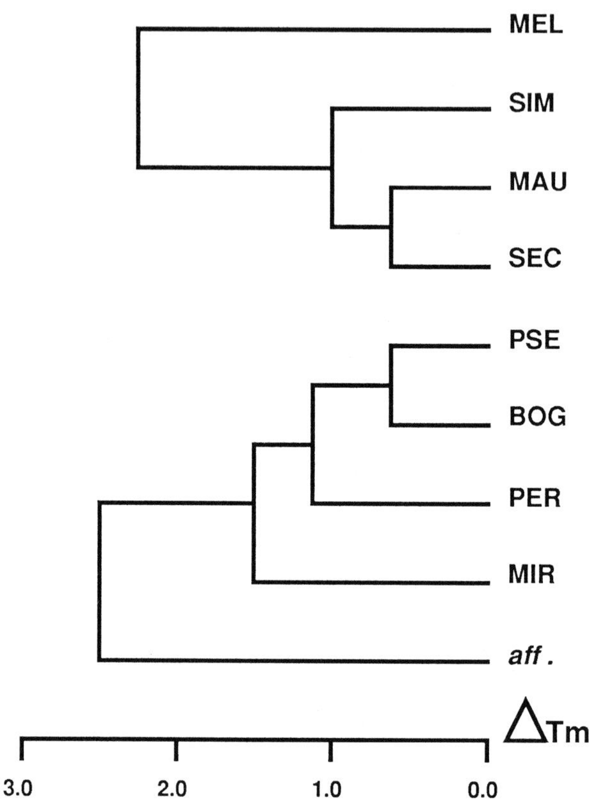

Figure 4. UPGMA dendrograms based on ΔTm for the *Drosophila melanogaster* subgroup on top and the *obscura* group on the bottom. The ΔTm scale is the same for both. Abbreviations are: MEL = *melanogaster*, SIM = *simulans*, MAU = *mauritiana*, SEC = *sechellia*, PS = North American *pseudoobscura*, BOG = bogotana subspecies of *pseudoobscura*, PER = *persimilis*, MIR = *miranda*, and aff = several species of the *affinis* subgroup. Details are in Caccone et al. (1988a) and Goddard et al. (1990).

Initially, using one strain of each species, we could not statistically reject a trichotomy for these three species. Our strategy in this case was to obtain more independently derived strains and perform replicate comparisons of the species. We obtained twelve independent interstrain measures through these nodes, each measure being replicated three to five times. Using both parametric and non-parametric statistics we were able to resolve this trichotomy into the bifurcations in Figure 4. The closests nodes are separated by a ΔT_m of 0.31°C (see Caccone et al. 1988a for details).

Our studies of the *obscura* group of *Drosophila* presented a similar problem, Figure 4. In this case we had five taxa which had a maximum ΔT_m of 2.5 between them. Furthermore, this is an extremely well-studied group and the phylogeny was quite well established before we did our study. We were able to statistically resolve all five taxa into a dendrogram consistent with all other data, i.e. morphology, chromosomes, allozymes, and ability to form hybrids. The minimum distance between nodes is 0.29 ΔT_m.

Thus in our experiments we have been able to resolve nodes down to a separation of about ΔT_m 0.3. This would represent about 0.3% or 0.5% nucleotide differences depending upon whether one uses the 1:1 or 1:1.7 conversion. Of course, the resolution is dependent upon the number of replicates performed through a node; ideally these should be independent measurements, e.g. using different individuals or strains. If one had ten individuals or strains of each taxa and did all pairwise comparisons, doubtless the resolution could be much smaller than 0.3°C. Of course this amount of work would be prohibitive in most cases. Our resolution to 0.3 was acheived with a moderate amount of effort.

What is the upper limit of measurement? Obviously there is one absolute upper limit, the limit imposed by the reasocciation conditions. If even the most conserved fraction of the genome is greater than 30% divergent, one cannot perform the experiment. Even if some of the genomes of distantly related species can hybridize, the amount may be too small to take advantage of the major strength of DNA-DNA hybridization. As a smaller portion of the genome is studied, the average divergence is across a smaller number of nucleotides. To be sure, even 10% of most eukaryote genomes contain at least five million nucleotide pairs, a number which would be very laborious to study directly by sequencing.

We have measured ΔT_m's as high as 17°C (Caccone and Powell, 1987) which is very close to the 30% cutoff if one uses the 1:1.7 conversion. However, in our experience we have found that measuring ΔT_m's above about 12°C becomes difficult and less accurate. Almost all our work has concerned taxa with ΔT_m's less than twelve and therefore our experience in this upper range of applicability is limited.

We need to make explicit that these ranges of resolution are for ΔT_m using the TEACL method. Greater delta's can be reached if one uses $\Delta T_{50}H$. For example, if one measures a ΔT_m of 15 between the fraction of the genomes which did hybridize, one might add another ten degrees or so due to the lower NPR (see Werman et al. 1990).

Statistical Considerations

Elsewhere (Powell and Caccone, 1990) we document the two points we will make in this section. In statistically analyzing any distance data, and ΔT_m in particular, it is important to know if the error on the estimates increases with increasing distance. We analyzed three large data sets and found that, up to a ΔT_m of 12°C, there is no significant increase in error or only very slight increase of an order which would not affect statistical tests. Our very limited data over ΔT_m 12 indicates the error may significantly increase at these large distances.

Can one use parametric tests on DNA-DNA hybridization results? In order to use parametric methods, the error of measurement of the variable should follow a normal distribution. We plotted the error distribution around the measurement of T_m's and found it not to deviate from the expected normal (Powell and Caccone, 1990). Because ΔT_m is formed by subtracting one normal distribution from another normal distribution, it will also be normal. Ahlquist (personal communication) has found a similar pattern for T_m's derived from the HAP technique. Thus there is justification for using parametric statistics on results from DNA-DNA hybridization experiments utilizing both techniques.

Finally in this section we cite the work of Felsenstein (1987) who has done statistical analyses of hybridization data especially on the problem of competing dendrograms. We have presented methods for deriving standard errors on ΔT_m's including cases where the numbers of replicates are unequal and reciprocal measurements are available (Caccone et al. 1988b; Caccone and Powell, this volume).

Conclusions

DNA-DNA hybridization is a powerful method for determining the overall single-copy DNA divergence between two taxa. For phylogenetic reconstruction it offers some advantages over other data such as a DNA sequence of a single allele of a single gene. The technique has a long history of use and most of the potential technical problems have been solved. The accuracy of the method, the experimental errors involved, and the range of applicability have all been documented. Several tree-building algorithms appropriate for DNA-DNA hybridization data are available. While not the major subject of this conference, we also note that one can gain insight into patterns of genome evolution using this technique (Powell and Caccone, 1989; Caccone and Powell, 1990) as well as relate overall genetic divergence with other factors such as morphology, ability to hybridize, and chromosomal evolution (Goddard et al., 1990).

References

Benveniste RE (1985) The contribution of retroviruses to the study of mammalian evolution. In: MacIntyre RJ (ed) Molecular evolutionary genetics. Plenum, New York, pp 359-417

Bledsoe AH, Sheldon FH (1989) The metric properties of DNA-DNA hybridization dissimilarity measures. Syst Zool 38(2):93-105

Britten RJ (1986) Rates of DNA sequence evolution differ between taxonomic groups. Science 231:1393-1398

Britten RJ, Cetta A, Davidson EH (1978) The single-copy DNA sequence polymorphism of the sea urchin *Strongylocentrotus purpuratus*. Cell 15:1175-1186

Brownell E (1983) DNA/DNA hybridization studies of muroid rodents: Symmetry and rates of molecular evolution. Evolution 37:1034-1051

Caccone A, Powell JR (1987) Molecular evolutionary divergence among North American cave crickets. II. DNA-DNA hybridization. Evolution 41:1215-1238

Caccone A, Powell JR (1989) DNA divergence among hominoids. Evolution 43:925-942

Caccone A, Powell JR (1990) Extreme rates of heterogeneity in insect DNA evolution. J Mol Evol 30:273-280

Caccone A, Powell JR Technical aspects of the TEACL method of DNA-DNA hybridization. (this volume)

Caccone A, Amato GD, Powell JR (1987) Intraspecific DNA divergence in

Drosophila: a study on parthenogenetic *D. mercatorum*. Mol Biol Evol 4:343-350

Caccone, A, Amato GD, Powell JR (1988a) Rates and patterns of scnDNA and mtDNA divergence within the *Drosophila melanogaster* subgroup. Genetics 118:671-683

Caccone A, DeSalle R, Powell JR (1988b) Calibration of the change in thermal stability of DNA duplexes and degree of base pair mismatch. J Mol Evol 27:212-216

Crothers, DM, Kallenback NR, Zimm BH (1965) The melting transition of low-molecular-weight DNA: theory and experiment. J Mol Biol 11:802-820

Felsenstein J (1987) Estimation of hominoid phylogeny from a DNA hybridization data set. J Mol Evol 26:123-131

Felsenstein J (1988) Phylogenies from molecular sequences: inference and reliability. Ann Rev Genet 22:521-565

Fitch WM, Margoliash E (1967) Construction of phylogenetic trees. Science 155:279-284

Goddard, K, Caccone A, Powell JR (1990) Evolutionary implications of DNA divergence in the *Drosophila obscura* group. Evolution (in press)

Kirsch JAW, Springer M, Krajewski C, Archer M, Aplin K, Dickerman AW (1990) DNA-DNA hybridization studies of the carnivorous marsupials I. The intergeneic relationships of bandicoots (Marsupialia: Perameloidea). J Mol Evol (in press)

Pamilo P, Nei M (1988) Relationship between gene trees and species trees. Mol Biol Evol 5:568-583

Powell JR, Caccone A (1989) Intra- and interspecific genetic variation in *Drosophila*. Genome, 31:233-238

Powell JR, Caccone A (1990) The TEACL method of DNA-DNA hybridization: technical considerations. J Mol Evol 30:267-272

Saitou N, Nei M (1987) The neighborhood-joining method: a new method for reconstructing phylogenetic trees. Mol Biol Evol 4:406-425

Sarich VM, Schmid CW, Marks J (1989) DNA hybridization as a guide to phylogenies: A critical analysis. Cladistics 5:3-32

Schulze DH, Lee CS (1986) DNA sequence comparison among closely related *Drosophila* species in the *mulleri* complex. Genetics, 113:287-303

Sheldon FH, Bledsoe AH (1989) Indices to the reassociation and stability of solution DNA hybrids. J Mol Evol 29:328-343

Sibley CG, Ahlquist JA (1987) DNA hybridization evidence of hominoid phylogeny: Results from an expanded data set. J Mol Evol 26:99-121

Sokal RR, Michener CD (1958) A statistical method for evaluating systematic relationships. Univ Kans Sci Bull 38:1409-1438

Swofford DL, Olsen GL (1990) Phylogeny reconstruction. In: Hillis DM, Moritz C (eds) "Molecular Systematics" pp 411-501, Sinauer Associate Publishers, Sunderland, Massachusetts, USA

Werman SD, Springer MS, Britten RJ (1990) Nucleic acids I: DNA-DNA hybridization. In: Hillis DM, Moritz C (eds) "Molecular Systematics", pp 204-249, Sinauer Associate Publishers, Sunderland, Massachusetts, USA

Satellite DNA

L. Bachmann, M. Raab, J. Schibel and D. Sperlich
Department of Population Genetics
University of Tuebingen
Auf der Morgenstelle 28
D 7400 Tuebingen/Germany

The genomes of eukaryotic organisms are regularly composed of three different classes of DNA sequences: Single copy, middle or moderately repetitive and highly repetitive DNA sequences, with copy numbers per genome of 1 to 10, 10 to 1000, and 1000 to 1000000 respectively. The highly repetitive sequences can be separated easily from the rest of the genomic DNA by density centrifugation if it happens that they deviate in base composition from the genome average. This deviation causes a higher (GC-rich) or lower (AT-rich) buoyant density and gives rise to the formation of discrete "satellite bands" of DNA in usual CsCl gradients at different densities than the "main band" DNA. Yet, conspicious departure in base composition is not observed in all families of repetitive sequences so that the repetitive DNA remains hidden in the bulk of the main band DNA as a so called "cryptic satellite".

Satellite DNA can be localized predominantly in the heterochromatic regions of the chromosomes around the centromeres but sometimes also near the telomeres or even dispersed in the euchromatin of all or some chromosomes of the set. The proportion of satellite DNA of the total genomic DNA of a species can often exceed 50 per cent or more and might reach in some amphibians, fishes or flowering plants even more than 90 per cent. The typical organization pattern of satellite DNA in the genome is a tandem array of identical basic repeats in clusters or blocks. Each of the tandem clusters can be composed of several hundred to several hundred of thousands copies and more than one cluster of the same repetitive unit might be present in the genome. For a proper analysis of satellite DNA it is first of all important to find and clone sequences that are representatives of the basic units of the various clusters. There are several strategies available to succeed in that matter:

1) If density centrifugation of total genomic DNA separates one or more clearly distinguishable satellite bands, these can be isolated and used for cloning

2) satellite DNA sequences reassociate after heat denaturation much quicker than single and middle repetitive sequences due to their high copy numbers and can be separated at proper

NATO ASI Series, Vol. H 57
Molecular Techniques in Taxonomy
Edited by G. M. Hewitt et al.

cot-values (depending on temperature and salt concentrations) as double stranded molecules from single stranded unique or low repetitive DNA by an hydroxylapatite column that preferentially retains duplex DNA. This DNA fraction can be eluted afterwards, purified and used for cloning.

3) So called restriction satellite bands become sometimes visible when total genomic DNA is digested with a restriction enzyme and separated on agarose or polyacrylamide gels. After staining of the gel with ethidium bromide the restriction satellites appear as conspicious bands in the background smear of the genomic DNA. These bands are composed of a great number of restriction fragments of the same length and are therefore in most instances basic repetition units of tandem clusters of the same satellite DNA sharing a conserved restriction site (mtDNA might give a similar pattern). Elution of DNA fragments of these bands with inserted filter papers or by electroelution gives mainly sequences of satellite DNA and can be used for cloning.

4) a genomic library of a species can be screened with labeled total genomic DNA of the same species. Only those colonies will give signals that carry an insert sequence that is present at high copy number in the total DNA.

Because of its tandem organization, satellite DNA gives very often a typical "ladder-like" hybridization pattern when a cloned satellite probe is hybridized to filterbound total genomic DNA that has been digested with a restriction enzyme before electrophoretic separation (e.g. see fig.5). This pattern results from the tandem organization of the satellite clusters. Since the pattern is observed after complete digestion it must be the result of an existing cleavage site variability between the copies of a cluster (fig.1).

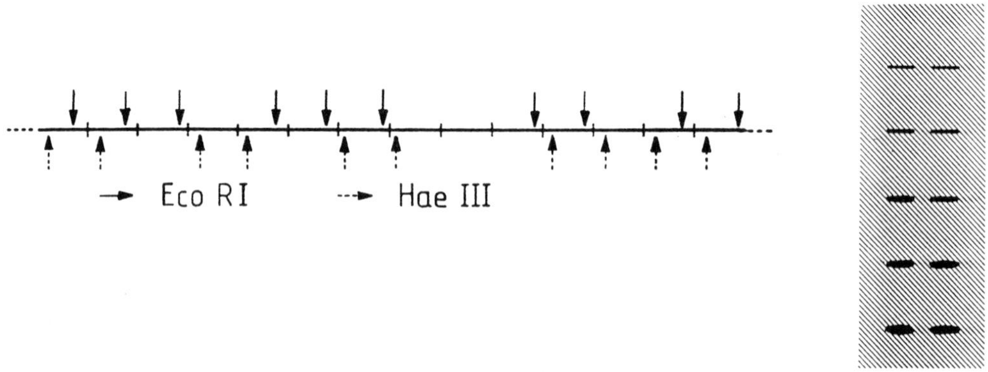

Figure 1 A "ladder-like" hybridization pattern appears if satellite repeats are tandemly arranged and if variation exists for the presence or absence of restriction sites. The pattern can be the same for different restriction enzymes.

Satellite basic sequences have been investigated in a number of animal and plant species (for summary see Miklos, 1985; Beridze, 1986; Verma, 1988). In *Drosophila* species of the *D. melanogaster* group (Lohe and Brutlag, 1987) and in the *D. virilis* group (Gall and Atherton, 1974) most of the satellite DNA is composed of simple repeat units of a length of 5 - 10 bp only. Yet, a more complex 359 bp repeat unit has been found in *D. melanogaster* too (Hsieh and Brutlag, 1979). As we shall show later, the presence of complex repeat units is typical for species of the *Drosophila obscura* group (Bachmann et al., 1989), for Y-specific repeats of *D. hydei* (Wlaschek et al., 1988), for the "Cla element" repeats of the dipteran genus *Chironomus* (Schmidt, 1984), for the grasshopper *Caledia* (Arnold et al., 1986) and other insects. In vertebrates, studies have been made mainly on amphibians (*Xenopus*: Lam and Carrol, 1983; *Rana*: Wu et al., 1986) and mammals (e.g. mouse: Hörz and Altenburger, 1981; primates: Singer, 1979). A proportion of 60-70 per cent of the satellite DNA of the house mouse is composed of a 234 bp repeat that has originated most probably from a dimer of two smaller repeats, that themselves might be explained originally as an association of simple 9 bp subunits.

It is generally agreed that satellite DNA is varies little between individuals of the same species but can be drastically different between even closely related species. Different assumptions can be made to explain this general observation. A direct selection pressure against mutational substitutions or other changes in the repeats appears unlikely, since no clear and convincing reports about any direct function of satellite DNA have been published so far. There are quite a number of known effects that are directly or indirectly related to the presence or absence of satellite DNA (e.g. crossing over frequency, length of cell cycles, morphological variation, formation of meiotic chromosome pairing etc.; for a summary see Verma, 1988) but all these phenomena are probably only side effects. It has been proposed therefore (Doolittle and Sapienza, 1980; Orgel and Crick, 1980) that satellite DNA is "selfish or parasitic DNA" that has acquired a mechanism for an increased replication rate. However, if this is so, it becomes very difficult to understand the high homogenity of the repeats in the same species opposed to the sometimes high divergence between species. Different explanations have been proposed. The most frequently discussed hypothesis is the assumption of unequal crossing over between the tandemly arranged repeats of the same cluster (Smith, 1976). It can be shown easily that such a mechanism of unequal crossing over can increase or decrease the copy numbers in a cluster. It has been also demonstrated by computer simulations that this mechanism has a strong homogenizing effect reducing the intraspecific variability between the repeats of satellite DNA (Ohta, 1983; Stephan, 1989).

Whatever the case might be, the existence of a considerable amount of highly repetitive satellite DNA in the genomes of most organisms should have significant implications for taxonomy and evolutionary biology. Because of the great quantity of copies available it is usually easy to use satellite DNA for molecular studies. Manifold comparisons for sequence homology or divergence can be made: between copies of the same cluster of the same chromosome, between copies of different clusters on other chromosomes, between copies of individuals from the same or from different populations of the same species or between copies from individuals of different species and so on. From all that, interesting phylogenetic relations may be deduced sometimes. Species specific satellite DNA, on the other hand, not helpful for phylogenetic considerations, may be useful as a reliable taxonomic tool, especially in those situations where morphological or other molecular techniques are not applicable. To demonstrate this, two examples of recent investigations of our group in Tuebingen might be mentioned briefly here.

The first example deals with two different triads of sibling species of the *Drosophila obscura* group: a) *D. subobscura*, *D. guanche*, *D. madeirensis* and b) *D. obscura*, *D. ambigua*, *D. tristis.*

From the species of the first triad only *D. subobscura* is distributed all over Europe. *D. guanche* is endemic to Canary Islands, *D. madeirensis* to the island of Madeira. Morphological discrimination is extremmely difficult and intercrosses between *D. madeirensis* with *D. guanche* or *D. madeirensis* with *D. subobscura* (Krimbas and Loukas, 1984) are possible. By use of molecular, and by specific fluorescent staining techniques of mitotic chromosomes, however, a marked difference with respect to the composition of satellite DNA could be detected.

For *D. guanche* a species specific satellite DNA could be found (Bachmann et al., 1989). An AT-rich, tandemly arranged 290 bp DNA-sequence (pGH 290) could be identified to be the basic repetition unit of this satellite DNA (fig.2). Approximately 80 000 copies of the pGH 290 sequence family are contributing roughly 16 per cent of the haploid genome of *D. guanche*. The presence of large DAPI-sensitive heterochromatic regions at the centromeres of the mitotic chromosomes of *D. guanche* are in good agreement with the molecular data (fig.3). Yet, no DNA-sequences homologue to the pGH 290 family could be detected in the genomes of *D. subobscura*, *D. madeirensis* or any other specie of the *D. obscura* group. On the other hand the species of this triad have other satellite sequences in common, that discriminate them clearly from all other obscura group species (Bachmann, unpubl.).

pGH290

```
        10         20         30         40         50         60
CTTTTCCGAG ACGGTAGGCC AAAACCACAC AGAGTTTGGA TCCATTTTAT AGCAGTATCA
        70         80         90        100        110        120
ATCAACTTTT CAGATAATCA AAGTTTTTTA CAACCAGATC TAAGATACTT AAGTTATGTG
       130        140        150        160        170        180
AAAAAACAAA CCGAAACCTC TGAAAAAACC TTAAACTAAA AATTACAAAA TTAATAGTTC
       190        200        210        220        230        240
TTAAAAAACA AGTTCTGCCA TAGAAAATAA AACCGACTTT TCCCGACTTT TTAAAGTCAA
       250        260        270        280        290
AATTTCGAAA AAAAAAATCT CACAACAATC AACATTTTTG CATTTCAAAG
```

Figure.2 The consensus sequence of pGH 290 repeats of *D. guanche* shown as a Hind III fragment. The sequence was derived from sequencing of six different copies of the pGH 290 sequence family.

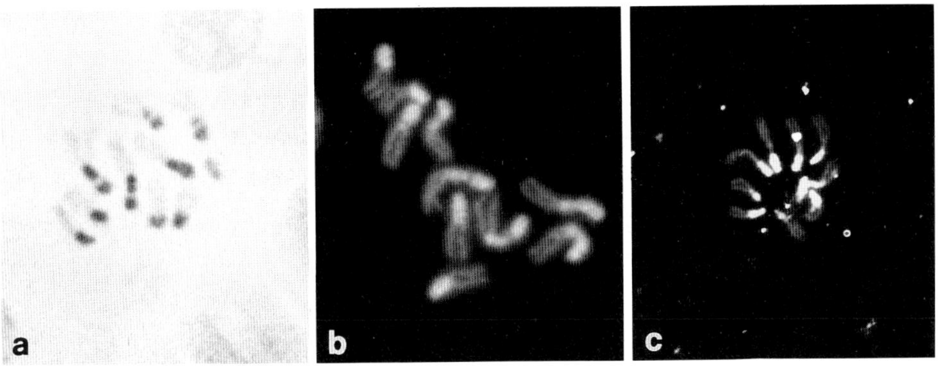

Figure.3 Mitotic chromosomes of *D. guanche.* a) C-banding showing the great amount of heterochromatin on all chromosomes. b) DAPI staining indicates that the heterochromatic regions are AT-rich. c) In situ hybridization with the pGH 290 DNA sequence (see fig.2) proves that this AT-rich sequence constitutes the major part of heterochromatin in this species.

A comparable situation was found in the other triad of species. The specific staining of mitotic chromosomes of *D. tristis* with the C-banding technique revealed conspicious heterochromatic regions at the telomeres of all chromosomes. A similar heterochromatic region could not be found in *D. ambigua*, *D. obscura* or any other *obscura* group species. A molecular analysis identified a tandemly arranged 181 bp DNA-sequence (pTET 181 see fig.4) to be the basic repetition unit of this satellite DNA forming the telomere associated heterochromatin (Bachmann et al., 1990). Homologue sequences to pTET 181 could be detected later with a low copy number in other species, too.

pTET181

```
        10         20         30         40         50         60
GAATTCCAAT TCGCATTTTG ATTGTGGTAT TGCGGATTAG AATTGCAGAT TATTGTGCCA
        70         80         90        100        110        120
TATATACATT TTGCAGCAGG AGATAGGTGG CTAGGGCCTG CCAGGAAATG TGTCGCCAGA
       130        140        150        160        170        180
ATATATGTCG CTATGCCTGC AGGGCATATT AAACTGCAAT CTTGATCCCA AATTTCAATCG
```

Figure 4 The consensus sequence of telomere associated pTET 181 repeats of *D. tristis* shown as an Eco RI fragment. The sequence was derived from sequencing of four different copies of the pTET 181 sequence family.

That means that the copies of pTET 181 are species specifically amplificated in the genome of *D. tristis* to a copy number of 82 000 per haploid genome, i.e. 10 per cent of the entire genome. In contrast to this species specific amplification of pTET 181 in D. tristis another satellite DNA located near the centromeres (Bachmann, unpubl.) illustrates the close phylogenetic relationship of *D. obscura, D. ambigua* and *D. tristis*. This specific repeat can be found only in these three species, separating them as a phylogenetic triad from the other species of the *D. obscura* group (fig.5). This is in good accordance with the results from other studies using other traits.

The second example for the use of satellite DNA for taxonomy are studies on filarial parasites (see also the contribution of R. Post)

Together with the Institute of Tropical Medicine in Tuebingen one of us (J. Schibel) is studying "The epidemiology and molecular biology of *Onchocerca volvulus* (Nematoda: Filarioidea)" in North Cameroon (CEC-Project by Dr. A. Renz and G. Wahl). This parasite is causing human onchocerciasis or river blindness, which occurs mainly in the tropical Africa.

The adult female worms of the parasite are living in subcutaneous nodules, where they are producing the first larval stages, the microfilariae. In a highly infected person there may be 50 to 200 millions of microfilariae distributed in the dermis. If the insect vector, a female of species of the *Simulium damnosum*-complex, takes a bloodmeal on an infected person which it needs for the maturation of eggs, some of the microfilariae are ingested. After about seven days they have passed two moults in the thoracic muscles of the blackfly, and as infective stage larvae (L_3) they enter the proboscis of the vector.

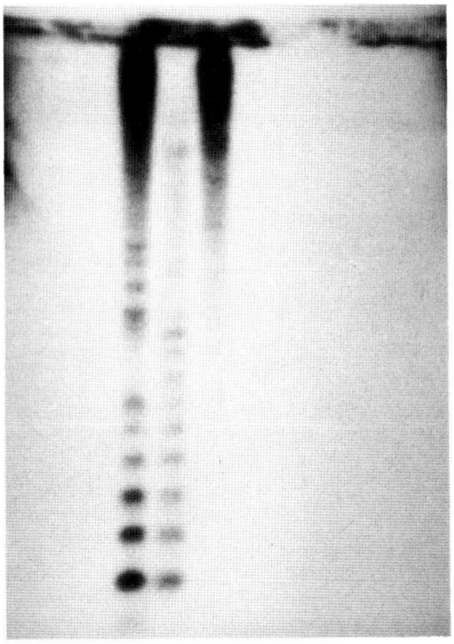

Figure.5 Filter hybridization of Eco RI fragments of total genomic DNA of the *Drosophila* species indicated above with a cloned satellite repeat of *D. ambigua.* A ladder-like hybridization pattern appears for the closely related species *D. ambigua, D. tristis* and *D. obscura.* No hybridizations can be seen in any of the other species. Lane 1: *D. melanogaster*; lane 2: *D. azteca*; lane 3: *D. pseudoobscura*; lane 4: *D. ambigua*; lane 5: *D. tristis*; lane 6: *D. obscura*; lane 7: *D. bifasciata*; lane 8: *D. subsilvestris*; lane 9: *D. subobscura*; lane 10: *D. madeirensis*; lane 11: *D. guanche.*

In Cameroon, significant differences between the clinical manifestations of onchocerciasis in savanna and rainforest have first been observed (Duke et al., 1966). In the savanna the disease is more serious and leads to ocular lesions resulting in a high blindness rate, whereas in the rainforest onchocerciasis is much less severe and does rarely lead to eye damage. Yet, it is not demonstrated whether these differences are caused by different strains of the parasite (Duke and Anderson, 1972) or by different vectors (Dunbar, 1966). In order to measure the risk of the local human population to get infected with *O. volvulus*, the vector flies have to be examined for filarial infections. Since female blackflies of species of the S. *damnosum*-complex are taking their bloodmeals also on warmblooded animals as cattle and antilopes, consequently non-*O. volvulus*-infective larvae can be found in man-biting *S. damnosum* s.l., too. Developing stage larvae and third stage larvae of the cattle filariae *O. ochengi* and *O. dukei*, both known to be transmitted by the same vector (Wahl et al., 1990), are, however, morphologically indistinguishable from *O. volvulus*. In order to be able to differentiate these larvae, our and other groups have tried to isolate species specific DNA sequences for *O. volvulus*, *O. gibsoni* and *O. armillata* (Murray et al., 1988; Harnett et al., 1989; Meredith et al., 1989) in the last few years. They have been tested by means of dot-blot hybridizations to work as specific probes and to be sensitive enough to be able to detect less than 100 pg of DNA (one infective larvae contains 1-2 ng). Dr. W. Harnett of the National Institute for Medical Research in London isolated a 154 bp sequence, that is highly repetitive in the genome of *O. volvulus*. He kindly gave some DNA of this probe (C1A1-2) to us. In Tuebingen the sequence was used successfully for dot-blot hybridizations of single infections (L_1, L_2 and L_3) from wild-caught flies of two breeding sites in North Cameroon.

	A	B	C
1	3 L_3	1 L_3 type D	1 L_3
2	1 L_3 type D	8 L_3	6 L_3
3	1 L_1	6 L_3	6 L_1
4	1 L_3	3 L_3	1 L_1
5	1 L_3	3 L_3	4 L_1
6	4 L_1	1 L_3	-

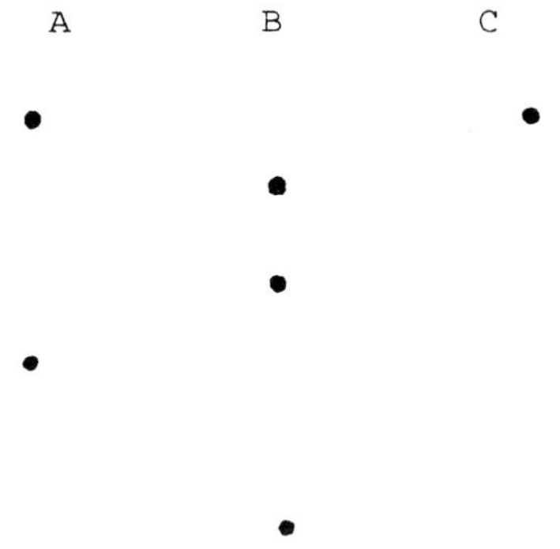

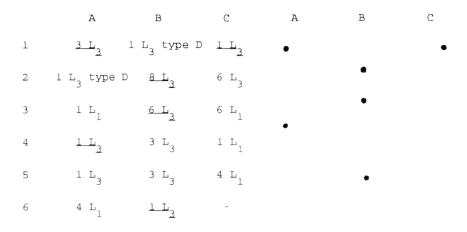

Figure.6 Dot-blot hybridizations of a labeled DNA sequence that is highly repetitive in *Onchocerca volvulus* and specific for that species to DNA of infections of black flies from two localities in Cameroon (Campement du Syrien: lanes A and B; Wakwa: Lane C). The parasitic larvae used are indicated on the left side (e.g. 3 L_3 means 3 third stage larvae were blotted). The corresponding hybridization pattern is shown on the right side. Only 6 larvae proved to be *O. volvulus*.

As can be seen from fig.6 infections with *O. volvulus*-stages are hybridizing strongly, whereas larvae of presumably animal origin do not react. Yet, the practical use of this DNA probe for the field work in Cameroon is still severely limited as long as a specific probe for *O. ochengi*, the most common cattle filaria in that region is still missing. In collaboration with Dr. S.E.O. Meredith of the Royal institute for Tropical Hygiene in Amsterdam, the highly repetitive DNA sequences of *O. volvulus* and *O. ochengi* have been therefore amplified using the PCR method. After purification and characterization it is intended to clone and to sequence the amplified *O. ochengi* DNA in order to find species specific sequence.

With two specific probes, one for *O. volvulus* and one for *O. ochengi*, it should be possible to bring more light into the epidemiology of human and cattle onchocerciasis in North Cameroon.

References

Arnold ML, Appels R, Shaw DD (1986) The heterochromatin of grasshoppers from *Caledia captiva* species complex I. Mol Biol Evol 3: 29 - 43

Bachmann L, Raab M, Sperlich D (1989) Satellite DNA and speciation: A species specific satellite DNA of *Drosophila guanche*. Z zool Syst Evolut-forsch 27: 84 - 93

Bachmann L, Raab M, Sperlich D (1990) Evolution of a telomere associated satellite DNA in the genome of *Drosophila tristis* and related species. Genetica: in press

Beridze T (1986) Satellite DNA. Springer, Berlin Heidelberg New York

Doolittle G, Sapienza C (1980) Selfish genes, the phenotype paradigm and genome evolution. Nature London 284: 601 - 603

Duke BOL, Anderson J (1972) a comparison of the lesions produced in the cornea of the rabbit eye by microfilariae of the forest and Sudan savanna strains of *Onchocerca volvulus* from Cameroon I. The clinical picture. Tropenmed Parasit 23: 354 - 368

Duke BOL, Lewis DJ, Moore PJ (1966) *Onchocerca Simulium* complexes I. Transmission of forest and Sudan savanna strains of *Onchocerca volvulus*, from Cameroon, by *Simulium damnosum* from various West African bioclimatic zones. Ann trop Med Parasitol 60: 318 - 336

Dunbar AW (1966) Four sibling species included in *Simulium damnosum* Theobald from Uganda. Nature London 209: 597 - 599

Gall JG, Atherton DD (1974) Satellite DNA sequences in *Drosophila virilis.* J Mol Biol 85: 633 - 664

Harnett W, Chambers AE, Renz A, Parkhouse RME (1989) An oligonucleotide probe specific for *Onchocerca volvulus.* Mol Biochem Parasit 35: 119 - 125

Hörz W, Altenburger W (1981) Nucleotide sequence of mouse satellite DNA. Nucleic Acids Res 9: 683 - 696

Hsieh T, Brutlag DL (1979) Sequence and sequence variation within the 1.688 satellite DNA of *Drosophila melanogaster.* J Mol Biol 135: 465 - 481

Krimbas CB, Loukas M (1984) Evolution of the obscura group *Drosophila* species. I. Salivary chromosomes and quantitative characters in *D. subobscura* and two closely related species. Heredity 53: 469 - 482

Lam BS, Carroll D (1983) Tandemly repeated DNA sequences from *Xenopus levis* I. Studies on sequence organization and variation in satellite I DNA. J Mol Biol 165: 567 - 585

Lohe AR, Brutlag DL (1987) Identical satellite DNA sequences in sibling species of *Drosophila.* J Mol Biol 194: 171 - 179

Meredith SEO, Unnasch TR, Karam M, Piessens WF, Wirth DF (1989) Cloning and characterization of an *Onchocerca volvulus* specific DNA sequence. Mol Biochem Parasit 36: 1 - 10

Miklos GLG (1985) Localized highly repetitive DNA sequences in vertebrate and invertebrate genomes. In: MacIntyre RJ (ed) Molecular evolutionary genetics. Plenum Press New York London, p 241

Murray KA, Post RJ, Crampton JM, McCall PJ, Kouyate B (1988) Cloning and characterization of a species-specific repetitive DNA sequence from *Onchocerca armillata.* Mol Biochem Parasit 30: 209 - 216

Ohta T (1983) Theoretical study on the accumulation of selfish DNA. Genet Res 41: 1 - 15

Orgel LE, Crick FHC (1980) Selfish DNA: The ultimate parasite. Nature London 284: 604 - 607

Pech M, Igo-Kemenes T, Zachau HG (1979) Nucleotide sequences of a highly repetitive component of rat DNA. Nucleic Acids Res 7: 417 - 432

Schmidt ER (1984) Clustered and interspersed repetitive DNA sequence family of *Chironomus*. J Mol Biol 178: 1 - 15

Singer MF (1982) Highly repeated sequences in mammalian genomes. Int Rev Cytol 76: 67 - 112

Smith GP (1976) Evolution of repeated DNA sequences by unequal crossing over. Science Wash DC 191: 528 - 535

Stephan W (1989) Tandem repetitive noncoding DNA: Forms and forces. Mol Biol Evol 6: 198 - 212

Verma RS (1988) Heterochromatin. Cambridge Univ Press Cambridge New York New Rochelle Melbourne Sydney

Wahl G, Ekale D, Enyong P, Renz A (1990) The development of *Onchocerca dukei* and *O. ochengi* microfilariae to infective stage larvae in *Simulium damnosum* s.l. and in members of the *S. medusaeforme* group, following intra-thoracic injection. Ann trop Med Parasit: submitted

Wlaschek M, Angulewitsch A, Bünemann H (1988) Structure and function of Y chromosomal DNA I. Sequence organization and localization of four families of repetitive DNA on the Y chromosome of *Drosophila hydei*. Chromosoma 96: 145 - 158

Wu Z, Murphy C, Gall JG (1986) A transcribed satellite DNA from the bullfrog *Rana catesbeiana*. Chromosoma 93: 291 - 297

Acknowledgements: The experimental investigations have been sustained by DFG grants Sp 146/6-6 and Sp 146/10-2

DNA FINGERPRINTING

David T Parkin and J H Wetton
Department of Genetics
School of Medicine
Queen's Medical Centre
Nottingham NG7 2UH
U K

Introduction

Evolution by natural selection takes place through the success and failure of individuals to survive and reproduce. It thus involves the lowest level of taxonomic structure - the individual. To understand many of the processes of evolution at higher taxonomic levels, it is necessary to study and understand the behaviour of populations and individuals in nature and the laboratory: to record their survival and measure their fitness. This study of the interaction between individual fitness and genetic variation lies within the field of population genetics.

A major discovery of theoretical and experimental population genetics has been the understanding of the relative importance of random and structural forces in the change of allele frequencies from generation to generation. Change in allele frequency is one of the ways that populations can evolve and change, and a significant contribution to evolutionary theory has been the appreciation that higher order taxonomic differentiation may have its roots in the analysis of single loci in individual populations. It is no surprise that authors such as Mayr (1963), Stebbins (1984) and Futuyma (1979) devote part of their books to simple population genetics.

This led to the debate over the actual structure of natural populations, and to the extent of genetic variation within them. On the one hand, the 'classical' school typified by Muller (1950) suggested that there was relatively little variation, and that the majority of loci were homozygous for a 'best' allele in most populations. The proponents of the 'balance' theory, such as Wallace (1958) disputed this. They argued that many loci were variable, alleles being held in the population by normalizing or balancing selection such as heterozygous advantage or frequency-dependence. In this unsatisfactory state, the case rested until the application of gel electrophoresis of proteins by Harris (1966) and Lewontin and Hubby (1966). They showed that this technique could be used to identify genetic variants and individual loci, and the period 1965-80 saw a rapid analysis of populations in a wide variety of species (Nevo, 1978). This 'golden age' allowed detailed study of evolutionary differentiation, both above the species level

NATO ASI Series, Vol. H 57
Molecular Techniques in Taxonomy
Edited by G. M. Hewitt et al.

(e.g. Ayala, 1975) and among populations from individual taxa (e.g. Barrowclough, 1980). Evidence for selection and neutrality of the loci was sought, and sometimes found (Lewontin,1974), and a whole theory of molecular population genetics emerged (e.g. Nei, 1975).

However, one particularly important facet of population genetics remained unapproachable. There was insufficient variation in even the most polymorphic species to permit the unequivocal recognition of individuality. For example, Wetton and Parkin (1989) showed that 11 loci were usually polymorphic in the House Sparrow, *Passer domesticus*. Such variation is not unusual in a vertebrate species, yet the probability that two individuals will have the same phenotype at all 11 loci simultaneously is 0.0023. Such a level of variation is insufficient for unique recognition or parentage assignment in even a modestly sized population, and so the system is unsuitable for the detailed demographic analysis of a single population. A more suitable system was needed. Such a system was provided by DNA fingerprinting.

DNA fingerprinting

Scattered through the genome are a series of 'families' of DNA known as hypervariable minisatellite or variable number of tandem repeats (VNTR). Each family has an essentially similar structure consisting of a short (usually 15-35 bp) core region which is constant across members of the family, and flanking DNA that varies between loci (see Fig. 1). The variation detected in a DNA fingerprint reflects not base substitutions, but different numbers of multiple tandem repeats of the entire unit of core plus flanking DNA (Jeffreys *et al.*, 1986b,c).

The individual steps in the visualization of a genetic fingerprint are detailed elsewhere in this book - generally by Carter () and in detail for each separate procedure. The essentials, however, are shown in Fig. 2. DNA is extracted and purified from an individual and cut with an appropriate restriction enzyme. The ideal enzyme cuts DNA frequently, but not within the sequence of DNA in the minisatellite under investigation. Preferred enzymes include Hae III, Alu I, Hinf I.

Restricted DNA is then assayed fluorometrically to equalize the concentration placed in the wells of an agarose gel. We normally use 6 μg per well. The electrophoresis run time depends upon the size of the DNA fragments to be retained in the gel, and the agarose concentration can also be varied to affect the fragment separation profile. The ideal is, however, to separate the fragments as widely and uniformly as possible to facilitate their subsequent resolution.

Transfer of DNA out of the agarose can be by vacuum blotting or electroblotting, but we routinely use Southern blotting. Especially if several gels are to be blotted simultaneously,

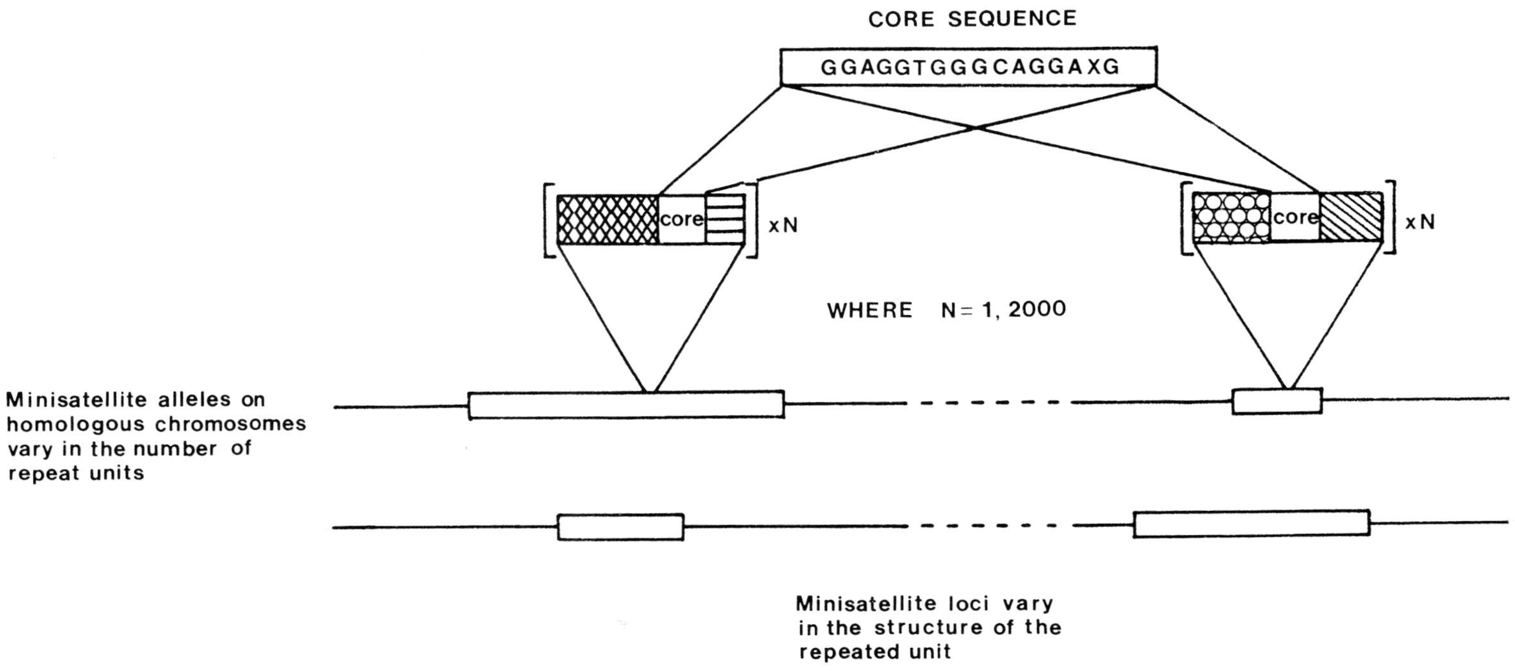

Fig 1 A generalised representation of the structure of hypervariable minisatellite DNA. See text for details.

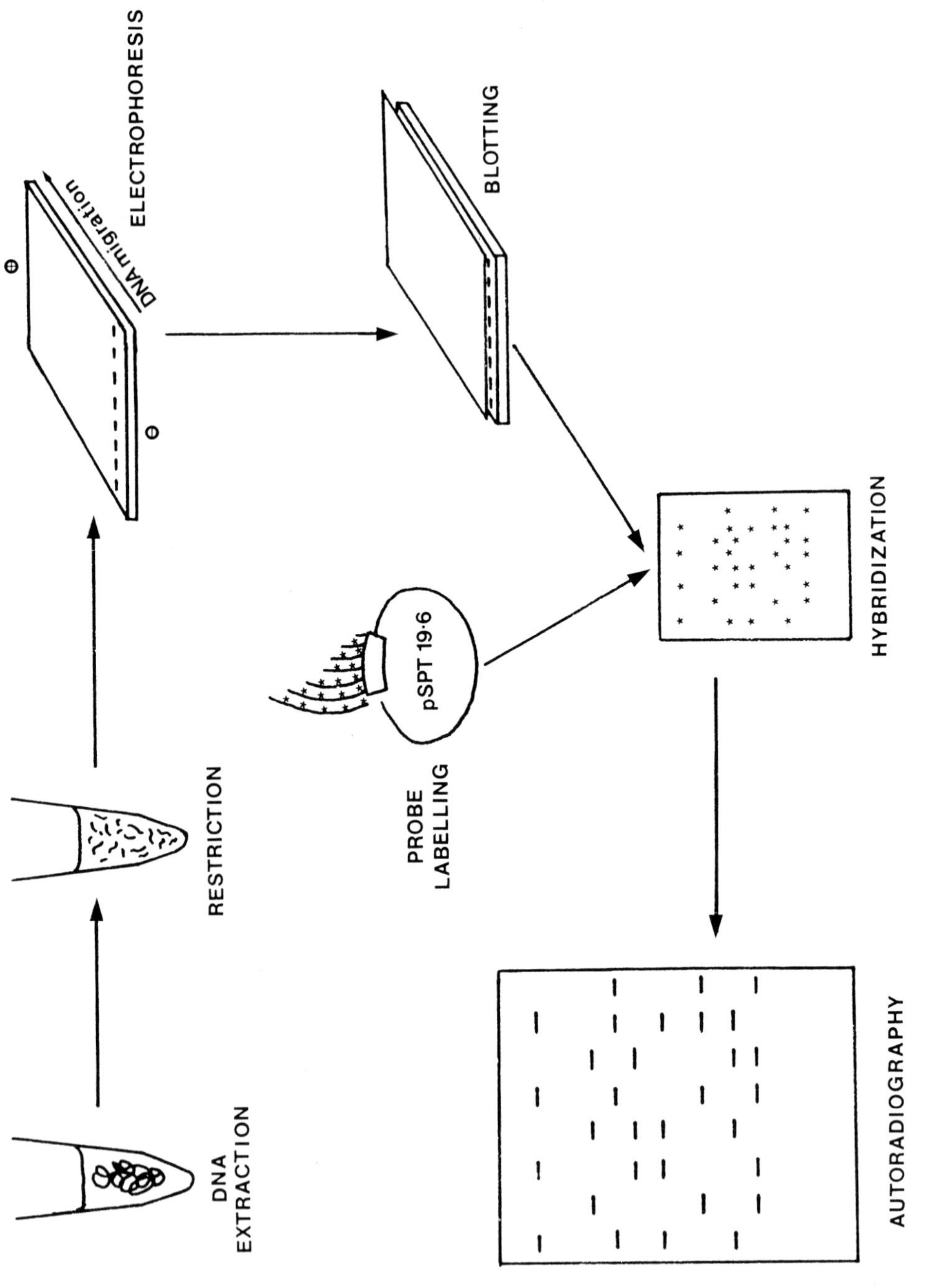

Fig 2 An outline of the procedure for visualising a DNA fingerprint. For details, see text and Carter.

we find it easier to do Southern transfers in parallel than the others in sequence. Finding a good membrane is a terrible problem. We have found difficulties and inconsistencies with the majority of those that are commercially available. Frequently, out of despair, we use nitrocellulose. It's more difficult to handle and cannot easily be reprobed, but it works consistently.

Several probes are available. We routinely use our own derivatives (Carter *et al*, 1989) of the 'Jeffreys' probes 33.6 and 33.15 which are available from Cellmark, or under agreement from other research laboratories around the world. These consist of multiple copies of the core sequence (Jeffreys *et al.*, 1985b,c) inserted into an appropriate vector, which can be amplified by culturing the host bacteria. Large quantities of the core sequence can then be purified and radiolabeled in vitro. The genomic DNA is dissociated into single strands prior to blotting, and rehybridization is permitted in the presence of the single stranded radioactive probe. The resulting fingerprints are revealed by autoradiography of the labelled blot onto Xray film. Sharpness of bands and hence resolution of the fingerprint is maximal with low concentrations of DNA. However, the obvious by-product of this is a weak signal, and thus long exposures. We now use RNA probes (Carter *et al.*, 1989). They can be more strongly labelled, and DNA/RNA binding is stronger giving a more powerful signal.

Interpretation

The number of bands revealed and the variation among individuals varies between species. The majority of our work has been directed towards avian subjects, but even here we find from c.10 (Sparrowhawk) to c.20 (House Sparrow). Unrelated adults of most wild populations show a high degree of variability (Wetton *et al*, 1987; Burke, 1989). A typical autoradiograph of unrelated House Sparrows will show about 60 bands of sufficient distinctness to be safely scored. The average individual will have about 20, of which about 10% are shared with unrelated birds. It is very unusual to find a single band present in every one of 16 unrelated birds on a standard blot. This indicates a very high incidence of heterozygosity that we can estimate from the Hardy-Weinberg law (Jeffreys *et al.*, 1985b). The frequency of three genotypes, AA, AB, BB are given by p^2, $2pq$, q^2 where p and q are the frequency of the two alleles. A single fingerprint band could be present as the product of a heterozygous or a homozygous locus. Its frequency is therefore expected to be $2pq + q^2$, if q is the frequency of the allele producing that minisatellite fragment. Assuming all alleles are equally abundant, the frequency of a single band, equals the amount of band-sharing, x. Thus,

$$2pq + q^2 = x$$

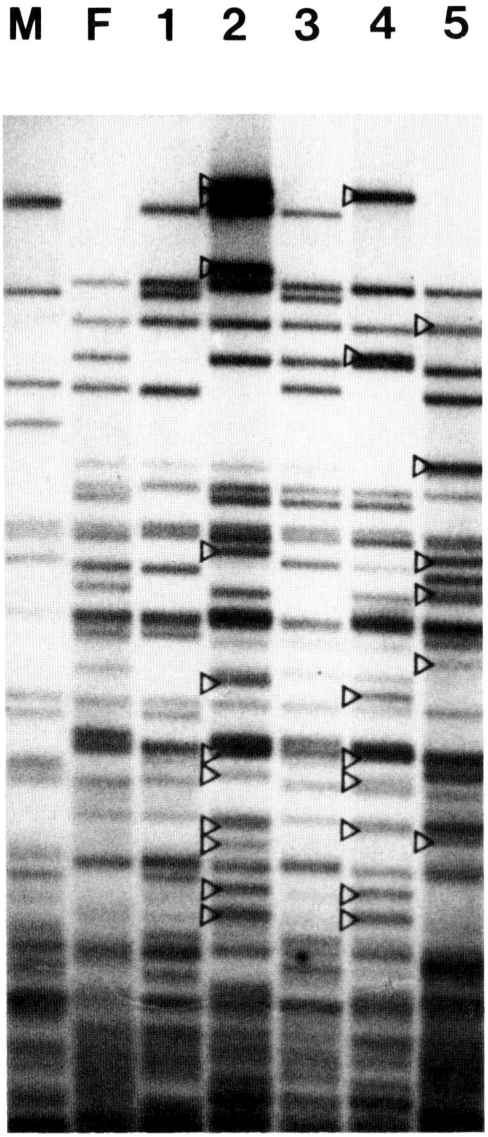

Fig 3. The DNA fingerprint of a 'family' of House Sparrows consisting of two adults and five progeny. Bands that are absent from the putative parents are indicated.

Since band-sharing is so low, it is unlikely that any homozygotes exist, and since p+q=1, this can be simplified to

$$2pq = x \quad \text{or} \quad x = 2q - q^2$$

With a band-sharing of 10%, this suggests q = 0.053, and that there are $\frac{1}{q}$ or about 20 alleles per minisatellite locus, assuming the allele frequencies are uniformly distributed. Such extensive variation is sufficient to allow individual recognition: a conservative estimate of the probability that two unrelated individuals will share the same constellation of bands is given by the band-sharing raised to the power of the average number of bands, 0.1^{20}, many orders of magnitude less than that obtained using enzyme data.

Most species that have been investigated show a high level of variability. Even populations that have been subject to high levels of inbreeding or severe bottlenecks, such as Red Kites (*Milvus milvus*) from Wales, or captive stocks of Rothschild's Mynah (*Leucopsar rothschildi*), are sufficiently variable to allow individual recognition within their modest populations (May *et al.*, in prep; Ashworth and Parkin, in prep.).

The fingerprints appear to be inherited in a relatively straightforward manner. Extensive family data are available from a variety of species, and the general pattern seems to be that every band present in an individual has been inherited from one (or possibly both) parents. Occasional bands may be seen that mismatch with the parents: usually all the other bands can be traced back to the male or female, and the single mismatch is assigned to a new mutation. A pattern of multiple mismatches is more likely to be due to mis-identified parentage, and is much more dramatic. Here, one adult matches as usual with about 50% of the bands being shared with its progeny, while the other adult shows little band-sharing at all - certainly no more than that expected from unrelated individuals.

Analysis

The statistical analysis of multi-band fingerprints is still under development (Jeffreys *et al.*, 1985a; Lynch, 1988; Brookfield, 1989). The most comprehensive review is that of Brookfield (1989) which we have discussed using data from our study population of sparrows (Parkin, Wetton & Brookfield, in prep.). The essentials of the analysis are to record the average number of bands per individual, band sharing between unrelateds, and the mutation rate. Then, for the individual,under investigation, the number in each track, and the number shared between all tracks is calculated. Thus, for five young House Sparrows and the adults that fed them in Fig.3 , the data were as follows:

Band-sharing in a family of House Sparrows (M = Male, F = Female, P = Progeny)

	Bands Present only in						
Progeny No.	**M**	**F**	**P**	**MF**	**MP**	**FP**	**MFP**
1	6	7	0	0	6	10	3
2	10	4	11	0	2	13	3
3	6	5	0	0	6	12	3
4	11	4	8	0	1	13	3
5	11	4	9	1	1	13	2

Taking each progeny in turn, there are four possible hypotheses: (a) the two adults are the parents, (b) the male alone is a parent, (c) the female alone is a parent, (d) neither adult is a parent. Brookfield's statistics allow the calculation the probability that the young House Sparrow will possess its particular array of bands given the patterns of the two adults under each of these separate hypotheses, and assuming normal Mendelian inheritance, particular levels of band sharing and mutation, and an absence of linkage, etc. These probabilities are shown below:-

Probability of observing the two adult fingerprints in association with each progeny fingerprint under four different hypotheses of relatedness

Progeny No.	Both adults are parents	Male only is a parent	Female only is a parent	Neither adult is a parent
1	$\underline{8.9 \times 10^{-10}}$	3.7×10^{-21}	2.2×10^{-17}	3.4×10^{-28}
2	3.6×10^{-45}	2.7×10^{-40}	$\underline{4.7 \times 10^{-27}}$	9.8×10^{-38}
3	$\underline{9.8 \times 10^{-10}}$	1.1×10^{-24}	3.0×10^{-18}	4.2×10^{-30}
4	1.5×10^{-35}	2.0×10^{-36}	$\underline{5.5 \times 10^{-22}}$	6.4×10^{-34}
5	2.8×10^{-38}	9.4×10^{-37}	$\underline{2.6 \times 10^{-23}}$	6.4×10^{-34}

It is at once obvious that all of these are very small, but there are many thousands of possible combinations of the adult bands in an individual progeny. However, for each young sparrow, there is one value (underlined) that is many times larger than the rest, and this is taken to be the 'most likely'. If it has a probability more than 100 times greater than another, it is assumed to be significantly more probable.

In this particular case, it is clear that all the young birds are the offspring of the female, but only two of the male. These two are thus full siblings, but what of the other three? They are clearly half siblings of 1 and 3, but do they all share the same father, or are the product of extra-pair fertilizations with separate males?

Again, Brookfield's (1989) statistics can help. We compare the progeny two at a time with their mother. They must be full or half siblings, depending upon whether or not they share a father. Counting bands reveals the following results:

Number of bands present Female and Progeny A and B

Progeny	Bands present in						
	Female only	A only	B only	Female and A	Female and B	A and B	All three
1 2	2	6	13	2	5	0	11
1 3	2	2	2	3	5	4	10
1 4	1	6	9	3	6	0	10
1 5	4	6	10	1	3	0	12
2 3	3	13	6	2	1	0	14
2 4	3	7	3	1	1	6	15
2 5	1	11	8	4	3	2	12
3 4	2	5	8	2	3	1	13
3 5	0	6	10	5	5	0	10
4 5	1	9	10	4	3	0	12

It is now possible to calculate the probability of two individuals sharing such an array of bands, given that they have the same mother, under the two rival hypotheses. The results are reasonably unequivocal.

Progeny being compared	Probability under alternative hypotheses	
	FULL SIBLING	HALF SIBLING
1 2	2.1×10^{-11}	$\underline{9.5 \times 10^{-8}}$
1 3	$\underline{3.9 \times 10^{-9}}$	$5.6. \times 10^{-14}$
1 4	6.8×10^{-10}	$\underline{2.1 \times 10^{-7}}$
1 5	4.3×10^{-10}	$\underline{2.1 \times 10^{-7}}$
2 3	2.8×10^{-11}	$\underline{1.3 \times 10^{-7}}$
2 4	$\underline{8.5 \times 10^{-11}}$	6.5×10^{-16}
2 5	8.0×10^{-12}	$\underline{4.1 \times 10^{-10}}$
3 4	2.1×10^{-9}	$\underline{1.4 \times 10^{-8}}$
3 5	2.4×10^{-10}	$\underline{1.6 \times 10^{-7}}$
4 5	3.2×10^{-11}	$\underline{1.5 \times 10^{-7}}$

Again, the most likely result is underlined. Progeny 1 and 3 are full siblings: we knew this already for they shared a father. Progeny 2 and 4 are also full siblings, but progeny 5 is half sibling to the rest. Thus, we seem to have three males involved with this female. Her mate who sired 1 and 3, and two others who sired 2 and 4, and 5,outside the pair bond.

This result could never have been reached without fingerprinting. There is no way that we could have recognized the mixed parentage, nor identified the sibling groups. The value of the methodology for determining individual fitnesses is clear, and of major significance to the demographic study of populations.

It is theoretically possible to take the analysis a stage further and compare the progeny two at a time in isolation from the adults. Here, there are three simple hypotheses: full sibling, half-sibling and unrelated. Again, Brookfield (1989) has produced statistics that allow the calculation of probabilities associated with the three hypotheses. Unfortunately, it is not often possible to differentiate between these (May *et al.*, in prep.): band-sharing is too high, and the observed number of bands too few, for unequivocal differentiation between the models.

This is unfortunate for there are many situations where an ecologist may have a group of similarly aged animals, and would like to know their relationships. Are several clusters of insect eggs on a single plant laid by the same adult, for example. Fingerprint analysis in the absence of the adult could help to show whether the progeny formed a single sibship (in which case a single female would be responsible) or a series of separate sibships (indicating several females). Formal significance may be hard to achieve, although an indication may be gained by simple observation of the results.

Single locus probes

The next stage in the technology of DNA fingerprinting will be the widespread development and use of single locus probes (SLPs). The probes presently in use are derived from the invariant core region of the minisatellite subunit. Used at a relatively low stringency, these hybridize with a wide range of fragments, binding to the core region with which they are homologous. Initially with man, and now with other species, laboratories are producing probes that contain not just this core region, but the flanking DNA as well. Used at low stringency, these will also behave as a multilocus probe, but when the stringency is increased, they hybridize only with the minisatellite locus from which they were derived. Consequently, under these conditions, they produce an autoradiograph pattern containing one or two bands, depending upon the homozygosity or heterozygosity of the locus involved. The advantage of these is that each SLP will produce a 'genotype' for that minisatellite, and some of them are extremely polymorphic (Wong *et al*, 1986). A relatively small number of such SLPs detecting highly polymorphic loci will be sufficient to recognize individuality as reliably as a multilocus probe.

A disadvantage is that several probes will be necessary, and several hybridizations may be needed - increasing both cost and time. Consequently, it seems likely that for rapid screening of extensive families, the multilocus system will retain its popularity. The real advantage of SLPs should be in the search for a true parent following the discovery of a mismatch. The progeny genotype can be determined at a series of loci, and the unknown parental gametic genotype deduced by comparison with the known parent. This gamete is then compared with the genotypes of all adults of that sex, and possible parents identified. Confirmation, were that necessary, can then be undertaken by a standard multilocus comparison of the putative family.

References

Ayala, F.J. 1975 Genetic differentiation during the speciation process. in *Evolutionary Biology* vol 8 (eds T. Dobzhansky, M.K.Hecht & W.C.Steere) Plenum, New York.

Barrowclough, G.F. 1980. Gene flow, effective population sizes and genetic variance components in birds. *Evolution*, 34, 789-798.

Brookfield, J.F.Y. 1989 Analysis of DNA fingerprinting data in cases of disputed paternity. *J. Maths Appl. Med. & Biol.*, 6,111-131.

Burke, T. 1989 DNA fingerprinting and other methods for the study of mating success. *Trends in Ecol. & Evol.*, 4, 139-144.

Carter, R.E., J.H.Wetton & D.T.Parkin 1989 Improved genetic fingerprinting using RNA probes. *Nucl. Acids Res.* 17, 5867.

Futuyma, D. 1979. *Evolutionary Biology.* Sinauer, Sunderland, Mass.

Harris, H. 1966. Enzyme polymorphisms in man. Proc. Roy. Soc. Ser. b, 164, 298-310.

Jeffreys, A.J., J.F.Y.Brookfield & R. Semeonoff 1985a Positive identification in an immigration test-case using human DNA fingerprints. *Nature*, 317, 818-819.

Jeffreys, A.J., V. Wilson & S.L.Thein 1985b Hypervariable 'minisatellite' regions in human DNA. *Nature*, 314, 67-73.

Jeffreys, A.J., V. Wilson & S.L.Thein 1985c. Individual-specific 'fingerprint' of human DNA. *Nature*, 316, 76-79.

Lewontin, R.C. & J.L.Hubby 1966. A molecular approach to the study of genic heterozygosity in natural populations. II Amount of variation and degree of heterozygosity in natural populations of *Drosophila pseudoobscura. Genetics*, 54, 595-609.

Lewontin, R.C. 1974 *The genetic basis of evolutionary change.* Columbia U.P., New York.

Lynch, M. 1988 Estimation of relatedness by DNA fingerprinting. *Mol. Biol. Evol.*, 5, 584-599.

Mayr, E. 1963 *Animal species and evolution.* Harvard U P, Cambridge, Mass.

Muller, H.J. 1950. Our laod of mutations. *Am. J. Hum. Genet.*, 2, 111-176.

Nei, M. 1975 *Molecular population genetics and evolution.* Elsevier, New York.

Nevo, E. 1978 Genetic variation in natural populations: patterns and theory. *Theoret. Pop. Biol.*, 13, 121-177.

Stebbins, G.L. 1974 *Flowering plants: evolution above the species level.* Harvard U P, Cambridge, Mass.

Wallace, B. 1958. The role of heterozygosity in Drosophila populations. *Proc. Int. Cong. Genet. 10th,* 1, 408-419.

Wetton, J.H., R.E.Carter, D.T.Parkin & D.Walters 1987 Demographic study of a wild House Sparrow population by DNA fingerprinting. *Nature*, 327, 147-149.

Wetton, J.H. & D.T Parkin 1989 DNA fingerprinting of house sparrows. in *Electrophoretic studies on agricultural pests* (eds H.D.Loxdale & J.den Hollander) Syst. Ass. Special Vol., Oxford U.P.

Wong, Z., V.Wilson, A.J. Jeffreys & S.L.Thein 1986 Cloning a selected fragment from a human DNA 'fingerprint': Isolation of an extremely polymorphic minisatellite. *Nucl. Acids Res.*, 14, 4605-4616.

THE STATISTICAL INTERPRETATION OF HYPERVARIABLE DNAs

J.F.Y. Brookfield
Department of Genetics
University of Nottingham
Queens Medical Centre
Nottingham
NG7 2UH

Introduction

Hypervariable DNA sequences are sequences which show high (80%+) levels of heterozygosity in wild populations. Using standard techniques of DNA extraction, restriction digestion, gel electrophoresis, Southern blotting, hybridisation and autoradiography, hypervariable sequences reveal very great diversity between individuals (Jeffreys, Wilson and Thein, 1985a,b; Vassart *et al.*, 1987). This high variation allows a comparison to be made to see if two DNAs are derived from the same individual. The hypervariable minisatellite regions discovered by Jeffreys are the most variable and therefore powerful hypervariable sequences known in man. Using Jeffreys' two minisatellite probes 33.15 and 33.6 sequentially the patterns revealed ('DNA fingerprints') vary so greatly that they can be calculated to be individual-specific, with the sole example of identical twins. In other species a wider range of probes may be available. The very high intrapopulation variation in *Drosophila melanogaster* in the chromosomal positions occupied by transposable elements (Montgomery and Langley, 1983) predicts that a restriction-digested DNA filter probed with a transposable element probe will also give an individual-specific pattern. 'DNA profiling' has been suggested as an alternative term to 'DNA fingerprinting' for studies of this kind since it avoids confusion with conventional fingerprinting in its forensic applications, and it leaves open the question of whether the banding patterns revealed are individual-specific.

The variation between individuals in such DNA profiles is so great that we are unlikely, in any reasonable sample size, to see the same pattern duplicated in unrelated individuals. Thus we cannot arrive empirically at an estimate of the low probability of two DNA profiles matching by chance. We have to use statistical models to estimate this probability, based on reasonable genetic assumptions and measurements that can be derived from the available data. The quantities normally used to measure DNA fingerprint information are x, the bandsharing, which is the proportion of bands shared by unrelated individuals in the population, and n, which is the mean number of scorable bands detected in the profiles. An inaccurate but generally statistically conservative assumption that is often made is that x is constant for all

NATO ASI Series, Vol. H 57
Molecular Techniques in Taxonomy
Edited by G. M. Hewitt et al.

bands. This allows the calculation of another useful quantity, q, ($=1-\sqrt{1-x}$), which is the probability that a given band is included in a random gamete, and which is again assumed to be constant for all bands. Some authors (Evett *et al.*, 1989) have used the statistic r, ($=q/x$) to represent the probability that an individual possessing a band passes it on to an offspring.

In addition to their use in establishing whether two DNAs have come from the same individual, DNA profile bands are inherited (with mutations occurring at an average rate of around 1 in 300 per band per generation, but being highly locus-dependent). This means that DNA profile bands can be used as genetic markers to establish the relatedness between individuals.

The initital 'DNA fingerprints' were created using hybridisation washes which allowed hybridisation at low stringency, i.e. even when the base sequences of the target DNAs on the filter were considerably diverged from those of the probe. The resulting autoradiograms show arrays of bands of high molecular weight derived from very many loci scattered around the chromosomes. The probes also detect a smear of low molecular weight bands that are not individually distinguishable. A typical minisatellite locus will yield zero or one bands in the scorable region of the autoradiogram. For this reason the individual bands in a DNA fingerprint, while heterozygous, are not usually allelic to any other bands in the scorable region. Allelism of bands in DNA fingerprints is thus generally ignored in the statistical interpretation. Such DNA profiling is now often referred to as using multi-locus probes, or MLPs. More recently, the trend has been towards the isolation of single-locus probes (SLPs). By screening a genomic library at low stringency with the multi-locus probe, one can isolate an array of different cloned minisatellites from diverse locations. By screening a filter of a DNA gel at high stringency, these probes detect the bands derived from a single genetic locus (Wong *et al.*, 1986). The pattern now produced is of two bands, unless the individual is homozygous at the locus. Differing minisatellite loci revealed in this way differ in their mutation rates and heterozygosities (Wong *et al.*, 1987). As would be expected from the population genetics of neutral alleles (Kimura, 1983), these quantities are strongly positively correlated (Jeffreys *et al.*, 1988). In simple matching between DNA profiles, the most variable loci are examined, since they give the information with the greatest statistical power. In paternity testing, these loci are generally also used, although in this application the high mutation rates associated with these loci are a disadvantage. Since all the bands visible in MLP DNA profiles, including those from the smeared unscorable region, are potentially obtainable by the exhaustive sequential probing with different SLPs, this technique can be made almost arbitrarily powerful. The only limit comes when all variable loci have been tried, or when the finiteness of the total genetic map length means that extra loci studied are correlated with earlier ones.

Hypervariable DNAs in forensic testing

The principles of the application of statistics to DNA profile information can be usefully illustrated by a consideration of forensic cases (Gill, Jeffreys and Werrett, 1985). In forensic situations the usual question that has to be applied to two DNA samples is whether they come from the same or from different individuals. Thus there could be a blood or semen stain at the scene of a crime and a blood sample donated by a suspect. These DNA samples can be compared and may be found to be indistinguishable. Clearly, this constitutes evidence that the suspect is the source of the scene-of-crime DNA, and thus probably guilty. The important question concerns the strength of this evidence. One advantage of the questions that DNA profiling is used to answer is that they generally consist of a choice between two statistically well-defined alternative hypotheses, which collectively have a total probability of one. This allows the statistical interpretation of the data using Bayes theory. If we have two statistically well-defined alternative hypotheses, then we are not constrained solely to calculate the significance of a deviation from a null hypothesis, but rather the ratio of two likelihoods calculated from our two alternative hypotheses. Suppose we find a match between two DNAs. If the DNAs came from the same source, then the probability of finding such a match is one (assuming there is no experimental error - see below). If they came from different sources, then the probability of a match is the probability that a random person shows the same DNA profile as that of the scene-of-crime DNA. This latter probability is low and has been calculated in a number of ways. One simple way is just to calculate the probability that all i bands seen in the scene-of-crime DNA are in the profile of a random man. This, given an assumption of band independence, is x^i, where x is the bandsharing between random individuals. This probability, of course, overstates the true probability of a match, since it represents only the probability all the scene-of-crime DNA bands are in a random man. The probability of a chance match is lower since we also require these to be the only bands in the suspect. We can estimate the number of bands in the population by n/x, with the result that the probability that we see the observed profile by chance is $x^i.(1 - x)^{(n/x-i)}$ (Jeffreys and Morton, 1987). The second term is obviously inaccurate, in that it involves raising a number to a power which includes a ratio between two poorly-measured sample means, and usually a simple x^i calculation is performed. The reason for this is that, with reasonable human estimates of x = 0.2 and n = 60, then, even if i was as low as 40, we would still arrive at values for a chance match of 10^{-28} and 7 x 10^{-54} using the two formulae. The important thing to realise here is that, whichever figure is used, it will be much smaller than any possible estimated probability of a chance match arising through human error, which must be 10^{-9} at least. Thus it is probably best to use x^i, since it contains fewer dubious statistical assumptions but will still give an abundantly significant result. It is unfortunate that great importance has been attached by some to potential errors in the use of these formulae (e.g. Cohen, 1990). Even if the hypothetical problems were real, we would still have a result which would be, for forensic testing, of unprecedentedly high significance.

It is the ratio between the two likelihoods that entirely encapsulates in a single number the strength of the evidence supplied by the DNA data in favour of one or other hypothesis. The likelihood ratio does not give the total probability of either hypothesis. The reason for this is that there may be other evidence for or against the suspect. This evidence could theoretically be expressed as a prior probability. This would be the probability that the man is guilty given all the evidence excepting the DNA. Such prior probabilities are not normally calculated, because the confidence that should be placed in some kinds of evidence, such as testimony of identification, cannot be easily quantified. Nevertheless, it will be true that, whatever the prior probability, the DNA evidence can be included in the following way:

$$\frac{\text{Posterior probability of guilt}}{\text{Posterior probability of innocence}} = \frac{\text{Prior probability of guilt}}{\text{Prior probability of innocence}}$$
$$\times \text{ Likelihood ratio from DNA evidence}$$

This standard Bayesian formula is a useful guide in situations in which the DNA evidence might be the same, but in which we might have very different confidence about the guilt of a suspect. As an illustration, consider the following two situations. In one case a suspect might be being tested at the DNA level as a result of evidence against him of other kinds, such as identification. In another situation a large number of individuals, perhaps identified as suspects only because they lived in the right town, could be tested. It could be that in each of these situations a match was found. The likelihood ratios would depend upon the particular profile of the scene-of-crime DNA but might well be the same. However, most people would agree that the evidence against the suspect in the latter case is weaker than in the former. This is, of course, correct and is reflected, not in the likelihood ratios from the DNA, but in the prior probabilities of guilt, which may be 50% in the first case and 1/5000 in the second. The prior probabilities of guilt and innocence are not generally calculable, and thus neither are the posterior probabilities, even with an accurate likelihood ratio. One unfortunate consequence of this is that it is a common error or, indeed, deception, for the likelihood ratio to be presented as the posterior odds, as in 'an innocent man has a one-in-a-million chance of producing a match, thus our suspect, who produces a match, has a one-in-a-million chance of being innocent'.

The use of single-locus probes has been complicated by the concept of a "match" between the suspect and the scene-of-crime DNA samples. The problem with this is that for reasons that are at least partially mysterious, the rate of movement of different preparations of the same DNA through an agarose gel may be different. This means that even if in fact the scene-of-crime sample came from the suspect the bands may be at different positions. Thus in seeking to see if two DNAs match what is usually done is to define a window of band mobilities and define any band that falls in the window as matching. The likelihood ratio depends on the probability of a chance match, which is the probability that an allele from a random individual will fall in this window. The concept of a match is one that can cause considerable difficulties. The main problem is that of the definition of the size of the window. One obvious approach would be to define the window size such that 95% of the time a true match will have the suspect's band fall within the window. This has the obvious and unfortunate consequence that 5% of the time a truly guilty person's band will fall outside the window and the person will be described as being eliminated by the DNA evidence. Furthermore, this 5% refers to a single minisatellite allele. As further bands are examined the proportion of guilty men being falsely eliminated will rise. The alternative is to increase the window size more and more, until maybe 99.9% of true matches will fall within the window. The consequence of this is that this increase out into the tails of the distribution increases hugely the probability that a random man's bands will fall within the window.

Some statistically illegitimate routes have been attempted out of this conundrum. If one apparently sees a match, one method is to measure the deviation between a band in the scene-of-crime stain and a band in the suspect and to calculate the probability that a random man would have a band that fell at least this close to the scene-of-crime band. This probability is now treated as the likelihood ratio. While initially appealing, this procedure is statistically illegitimate, as it consistently over-estimates the strength of the evidence against the suspect. Even if the suspect truly was the source of the scene-of-crime DNA it is not certain that the two bands would be as close as they are found to be. Thus the probability of the evidence given the hypothesis of a true match must be less than one. For an account of this and other statistical errors using single-locus probes in forensic testing, see Lander (1989).

For these various reasons it is suggested that the way forward is to discard the concept of the match altogether, and to use in its place precise probabilities of getting the observed deviation between the two bands (scene-of-crime and suspect) under the two hypotheses. In the case of the hypothesis of guilt, this probability would be calculated from a normal distribution determined by the measurement error. In the case of innocence, the probability would be calculated from the population frequency of bands of this mobility. The likelihood ratio would now be the ratio of two small probabilities, the probabilities of getting the data under the two hypotheses.

This problem is not so acute when using multi-locus probes, since in these situations bands can act as length standards for others. Furthermore, it seems probable that, even for single-locus probes, by improvements in technique such as mixing experiments these problems may be considerably reduced.

Biological applications in non-human species

While there may be forensic application involving non-human species, such as the identification of stolen birds of prey, the most important use of hypervariable DNAs is in the establishment of the relationships between organisms from wild populations (Burke, 1989). In humans there is now an established record in the use of DNA fingerprints in paternity testing (Jeffreys, Brookfield and Semeonoff, 1985; Brookfield, 1989a; Evett *et al.*, 1989; King, 1989). The normal situation is to have a mother, a child, and a putative father, and to try to establish the probability of the DNA data under two hypotheses, firstly, that the man truly is the father and, secondly, that he is not - a hypothesis that normally includes the assumption that he is not related to the child at all. Given the DNA profile of the mother and the putative father, it is easy to calculate the probability of the child showing its precise DNA profile under each of the two hypotheses. These probabilities will certainly both be very low, but what matters is their ratio. This likelihood ratio, which has been called the paternity index, and can be calculated for any kind of genetic variability (such as blood group or enzyme data), shows the strength of evidence given by the data in favour of one or other hypothesis. An alternative approach that has been adopted has been to calculate, given the child and the mother, the probability that the father has his DNA profile, under each of the two hypotheses. While the absolute values of these two probabilities will be, in any given case, different from those calculated by the other method, their ratio will still be the same as the paternity index. This method of calculating the probabilities of the father's profile, given the mother and child's, has the advantage that it leads easily into a useful approximation in which paternal-specific bands are identified in the child, and then the father is examined to see if he has some or all of these. If he truly is the father, then any paternal-specific bands in the child that he lacks must be due to new mutations.

A situation that is sometimes encountered is when only two individuals are available, such as a child and putative father. Paternal-specific bands in the child cannot now be identified and the only technique that can be adopted is to compare the two individuals merely on the basis of the numbers of bands that they share. Suppose we have two individuals with a and b bands respectively, of which c bands are shared between them. We can imagine two alternative hypotheses for their relationship, 1 and 2, such as parent-offspring and unrelated. We can calculate, from genetic arguments, the expected proportions of bands that the two individuals share. Let B_1 and B_2 represent the expected proportions of bands shared under the two hypotheses. If hypothesis 2 is no relationship, then $B_2 = x$. It is easy to show that the likelihood ratio under the two hypotheses is:

$$\frac{B_1^{c}.(1-B_1)^{a+b-2c}.(1-2x+xB_1)^{n/x-a-b+c}}{B_2^{c}.(1-B_2)^{a+b-2c}.(1-2x+xB_2)^{n/x-a-b+c}} \qquad (1)$$

The kinds of situations in which we might be interested in using these formulae are to measure how many offspring individual animals have as a means of measuring lifetime reproductive success or seeing if animals favour their relatives. Alternatively, we may be interested in female behaviour. We might find a single female and a collection of offspring and be interested in whether the offspring all have the same father. There might be 8 offspring, and we might find that offspring 1, 2, 4, 6 and 7 had an apparently different collection of paternal-specific bands to the other three. Thus we would like to calculate the probability of getting our DNA profiles under hypotheses in which the offspring have one or two fathers. The calculations of such likelihood ratios are straightforward, if tedious (Brookfield, 1989b). The only potential hazard is the need for care in the allocation of a prior probability to the two-father hypothesis. It may well be that we feel on the basis of what is known about the biology of the species that the collection of offspring have prior probabilities of one and two fathers that are equal. However, while there is only a single one-father hypothesis, the two-father hypothesis that has 1, 2, 4, 6 and 7 sharing the same father is just one of 127 ($= 2^{8-1} - 1$) two-father hypotheses, which differ in the distribution of the offspring between the two fathers. Thus, even if the data set is much more likely with this particular two-father hypothesis than with a one-father hypothesis, this has to be set against a 1 to 127 ratio of their prior probabilities.

Problems with Assumptions

The calculations which are used require certain assumptions to hold. Perhaps the most important of these is the independence of bands. In forensic cases where a match is found between a suspect's DNA profile and that of a scene-of-crime stain, the probability of a chance match is calculated as x^i. This probability being correct assumes that all the bands are independent of each other. In genetic terms, it requires that the bands are in linkage equilibrium. There is sometimes misunderstanding of the relationship between linkage and linkage disequilibrium. Linkage refers to the presence of different minisatellite bands on the same chromosome, such that they fail to assort independently of each other in a meiosis. Linkage disequilibrium, however, is a description of the association between bands in a population and means that the presence of one DNA band in an individual from a population makes it more likely that the individual has a specific other band. Thus linkage is neither a necessary nor a sufficient condition for linkage disequilibrium, although the concepts are connected in a subtle way. In a population with random mating on which there is no selection acting that is initially in linkage disequilibrium, linkage equilibrium is established very rapidly for loci which are genetically unlinked. For linked loci, however, linkage equilibrium is approached asymptotically and slowly over a number of generations, this rate of approach increasing with the genetic map distance. Linkage disequilibria are being generated constantly at minisatellite loci by the occurrence of new mutations which, being unique events, will initially be associated with a single chromosome from the population. (If separate mutations repeatedly generate indistinguishable alleles, then the linkage disequilibria expected will be less strong.) If a number of minisatellite loci are linked, then strong linkage disequilibria are likely to be found. There will thus be a tendency for linkage disequilibrium to increase with linkage. However, linkage disequilibrium may arise between unlinked loci if mating is non-random. If the population is divided into a number of races, these races may show collections of bands which, within an individual race, have high frequencies. However, in the population as a whole, these bands may have low frequencies. In the extreme case of racial subdivision, we can imagine that all the DNA profile variation is inter-racial and none intra-racial. We can consider a hypothetical example of such a situation in which the population is divided into 10 equally common races, and that all DNA profile bands are race-specific (i.e. found in only one race). Now the bandsharing x would be 0.1, and if we looked at 20 or 30 bands we would calculate, using x^i, probabilities of a random man matching at all bands as 10^{-20} or 10^{-30}. However, since there is no variation within races, there are, in fact, only 10 equally-frequent DNA profiles in the population - one for each race - and thus the true probability of a chance match is 10%. Thus, the statistical problems associated with undetected population subdivision are potentially major. The important issue, however, is the effect of real amounts of population subdivision on the accuracy of estimates based on x^i. We can imagine a situation in which bands are not entirely race specific but are found in at frequencies of 10% or 20% or so higher in one race than another. It can be shown (Brookfield, in prep.) that variation of this kind leaves estimates of the probability of chance matches based on x^i almost exactly correct. Furthermore, there are now direct estimates that show that there is very high variation in bands even between individuals within all tested races (Jeffreys, Turner and Debenham, submitted). The variation between human races is not, despite fears to the contrary (Cohen, 1990) sufficient to cause us to reasonably doubt conclusions based on the assumption of band independence in forensic or paternity cases. One important point which does emerge from these considerations, however, is that a new species being examined with DNA profiling for the first time might show more linkage disequilibrium than that found in humans. In some species the examination of large families has shown that bands are unlinked, which offers evidence that linkage disequilibria are unlikely (Burke and Bruford, 1987; Wetton *et al.*, 1987; Gyllensten *et al.*, 1989). In other species, however, segregation analysis has not been performed, and probability calculations assuming band independence may give grossly inaccurate results. In mice, for example, the *b* allele of the *MS 15-1* minisatellite locus yields 10 separate bands of *Hinf I* digestion, which are in complete linkage disequilibrium when recombinant inbred strains are compared (Jeffreys *et al.*, 1987).

Another set of assumptions concerns allelism. For reasons given above, it is assumed that bands of differing mobility in a fingerprint are not allelic to each other. This assumption may be incorrect but is conservative in that it almost always overestimates the probability of a chance match. If two bands are found in a scene-of-crime DNA the probability that they are both found in a suspect will normally be calculated as x^2, which assumes they are not alleles. If, however, these bands are heterozygous alleles, the probability that a random man would have both is $2q^2$. This will be less than x^2 if $q < 0.58$, which it always will be. There is a further assumption that the bands of equal mobility in different individuals are the same band, whereas it could be that they are bands at different loci which, by chance, have the same mobility. Inaccuracy of this assumption only has any effect if we are testing for relationship in a pedigree with inbreeding, and even in this situation it can be shown (Evett and Brookfield, submitted) that the effect is slight.

What kinds of relationship can be detected?

If two individuals from a wild population are to be examined and we wish to determine, using MLPs, their relationship, then formula (1) above allows us to calculate the likelihood ratio for their DNA profiles under two relationship hypotheses. The power of such a test relies on the extent to which B_1 and B_2 are different. As pointed out by Lynch (1989), however, when the more distant degrees of relationship are being compared, the expected bandsharing values become increasingly similar. This means that very large numbers of bands would be required before cousins and second cousins, for example, could be distinguished. With single-locus probes, one can theoretically screen individuals' DNAs with arbitrarily large numbers of probes. This is not a complete solution to the problem, however, since once the number of probes starts to approach the number of chromosome pairs, there will be an increasing probability of linkage between the loci being examined that will prevent the normal statistical assumptions used from being correct. It may be that with these very distant degrees of relationship it will be impossible to establish with certainty the relation between the individuals. This is not, however, merely a technical problem. If the total map length is short, and the difference between the coefficients of relatedness corresponding to the two relationship models is small, then it may well be that, for example, 2 second cousins have fewer genes in common than do 2 third cousins in a particular case.

A numerical illustration will show the lack of power of the technique. Suppose we have a pair of human cousins and that each has 60 detectable bands, which is also n (the average number of bands), using the combination of the two Jeffreys probes. A good estimate of x is 20%. From this, one can easily calculate that the expected bandsharing between cousins is 29.44%. The most probable number of bands shared for two such cousins is thus 18. If they were unrelated the most likely number of bands shared would be 12. If we have a = b = 60, and c = 18, we can use (1) above to calculate the likelihood ratio. The number resulting is 8.75. Thus, the most probable data set if the individuals are cousins is only 8.75 times more probable than it would be if they were unrelated. Therefore it will normally be impossible to establish with statistical confidence that cousins are related.

Conclusions

Hypervariable DNA sequences have revolutionised forensic science. They can reveal so much information that the probability of a chance match between two individuals is vanishingly small. They can also be used, with slightly less power, to demonstrate close relationships between individuals from a wild population. Their use in studying relationships in wild populations is limited in two ways. Firstly, their high mutation rates make them unsuitable for the study of distant relationships such as that between individuals from different populations. Secondly, it should always be remembered that the establishment of the relatedness between individuals in a wild population, even if it was possible, is not a satisfactory end in itself. We should always consider what biologically important questions of a general kind such information will help us to answer.

I thank Professor A.J. Jeffreys, F.R.S., for numerous helpful comments on an earlier draft of this paper.

References

Brookfield JFY (1989a) Analysis of DNA fingerprinting data in cases of disputed paternity. IMA Journal of Mathematics Applied in Medicine and Biology 6: 111-131

Brookfield JFY (1989b) How many fathers? Fingerprint News July 1989: 7-9

Burke T (1989) DNA fingerprinting and other methods for the study of mating success. Trends in Ecology and Evolution 4: 139-144

Burke T, Bruford MW (1987) DNA fingerprinting in birds. Nature 327: 149-152

Cohen JE (1990) DNA fingerprinting for forensic identification : potential effects on data interpretation of subpopulation heterogeneity and band number variability. Am J Human Genetics 46: 358-368

Evett IW, Werrett DJ, Buckleton JS (1989) Paternity calculations from DNA multilocus profiles. J Forensic Sci Soc 29: 249-254

Gill P, Jeffreys AJ, Werrett DJ (1985) Forensic application of DNA 'fingerprints'. Nature 318: 577-579

Gyllensten UB, Jakobsson S, Temrin H, Wilson AC (1989) Nucleotide sequence and genomic organisation of bird minisatellites. Nucl Acids Res 17: 2203-2214

Jeffreys AJ, Brookfield JFY, Semeonoff R (1985) Positive identification of an immigration test-case using human DNA fingerprints. Nature 317: 818-819

Jeffreys AJ, Wilson V, Thein SL (1985a) Hypervariable minisatellite regions in human DNA. Nature 314: 67-73

Jeffreys AJ, Wilson V, Thein SL (1985b) Individual-specific 'fingerprints' of human DNA. Nature 316: 76-79

Jeffreys AJ, Wilson V, Kelly R, Taylor BA, Bulfield G (1987) Mouse DNA 'fingerprints': analysis of chromosome localization and germ-line stability of hypervariable loci in recombinant inbred strains. Nucl. Acids Res. 15: 2823-2836

Jeffreys AJ, Morton DB (1987) DNA fingerprints of dogs and cats. Animal Genet 18: 1-15

Jeffreys AJ, Royle NJ, Wilson V, Wong Z (1988) Spontaneous mutation rates to new length alleles at tandem-repetitive hypervariable loci in human DNA. Nature 332: 278-281

Kimura M (1983) The Neutral Theory of Molecular Evolution. Cambridge University Press

King MC (1989) Invited editorial: genetic testing of identity and relationship. Am J Hum Genet 44: 179-181

Lander ES (1989) DNA fingerprinting on trial. Nature 339: 501-505

Lynch M (1988) Estimation of relatedness by DNA fingerprinting. Mol Biol Evol 5: 584-599

Montgomery EA, Langley CH (1983) Transposable elements in Mendelian populations II Distribution of three *copia*-like elements in a natural population. Genetics 104: 473-483

Vassart G, Georges M, Monsieur R, Brocas H, Lequarre AS, Christophe D (1987) A sequence in M13 phage detects hypervariable minisatellites in human and animal DNA. Science 235: 683-684

Wetton JH, Carter RE, Parkin DT, Walters D (1987) Demographic study of a wild house sparrow population by DNA fingerprinting. Nature 327: 147-149

Wong Z, Wilson V, Jeffreys AJ, Thein SL (1986) Cloning a selected fragment from a human DNA 'fingerprint': isolation of an extremely polymorphic minisatellite. Nucl Acids Res 14: 4605-4616

Wong Z, Wilson V, Patel I, Povey S, Jeffreys AJ (1987) Characterisation of a panel of highly variable minisatellites cloned from human DNA. Ann Hum Genet 51: 259-288

A MULTIDIMENSIONAL APPROACH TO THE EVOLUTION AND SYSTEMATICS OF *DOLICHOPODA* CAVE CRICKETS

V.Sbordoni, G.Allegrucci and D.Cesaroni

Department of Biology
University of Roma "Tor Vergata"
Via Orazio Raimondo
00173 Roma
Italy

INTRODUCTION

In this paper we will address cave crickets as material, and molecular techniques and multivariate morphometrics as methods, to investigate problems of systematics (i.e.: classification and phylogenetic inference) at the species level, by means of a microevolutionary approach. This outlines a field laying at the interface between population genetics and taxonomy. In a recent article, Felsenstein (1988) complained of little communication existing between population genetists and systematists and pointed out that systematic studies can greatly benefit from facing microevolutionary approaches. Anybody having been trained in both fields can acknowledge that a tighter link should exist between the study of historical processes and the study of underlying mechanisms.

Here we attempt to fill this gap showing that the comparative study of geographic variation patterns of various characters is a wise tool to select characters meaningful for phylogenetic inference. Any character evolving at a reasonably constant rate over a given period of time can be regarded as a useful phylogenetic tracer at the time scale considered. Accordingly, characters subjected to large directional selection pressures or to other significant environmental effects affecting their variation are not such good tracers. On the other hand, characters whose variation reflects a prevailing role of the balance between mutation rate and genetic drift in isolation (i.e.: reduced or absent gene flow) can be regarded as meaningful estimators of historical evolutionary events.

NATO ASI Series, Vol. H 57
Molecular Techniques in Taxonomy
Edited by G. M. Hewitt et al.

Characters vary over time and in space as well. Analysis of variation over time is severely limited by obvious constraints. However, the study of geographic variation was a long recognized tool to investigate the relative roles of evolutionary factors. Numerous quantitative methods have been developed for analyzing patterns of spatial variation that also allow treatment of multivariate sets of data. We utilized some of these methods to investigate the behaviour of sets of characters which will subsequently be employed in systematic analyses and tree reconstruction. Since patterns of variation may change according to the scale considered, they were studied both at the within- population and the across-species level.

Although phylogeny is not the primary scope of our studies, we believe that careful sifting of characters on the basis of their biological properties does represent the stuff of phylogenetic inference. This is not to say that current controversies on superiority of methods to treat data (e.g: discrete character states vs. distance measures) or to construct trees (e.g.: parsimony, compatibility, phenetic clustering etc.) are worthless. Developments of these methodologies have largely contributed to the establishment of systematic biology as a science. But, attention to methodology should not obscure the differences existing among characters in their informational value, or even limit their choice on the basis of *a priori* assumed geometric or esthetic properties.

Dolichopoda cave crickets are currently the object of extensive research in our laboratory aimed at elucidating patterns of adaptation, population divergence and speciation by combining several methodologies and approaches. A simplifying feature of these cave dwelling organisms is that populations are usually isolated each other, and that species show vicariant allopatric ranges. Moreovever, in some cases, cladogenetic events can be dated by palaeogeographic evidence. A rather extensive deal of biological information on these crickets has been gathered over the last 15 years, for microevolutionary studies (Sbordoni et al., 1987). Basically our approach attempts to characterize the place of any population in the multidimensional space described by molecular, morphological and ecological variables and to compare the resulting ordinations for a given set of populations.

At this time, a sufficiently wide sample of populations belonging to different species, and a reasonably diversified array of descriptors have become available to compare, (a) patterns of variation at the micro- and macro-geographic scale from different sets of descriptors and, (b) trees for an assumed monophyletic sample of five species resulting from either molecular or morphological data sets.

Molecular descriptors here considered include allele frequencies at enzyme loci and DNA-DNA hybridization data. Morphological descriptors include morphometric variables describing the external morphology, the shape of the epiphallus (i.e.: male copulatory structure), and the patterns of egg chorion ultrastructure as well. Other character sets utilized for comparative purposes were related to the ecology and life cycle of populations studied.

SYSTEMATIC BACKGROUND, GEOGRAPHICAL DISTRIBUTION, AND ECOLOGY OF THE GENUS *DOLICHOPODA*

The cave cricket genus *Dolichopoda* belongs to the family Rhaphidophoridae and consists of around thirty species discontinuously distributed throughout the North Mediterranean region from the Pyrenees to Turkish Armenia and the Caucasus. Areas of high species diversity include insular and peninsular Italy and Greece. In Italy this genus is represented by nine species distributed in some minor Tyrrhenian islands and a large part of continental Italy with the exception of most of the Alpine region, the Po valley, the North -Western Apennines and peninsular Apulia.

An up to date map of the sampling localities and the ranges of the Italian species are reported in Figure 1. As indicated by the map the various species show a typical allopatric, vicariant distribution. However at least one instance of sympatric occurrence was revealed in a small cave

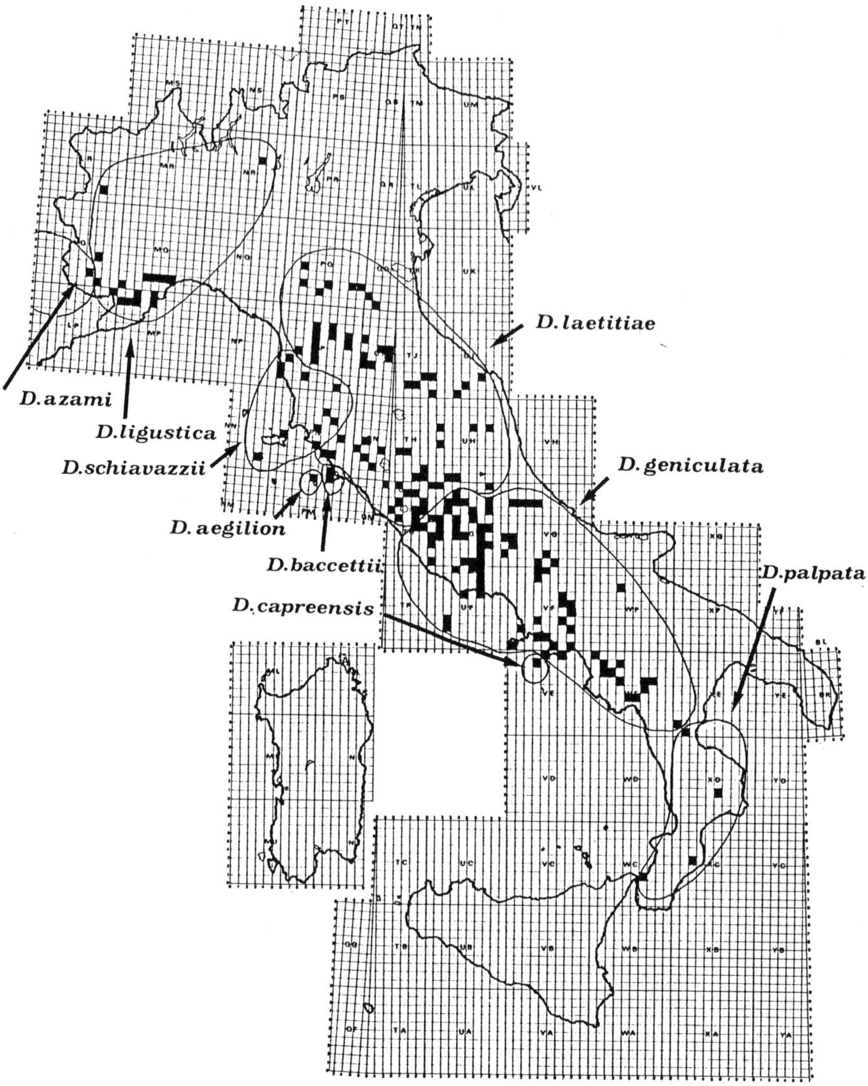

Figure 1. The genus *Dolichopoda* in Italy: sampling localities and species ranges.

on the Monte Argentario promontory in Tuscany, where the endemic *D.baccettii* coexists with *D.schiavazzii*. This last species had probably been unintentionally introduced by Passionist Fathers established in this area since the 18th century. Allozymic studies provided evidence that the two species hybridize at some extent but could'nt ascertain the occurrence of introgression (Allegrucci et al., 1982).

The taxonomic arrangement is still open to change

especially in the *laetitiae-geniculata* complex. This group includes *D.laetitiae laetitiae*, *D.laetitiae etrusca*, *D.geniculata geniculata*, *D.geniculata pontiana*, *D.capreensis*, *D.palpata*, which had been mainly separated by the morphology of the tenth tergite and the epiphallus shape (see Baccetti and Capra, 1970; Capra, 1967, 1968; Sbordoni et al., 1979, Baccetti, 1982). In some populations low morphological differentiation is associated with either identical or different chromosome number and the ranges of allozyme genetic distance values between morphologically conspecific and heterospecific populations show consistent overlap (Baccetti, 1982; Sbordoni et al., 1985; Allegrucci et al., 1987).

Most populations are strictly dependent upon caves, a few others may occur in cave-like habitats such as rock crevices and ravines in mesic or moist woods. In Italy, *Dolichopoda* populations often occur also in man made hypogean habitats, such as cellars, catacombs, Etruscan tombs and aqueducts. These environments differ from each other in several features. Natural caves are habitats established since a rather long time, relatively stable and predictable in climate, community patterns and trophic resources. On the contrary, artificial caves are younger habitats, often seasonally variable in climate, poor in trophic resources, and characterized by simpler and relatively unpredictable community patterns. Differences in population size, phenology, age structure and other life history traits appear to characterize populations living in either type of habitat (Carchini et al., 1983, 1990; Di Russo et al., 1987).

VARIATION OF ALLOZYMIC CHARACTERS

Within population variation

In *Dolichopoda* populations mean heterozygosity estimates based on 26 gene loci range from 0.081 to 0.198, with an across-species average of 0.149 (unpublished data). These values are of the same order of magnitude of values found in other cave cricket genera such as *Troglophilus*, *Hadenoecus*,

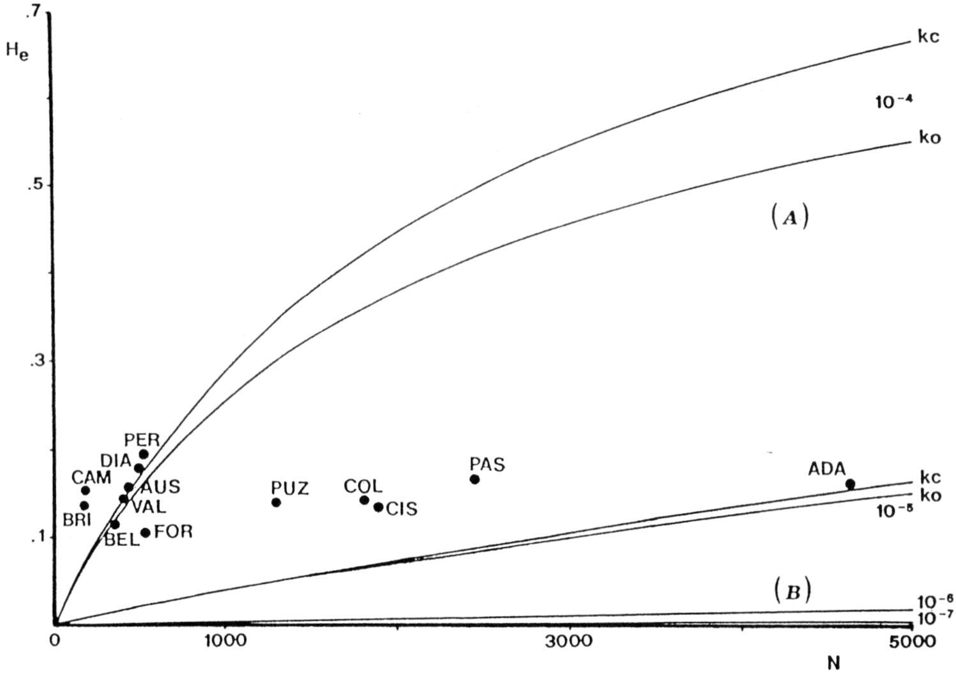

Figure 2. Empirical *versus* predicted relationships between heterozygosity at enzyme loci (He) and population size (N) in *Dolichopoda* populations. Experimental values are plotted against theoretical curves calculated at different mutation rates ($10^{-7} < \mu < 10^{-4}$) from the neutral evolution models by Kimura and Crow, 1964 (kc) and Kimura and Otha, 1973 (ko).

Euhadenoecus if compared at a common set of loci (Sbordoni et al., 1981, 1987; Caccone and Sbordoni, 1987). These values are also comparable to variability estimates from several other cave organisms (see reviews in Sbordoni, 1982; Culver, 1982), and show that even small cave populations can maintain considerable amounts of genetic variation.

Population size estimates available for several populations of *Dolichopoda* were in the class of magnitude $100 < N < 10{,}000$ (Carchini et al., 1982, 1983). From these data we tested the relationship between heterozygosity (He) and population size (N) predicted by the basic formulas of the "infinites sites" model (Kimura and Crow, 1964) and the "stepwise mutation" model (Kimura and Otha, 1973). Results from this study (Sbordoni et al., 1987) showed that to obtain the heterozygosity levels observed in *Dolichopoda*, by using the population size estimated, one should assume mutation rates two or three orders of magnitude higher than those commonly found in electrophoretic variants (Kimura and Otha, 1971; Nei, 1975; Nei and Graur, 1984). Figure 2 shows the curves theoretically expected from the two models, when the mutation rates assumed were (A) the best average value fitting to the distribution of experimental data points and (B) those commonly assumed for electrophoretic variants. A striking departure of experimental heterozygosities from theoretical ones was apparent in both cases. Higher levels of genetic variability in *Dolichopoda* populations did not necessarily correspond to larger population sizes. Although a departure of observed heterozygosities from theoretical ones is expected in default, because the possible influence of past bottlenecks, the neutral theory does not predict at all an excess of heterozygosity. In Figure 2 a notable example is represented by the CAM population of *Dolichopoda aegilion* restricted in a very small limestone area in the Giglio island (Tuscan Archipelago) and isolated by sea from its continental relatives since at least 700,000 years. Estimated population size for this population was N = 168 ± 150, a very low figure compared to its relatively high average heterozygosity (He = 0.147 ± 0.038).

These findings suggest that genetic drift and gene flow alone are not good predictors of levels of genetic variability in *Dolichopoda* populations and that some form of balancing selection could be involved.

Spatial patterns: microgeographic level

The peculiar structure of Cerveteri's Etruscan necropolis "La Banditaccia", near Rome, and the presence therein of a population of *Dolichopoda laetitiae*, gave us the opportunity of a detailed study on the population structure in relation to ecological parameters and topography as well.

The necropolis consists of a few hundred tombs, excavated in the tufa rock, varying in morphology, size, climate and community patterns. These tombs represent a peculiar habitat

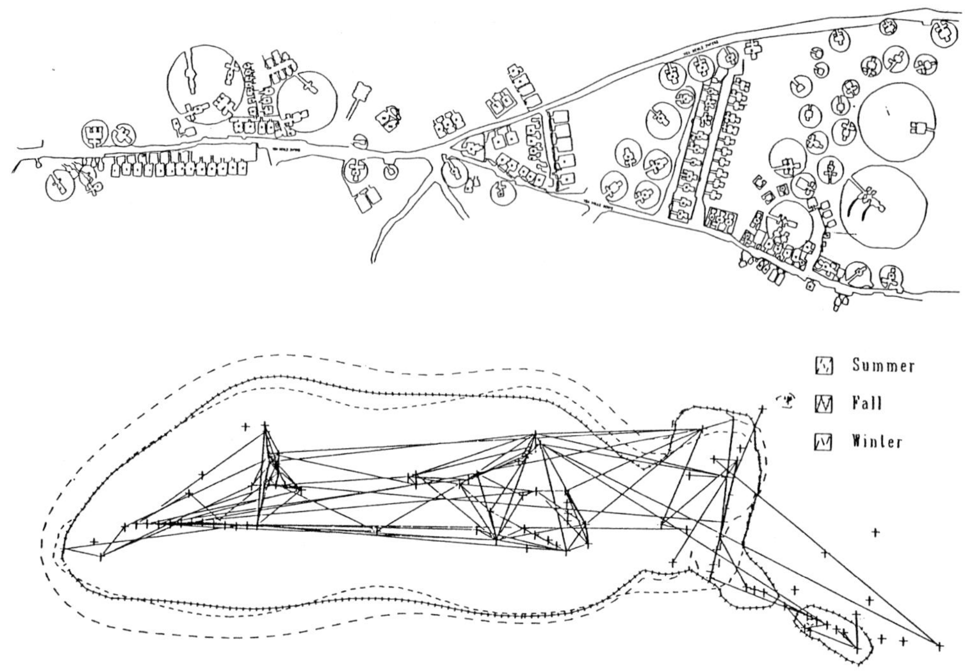

Figure 3. Map of Cerveteri's Etruscan necropolis (above) and plot of the movements network of individually marked *Dolichopoda* (below). The areas enclosed in the three polygons show an almost continuous utilization of the whole area in different seasons. The wider population home range revealed in fall was associated with a grater population size as estimated by mark-recapture data.

for several other cave organisms beside *Dolichopoda*. Most, yet not all, tombs are inhabited by varying number of crickets, representing more or less stable subdivided units.

The opportunity to quantify several parameters such as microclimate, community patterns, and dispersion patterns among tombs, led us to investigate whether any evidence of genetic structuring associated to habitat utilization was detectable within the overall *Dolichopoda* population. The ultimate question raised was: is there any evidence that allozyme polymorphism could be maintained by habitat choice?

Three thousand individually labelled crickets were the basis of a year long investigation of population and "unit" sizes, dispersion patterns, age and sex structure. A sub-sample of adult individuals was genotypically characterized through cellogel electrophoresis at 22 loci (Sansotta et al., 1989).

Figure 3 shows the map of the necropolis and the network of individual displacements across tombs. This distribution was tested for discontinuities and was proved to be not statistically different from a hypothetical distribution of uniform utilization based on population densities at each tomb (Kolmogorov test, $P < 0.05$). The areas enclosed in the three polygons revealed a similar preferential utilization of the necropolis in the three seasons analysed.

In order to quantify and summarize ecological differences within the necropolis, principal component analyses were carried out to study the ordination of tombs on the basis of several biotic and environmental variables measured in different seasons during the mark-recapture survey. Since temperature was well correlated to the remaining variables and accounted for over 70% of the overall variance, Cerveteri's necropolis was subdivided in five tomb groups with respect to topographical distance among the tombs and to average temperature within any group of tombs.

Each of 66 sexually mature individuals tested by electrophoresis, was assigned to one or another group on the basis of the tombs of recapture. The values of Wright's F, calculated for this five groups at the 9 polymorphic loci, showed some evidence of structure. Tested with X^2, all F_{ST} values were significant, suggesting a non random distribution of genotypes with respect to habitat diversity.

Due the small sample size these results are still preliminary and further research is now in progress. However, if these data will be confirmed, they would be particularly inspiring of optimal habitat choice as a possible mechanism involved in the maintainance of polymorphism in *Dolichopoda*. In fact, the high rate of migration of crickets between tombs and the non random distribution of adult genotypes, recaptured in tombs where they had actively migrated, seem to exclude other possible mechanisms for genotypic structuring, such as incomplete mixing. (For a general discussion on habitat choice see: Taylor, 1976, Taylor and Powell, 1978, etc.).

Spatial patterns: geographic level

Let us now look at the geographic variation of allozyme descriptors in *Dolichopoda* from a broader window. Twentyfive populations belonging to five morphologically inferred species from Central-Southern Italy (1 population of *D.aegilion*, 3 of *D.baccettii*, 6 of *D.schiavazzii*, 6 of *D.laetitiae*, and 9 of *D.geniculata*) have been electrophoretically studied for 15 gene loci (Sbordoni et al., 1985).

Interspecific mean values of Nei's genetic distance between *Dolichopoda* populations ranged from 0.107 to 0.405. Intra- and inter-specific distances were clearly distinct in all species except in comparisons between *D.laetitiae* and *D.geniculata*. Clinal variation of allele frequencies at a few loci, in particular at the *Pgm* locus, had been previously revealed within the *laetitiae-geniculata* complex (Sbordoni et al.,1976). Although this variation appeared to be associated to some environmental factors, such as altitude and temperature, and showed patterns of statistically significant spatial autocorrelation, different multivariate analyses showed that the overall geographic variation of allele frequencies is mainly influenced by lack of gene flow (i.e.: past or present occurrence of geographic barriers). In fact, peripheral populations or groups of populations adjacent to the Tyrrenian and Adriatic coast resulted genetically highly differentiated from each other and from the central inland group of populations. Actually most of these peripheral populations are strictly dependent upon caves because the dry Mediterranean "macchia", prevailing in these coastal areas, prevents occurrence of *Dolichopoda* outside caves. This eliminates gene flow among cave populations. The inland group was genetically more homogeneous, regardless of the geographic distance and even of their different taxonomic attribution to either *laetitiae* or *geniculata*.

Relationships within the *laetitiae-geniculata* complex have also been investigated by using the correspondence analysis of allelic frequencies and the kriging linear interpolation (Matheron, 1969, 1970). This last technique, widely employed in

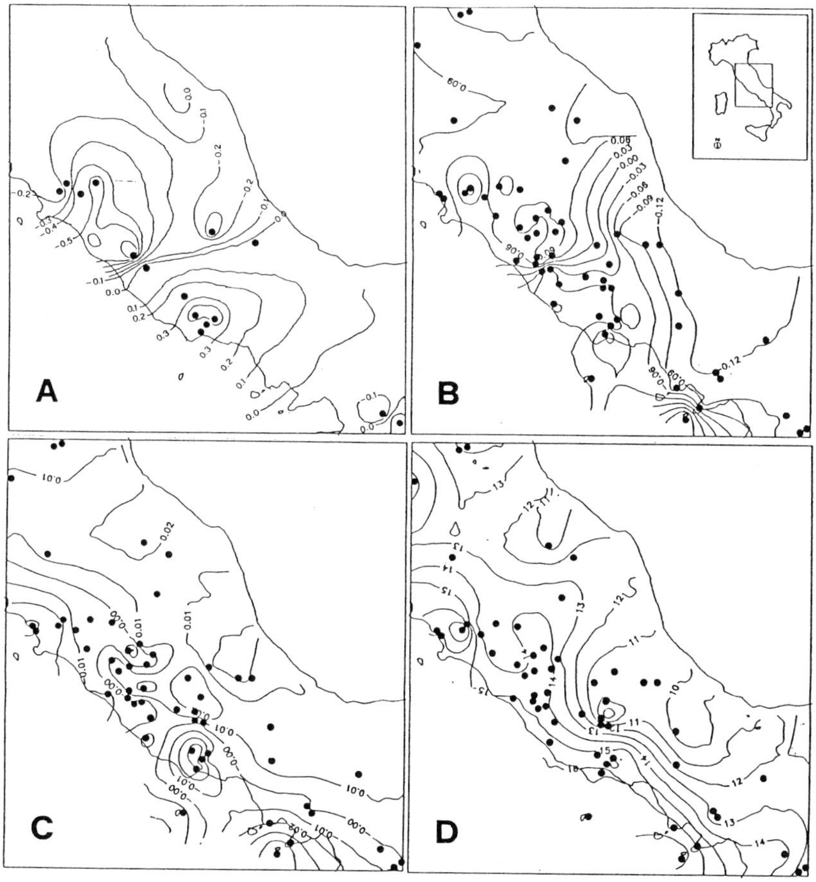

FIGURE 4. Kriging linear interpolations of different character sets; this technique utilizes weights estimated on the basis of spatial autocorrelation functions. A: interpolation of the first axis from a correspondence analysis performed on allele frequencies of 15 populations. B: analogous interpolation on epiphallus morphometric descriptors of 57 populations. C: analogous interpolation on body and appendage morphometric descriptors. D: interpolation of average cave temperature. Note that the isolines in A and B show a general NW-SE trend, with a strong discontinuity corresponding to the Tiber valley, whereas in C the isolines have a mainly SW-NE orientation, i.e. from low to high altitudes, similar to the isotherms in D.

geostatistics, utilizes weights estimated on the basis of spatial autocorrelation functions. In Figure 4A a kriging map of allozyme frequencies is represented, on the basis of the first principal axis of the correspondence analysis. Also in this case peripheral populations show levels of genetic differentiation higher than the inland group, but it is especially outlined a steep variation corresponding to the Tiber valley which appears to represent a notable barrier to gene flow.

Spatial autocorrelation analysis of allozyme frequencies led to similar results (Barbujani and Sbordoni, unpublished). With the exception of the previously discussed *Pgm* locus, a substantial lack of spatial autocorrelation was revealed, suggesting the absence of either constant gene flow, even very low, or selection pressures producing a geographic structure.

VARIATION OF MORPHOMETRIC CHARACTERS

Morphological investigations of geographical variation have progressed to understand the relative roles of gene flow, genetic drift, selection and phenotypic plasticity in shaping the patterns of intra- and interspecific phenotypic variability.

Within the *D.laetitiae-geniculata* complex, studies on two different morphometric character sets on 57 populations sampled from a variety of cave biotopes in Central and Southern Italy have been undertaken (Allegrucci et al., 1987; Cesaroni et al., 1989, and in prep.). A series of multivariate analyses performed on the two morphometric character sets, compared to a set of environmental descriptors, allowed to test the degree of matching of patterns within the geographical "window" utilized for this study.

Epiphallus morphology

The first set describes the morphology of the adult male epiphallus, which has long been considered as a major diagnostic character in the conventional taxonomy of this group. We assumed that epiphallus shape is a character that should not be under strong selective pressures because the allopatric, vicariant distribution of *Dolichopoda* populations and species. Therefore, besides its diagnostic values, it could be considered a "tracer" of the gene flow and genetic drift effects. The numerical description of the epiphallus shape was obtained utilizing the measurements of six sections along the epiphallus axis and the length of the axis itself.

Results of ordination analyses confirmed that the male epiphallus shape shows significant differences between species, and consequently a discontinuous trend in geographical variation (Cesaroni et al., 1989; and in prep.). The kriging linear interpolation based on the first principal axis of a correspondence analysis summarizes the geographical variation patterns (Figure 4B). Although the epiphallus morphology presents clinal variation, slightly associated with SW-NE environmental gradients, the major source of variability has a NW-SE direction. The strong discontinuity observed in this trend coincided again with the Tiber valley which therefore stands as the main biogeographic barrier in the studied area. A second discontinuity clearly reflects the outlier position of the Capri island population of *D.capreensis*. All that seems to substantiate the hypothesis that epiphallus shape is relatively independent of selective pressures and that it is thus a good tracer of the degree of genetic (reproductive) isolation of populations. That may be revealable also by comparing kriging interpolations on allozymic data (Fig.4A) and epiphallus shape data (Fig.4B). Notwithstanding the different number of population utilized, the resulting patterns are coarsely similar; note that in the allozyme study the Capri population had not been investigated.

Body and appendage morphology

The second morphometric character set describes body and appendage morphology of adult males. Five characters were measured: pronotum length and width, metafemur, metatibia and metatarsus length. A preliminary study, involving only 15 populations with a known pattern of allozymic variation, already suggested that variation of these descriptors would result from either selection or phenotypic plasticity, rather than being the expression of isolation of gene pools (Allegrucci et al., 1987). In fact, these morphometric indices and allele frequencies failed to show congruent geographic variation.

Multivariate analyses performed on the increased sample of 57 populations supported that hypothesis (Cesaroni et al., 1989, and in prep.). Body and appendage morphometric descriptors did not allow to discriminate among species, but showed a clinal variation across cave biotopes characterized by different bioclimate and vegetation types (from the Mediterranean zone to mountain areas). The spatial trend was well related not only to geographical structure of the sampling design, but also to the spatial configuration of ecological descriptors. Average cave temperature standed as the most important ecological factor associated with external morphometric variation. A canonical correlation analysis between body and appendage morphology *versus* environmental descriptors also confirmed the importance of temperature. Figures 4C and 4D show the kriging linear interpolations respectively of body and appendage morphometric descriptor and average cave temperatures. It is noticeable that body and appendage morphology are almost entirely under the control of environmental gradients, which develop eastward from warmer mediterranean to cooler mountain areas. Such findings suggest that body and appendage morphology is a good tracer of adaptive responses and/or the "mean" phenotypic plasticity within the populations studied.

The hypotheses on the relative value of different morphological descriptors, which we outlined in the introduction, are in part substantiated by the results of these analyses. Actually, the clinal variations of body and appendage morphology (Fig.4C) are not correlated with the epiphallus shape, but on the contrary the patterns in Figures 4B and 4C are orthogonal to each other.

A COMPARATIVE APPROACH TO THE SPECIES LEVEL RELATIONSHIPS

Since *Dolichopoda* cave crickets have been the object of extensive research, they are an excellent material to attempt to examine the degree of taxonomic congruence between different data sets, where the same group of taxa is examined using the same type of analytical method for both molecules and

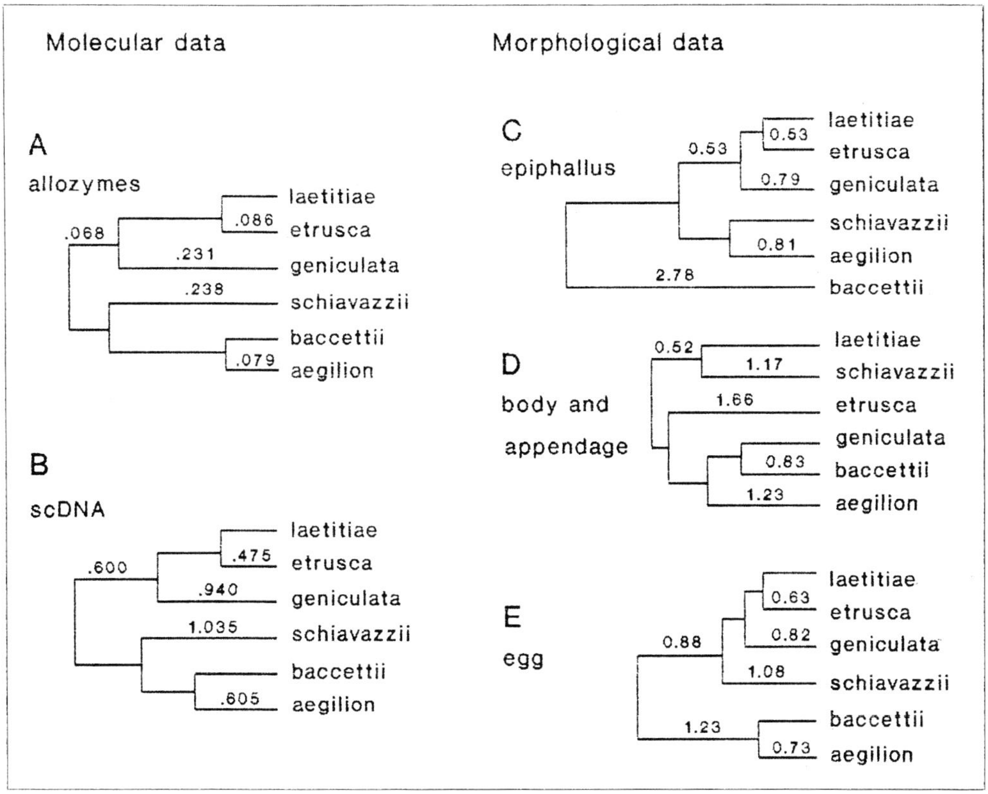

FIGURE 5. Comparison of UPGMA trees deriving from different molecular and morphological character sets in six allospecific populations. Note that the topology of the dendrogram is not the same for all types of data. Whereas both molecular data sets lead to similar tree topology, although with different branch lenghts, morphological data sets are more or less dissimilar from each other and from molecular ones. See text for further explanations.

morphology. Studies of five species of *Dolichopoda* cave crickets were available to compare molecular data (allozymes and single-copy DNA) with each other, with three different morphological data sets (body and appendage morphology, epiphallus morphometry and egg chorion ultrastructure), and finally with a series of ecological descriptors of the populations studied (Allegrucci et. al., 1989, and in prep.).

The species selected were: *D.schiavazzii*, *D.baccettii*, *D.aegilion*, *D.geniculata* and *D.laetitiae*. In order to make the picture clear we utilized only one population for each species, with the exception of *D.laetitiae* where one population for each subspecies, *D.l.laetitiae* and *D.l.etrusca*, was considered. The populations were the same six for each character set. For comparative purpose the clustering algorithm used was always the UPGMA (Sokal and Michener, 1958). It is well established that dendrograms constructed by this phenetic clustering method can be interpreted reasonably in a phylogenetic sense when rates of evolutionary divergence are relatively homogeneous across phyletic lines (Farris, 1972). This indeed occurred in our molecular data sets, where trees constructed by alternative procedures as distance Wagner analysis (Farris, 1972), least-square methods (Fitch and Margoliash, 1967; Cavalli Sforza and Edwards, 1967), and neighbor-joining method (Saitou and Nei, 1987), produced similar topologies and corresponding branch lengths. Therefore, molecular generated trees were used as standards for comparison with other UPGMA trees based on morphology.

Molecular data

Allele frequency data for 26 enzyme loci were the basis for computing genetic distances. Unlike previous work carried out in our laboratory, cellulose acetate was used as support medium rather than starch gel. Cellulose acetate offers several technical advantages and a better resolution with respect to starch gel.

Figure 5A shows the UPGMA dendrogram constructed on unbiased genetic distances (Nei, 1978). Allozyme data separate two clusters: *D.laetitiae-D.geniculata* and *D.schiavazzii-D.baccettii-D.aegilion*. However, the internodal distance separating these two clusters is relatively short while the branch lengths leading to each species are comparatively longer. Actually, the standard errors associated with branching points (Nei et al., 1985) indicate that nodes within groups are statistically significantly different, whereas nodes at higher levels of genetic distance are not. This result suggests that allozymes well differentiate the terminal tips of a branching pattern, but they are less sensitive to higher levels of divergence.

The overall genetic divergence at nuclear genes in *Dolichopoda* was also estimated by single-copy DNA hybridization (Allegrucci et al., 1989, and in prep.). The technique relies on the fact that thermal stability of DNA duplexes is determined by the fidelity of base-pair matching, A with T and G with C. The relationship between thermal stability and base-pair mismatch is thought to be quantitative and linear (Kohne, 1970; Britten et al., 1974). The methods are described elsewhere in this book (Caccone and Powell, 1990; Powell and Caccone, 1990). The median difference in thermal stability between the median melting temperature of the homo- and hetero-duplex molecules, ΔTm, was calculated for each comparison. ΔTm measures the sequence divergence between the two taxa compared.

Figures 5B illustrates the relationships among the six populations studied based on their ΔTm values. As the ΔTm's have associated standard errors ranging from 0.15 to 0.36°C, the nodes linking the two *D.laetitiae* populations to the *D.geniculata* population are not significantly different. The other nodes are all statistically significantly different from one another even using conservative tests such as the non-parametric Mann-Whitney U-test.

ΔTm values provides a measure of divergence of the entire single-copy genome, while Nei's D provide information on the divergence of a small sample of single-copy genes, i.e.: a subset of the genes coding for soluble enzymatic proteins. These two measures should be related and indeed they are. Figure 6 shows the relationships between D and ΔTm for all cave crickets studied, *Dolichopoda* (Allegrucci et al., 1989), *Hadenoecus*, and *Euhadenoecus* (Caccone & Sbordoni, 1987; Caccone & Powell, 1987). Intergeneric comparisons with values of Nei's D > 1 and $\Delta Tm > 10$ were not reported because the reduced reliability of both measures at those levels. However, distance measures, which are within the reliability range of each technique, seem to be linearly related within each genus. Comparisons were performed with D values calculated on the basis of 26 gene loci common to all populations of the three genera. However, the steepness of the line is quite different

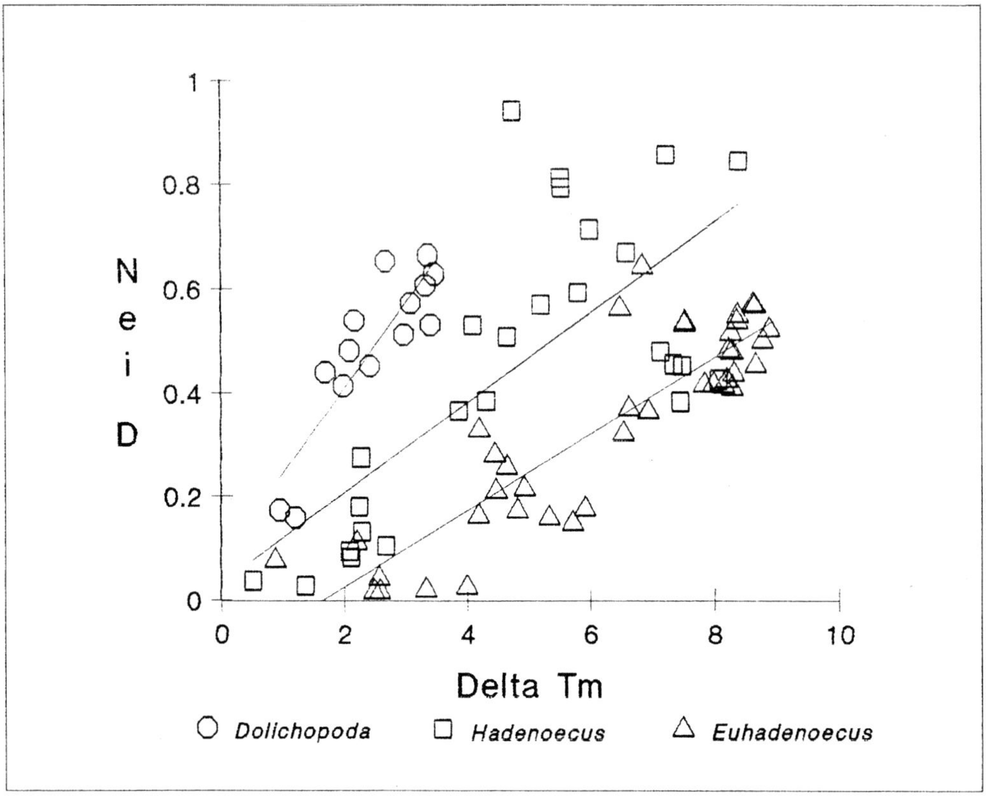

FIGURE 6. Relationships between measures of divergence from allozymic data (Nei D) and scDNA hybridization data (△Tm) for interpopulation comparison in three genera of cave crickets studied: Italian *Dolichopoda*, and North American *Hadenoecus* and *Euhadenoecus*. The steepness of the line is quite different from genus to genus, making it difficult to formulate a general rule.

from genus to genus, making it difficult to arrive at any generalization on the absolute numerical relation between Nei's D and △Tm values.

Allozyme distances and △Tm data produced highly congruent distance matrices (Mantel test, Table 1). The clustering analyses produced trees which have similar topologies but different branch lengths (Fig. 5A and 5B). In both cases the five *Dolichopoda* species are separated in two distinct clusters with *D.laetitiae* and *D.geniculata* belonging to one clade, and *D.baccettii*, *D.schiavazzii* and *D.aegilion* grouped in another clade. Within each group, both data sets arrive at similar relationships. However, the branch lengths are quite different, and the statistical confidence that all nodes are significantly different from one another varies between the two trees.

By using both techniques on the same taxa, we can ascertain with more confidence the genetic relatedness of different levels of divergence than we could have done using only one technique. The rationale behind this approach is straightforward; each technique is most sensitive and reliable to different degrees of divergence. By using a combination of molecular techniques one can arrive at a consensus tree, which uses the most reliable data at each hierarchical level.

Morphological data

Three sets of morphological characters studied by morphometric techniques were available comparatively for the six heterospecific populations of *Dolichopoda*: (1) epiphallus shape (Cesaroni et al., 1989, and in prep.), (2) body and appendage morphology (Allegrucci et al., 1987), and (3) egg chorion ultrastructure (Rampini and Saltini, 1988, and in prep.). The morphometric measurements used to describe quantitatively epiphallus shape, body and appendage morphology, and egg structure were first standardized, then transformed in Euclidean distances, which finally were used to construct dendrograms.

As previously outlined, the importance of the shape of the epiphallus as a taxonomically informative character relies on the assumption of its evolution being less constrained by selective forces than other morphological traits. Actually, the male epiphallus shape revealed significant differences between species, and discontinuities in its geographic variation. However the dendrogram based on the epiphallus shape reported in Figure 5C shows a notable difference compared to molecular data. While relationships within the *laetitiae-geniculata* group conform to the previous branching patterns, *D.baccetti* stands as an outgroup of two species groups; *D.laetitiae-D.geniculata*, and *D.schiavazzii-D.aegilion*. In particular, the UPGMA tree outlines the high level of divergence between *D.baccettii* and the other species. Whether such departure should be interpreted as the result of a relatively fast evolutionary change in response to genetic drift in a small isolated population is an open question. Uneven evolutionary rates can be expected to occur in polygenic characters controlled by a small number of genes, however patterns of heritability of the epiphallus are unknown. Therefore, we are inclined to consider this character as a biased tracer of phylogeny in spite of its diagnostic value and its pattern of geographic variation. (Note that the geographic sample studied did not include *D.baccettii*).

If body and appendage morphometrics are considered, the two *D.laetitiae* subspecies are separated into two different clusters (Fig.5D). *D.l.laetitiae* is grouped with *D.schiavazzii*, while *D.l.etrusca* becomes the outgroup of all the other species, with *D.baccettii* being closer to *D.geniculata* than to *D.aegilion*. It is not surprising that this tree is at odds with both traditional taxonomy and molecular data. As suggested by the analysis of geographic variation, these characters seem to be largely influenced by environmental effects.

In order to provide a direct test for this hypothesis we attempted to investigate the degree of correspondence between the morphometric tree and trees obtained from "ecological" descriptors of the populations studied. Two sets of such variables were considered, the first one describing life history traits of the population, the second one including habitat parameters. Population descriptors included age class structure, fecundity, individual mobility, population size, and aggregative tendency of individuals expressed as codified variables. Habitat descriptors were cave age and temperature, scores for bat guano abundance and scores for bioclimate and vegetation type outside caves. The trees in Figure 7 have been constructed by UPGMA clustering using complements of simple matching coefficients calculated on either set of descriptors, singularly taken or combined. The branching pattern of morphometric data resulted highly congruent with that of habitat descriptors (see also Mantel test in Table 1), supporting from another point of view what we basically

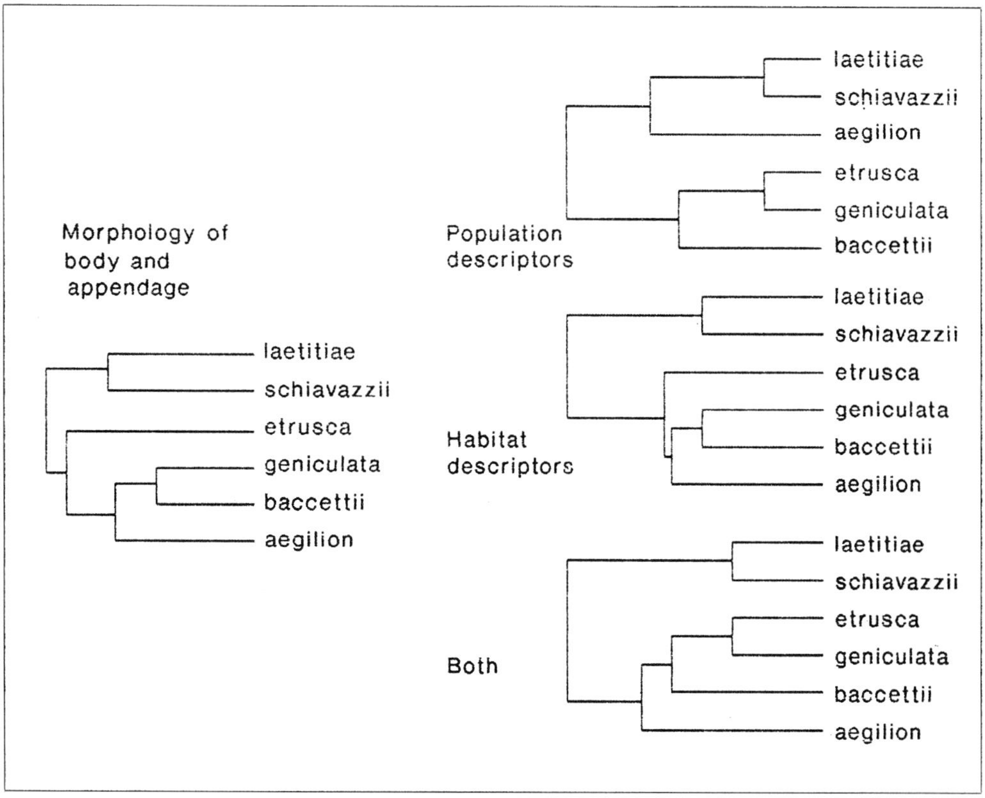

FIGURE 7. Comparison of UPGMA dendrograms deriving from body and appendage morphology with any of two purely ecological descriptors of the populations studied, habitat features and demographic and trophic features of population, and a combination of the two. The topology of the tree reconstructed from this morphometric character set appears to be close to that obtained from habitat descriptors.

inferred from the geographic structure of the data.

Therefore the species arrangement obtained from the body and appendage morphometrics is more likely to reflect phenotypic plasticity and/or short term adaptation to local environmental conditions than actual phylogenetic relationships.

Surprisingly enough, the character set describing egg chorion ultrastructure produced a picture very similar to molecular data. The dendrogram in Figure 5E differs from allozyme and scDNA data (Figg.5A and 5B) only because *D.schiavazzii* is clustered with *D.laetitiae-D.geniculata* instead of *D.baccetti-D.aegilion*. This character set was chosen

Table 1. Results of Mantel (1967) tests on distance matrices among six populations of *Dolichopoda*. Rogers (1972) genetic distances were employed for allozymes, ΔTm values for single copy DNA data, Euclidean distances for morphological data, and complements of simple matching coefficients for population and habitat descriptors.

	1	2	3	4	5
1 - allozymes	-	r= 0.858 t= 3.10 p= 0.001	r=-0.256 t=-0.98 p= 0.178	r= 0.006 t= 0.02 p= 0.417	r= 0.545 t= 2.03 p= 0.003
2 - single copy DNA		-	r=-0.380 t=-1.58 p= 0.061	r= 0.189 t= 1.24 p= 0.115	r= 0.632 t= 2.33 p= 0.002
3 - body and appendage morph.			-	r=-0.053 t=-0.17 p= 0.310	r= 0.057 t= 0.22 p= 0.378
4 - epiphallus morph.				-	r= 0.374 t= 1.44 p= 0.152
5 - egg chorion morph.					-
population descriptors			r= 0.316 t= 1.30 p= 0.123		
habitat descriptors			r= 0.572 t= 2.20 p= 0.009		

r (matrix correlation) = normalized Mantel statistic Z;
t = approximate Mantel t-test;
p = one tail probability of $|$random Z$| \geq |$observed Z$|$ out of 10,000 random permutations.

because of the hypothesis of reduced selective constraints on this structure compared with other morphological traits, and differences among species might be expected to evolve as a by-product of the overall divergence following geographic isolation.

Statistical comparisons

In the previous discussion we have broadly compared trees generated from widely differing character sets. For each tree, a matrix of ultrametric distance values has been computed in order to test the goodness of fit of the cluster analyses to the data (Rohlf and Sokal, 1981). The resultant coefficients of cophenetic correlation indicated very good fits ($r \geq 0.9$) for all tested trees with the exception of the tree based on body and appendage morphology ($r = 0.7$).

The problem arises when attempting to statistically evaluate the degree of congruence between trees from different data sets. Among several formal methods which have been developed, we used the Mantel test (1967). This is a non parametric procedure, based on randomization, which generates statistical significance levels for correlational measures of similarity between distance matrices. We selected this technique because of its flexibility and its direct dependence on original distance data rather than on distributional assumptions or clustering algorithms.

For the allozymic data, Rogers (1972) distances were used, since Mantel test requires metric distances, and Nei's D are not. Results are summarized in Table 1. As already seen, molecular data matrices showed a statistically significant correlation with one another. On the other hand, morphometric matrices were not congruent with one another. However, while body and appendage morphometrics and epiphallus shape were not correlated to any data matrix either morphological or molecular, egg chorion distances were statistically significantly correlated with both allozymic distances and ΔTm values.

CONCLUSIONS

It is now recognized that the use of different character sets may eventually lead to contrasting systematic results within a given sample of taxa. Attention has been particularly focused on the relative value of molecules *versus* morphology in inferring phylogenies, raising the question whether conflict or compromise should be taken as a guideline to interpret genealogical historical relationships. From the rather wide literature which has been rapidly accumulated on this topic, it appears that search for compromise is a preferred road. For example, in a recent review Hillis (1987) argues that conflicting results stemming from the use of different data sets need to be reconcilied, and that consensus procedures should be applied in order to ascertain a greater portion of phylogeny. This strategy is based on the assumption that, because the organisms studied have a single history, congruence between different sets of descriptors should in principle be expected. Conflict is regarded as an indicator of either "theoretical or procedural problems".

Our view is rather discordant. Since characters evolve at different rates, and are subjected to different forces producing either constant or variable evolutionary rates, conflictual results should be expected. In addition, as more and more comparative studies will become available, conflict is expected to occur as a common outcome even among sets of characters of the same nature, either morphological or molecular.

Actually, using homogeneous techniques, we have already showed that conflictual results occurred not only by contrasting molecular data to morphology but also by comparing each other different morphological data sets. On the other hand, we showed that both kinds of descriptors can eventually produce highly concordant topologies and statistically significant correlations between distance matrices. In particular, highly correlated distance matrices were obtained between such differing character sets as egg chorion ultrastructure and both allozymes and DNA-DNA hybridization data. However, each of these descriptors was at odds with a synthetic descriptor of body and appendage morphometrics, and produced discordant patterns also in respect of the epiphallus morphology.

In our opinion, any attempt to find a compromise, putting all these descriptors in the same pot, would obscure rather than enlighten genealogical evolutionary relationships among taxa considered, resulting in uneven taxonomic arrangements. However, consensus may be applied to compatible sets of data. Therefore, sifting of characters should precede search for consensus in order to optimize phylogenetic inference. From our data we argue that the pattern of geographic variation of allozyme frequencies is largely dependent on isolation effects rather than on environmental gradients. This finding combined with the very good agreement found between allozymic distances and ΔTm values from DNA-DNA hybridization, represents a strong hint of the value of these parameters as meaningful indicators of differences at the nuclear genome. In addition, several lines of empirical evidence indicate that both these parameters behave in a clocklike manner, even if the rate of change is taxon specific (Thorpe, 1982; Powell and Caccone, 1989; Sbordoni et al., 1990).

Hence, we can take allozymes and DNA-DNA hybridization data as guidelines to choice among morphological characters. On this basis, egg chorion structure stands as a very promising morphological indicator of phylogeny. This finding supports the increasing use of egg chorion morphology as a systematic tool in taxonomy of Orthoptera and other insects (Hinton, 1981; Mazzini, 1987). On the other hand, the temperature-dependent pattern of geographic variation of body and appendages morphometry, warns one against use of this character set as a reliable tool for evaluating systematic relationships in *Dolichopoda*.

We think we are still far away from depicting a "correct" phylogenetic tree for our cave crickets. In effect, the scope of this paper was to present a rather heterogeneous array of comparative approaches, shifting from population genetics to systematics, and to offer some methodological starting points useful to the study of evolutionary relationships among organisms at the population and species levels. Some emphasis on organisms might be wise even within a context, such as this meeting, dominated by molecular technologies.

ACKNOWLEDGEMENTS

We are grateful to all persons concerned, with the present authors, in this long term research program. In particular, we wish to thank G.Barbujani, A.Caccone, F.Corsi, E.Fresi, P.Matarazzo, M.Rampini, G.Saltini, A.Sansotta, and M.Scardi for their permission to include in this paper some data from still unpublished papers, which we are preparing with their collaboration. G.Carchini and S.Forestiero provided helpful comments on the manuscript.

The senior author is particularly grateful to G.M.Hewitt

for his kind invitation to give a lecture at the NATO ASI meeting.

These studies have been supported by grants from Italian National Research Council (CNR) and Ministry of University and Scientific and Technologic Research (MURST).

REFERENCES

Allegrucci G, Caccone A, Cesaroni D, Cobolli Sbordoni M, De Matthaeis E, Sbordoni V (1982) Natural and experimental interspecific hybridization between populations of *Dolichopoda* cave crickets. Experientia 38:96-98

Allegrucci G, Cesaroni D, Sbordoni V (1987) Adaptation and speciation of *Dolichopoda* cave crickets (Orthoptera, Rhaphidophoridae): geographic variation of morphometric indices and allozyme frequencies. Biol J Linn Soc 31:151-160

Allegrucci G, Caccone A, Cesaroni D, Sbordoni V (1989) Phylogeny of *Dolichopoda* cave crickets: a comparison of scDNA hybridization data with allozymes and morphometric distances. 2nd Congr Europ Soc Evol Biol, Roma 25-29 Sept 1989, Abstracts, p 7

Baccetti B (1982) Ortotteri cavernicoli italiani (Notulae orthopterologicae XXXVI). Lav Soc Ital Biogeogr Verona 1978 (n.s.) 6:195-200

Baccetti B, Capra F (1970) Notulae orthopterologicae. XXVII. Nuove osservazioni sistematiche su alcune *Dolichopoda* italiane esaminate anche al microscopio elettronico a scansione. Mem Soc Entomol Ital 48:351-367

Britten RJ, Graham DE, Neufeld R (1974) Analysis of repeating DNA sequences by reassociation. In: Grossman L, Moldave K (eds) Methods in enzymology, vol 29e. Academic Press, New York, p 363

Caccone A, Powell JR (1987) Molecular evolutionary divergence among North American cave crickets. II.DNA-DNA hybridization. Evolution 41:1215-1238

Caccone A, Powell JR (1990) A protocol for the TEACL method of DNA-DNA hybridization. In: GM Hewitt (ed) Molecular techniques in taxonomy, NATO ASI Series (this volume). Springer, Berlin Heidelberg New York

Caccone A, Sbordoni V (1987) Molecular evolutionary divergence among North American cave crickets. I.Allozyme variation. Evolution 41:1198-1214

Capra F (1967) Una nuova forma di *Dolichopoda* dell'Arcipelago Pontino (Orth. Rhaph.). Fragm Entomol 4:171-175

Capra F (1968) Una nuova *Dolichopoda* dell'isola di Capri (Orth. Rhaph.). Fragm Entomol 6:39-44

Carchini G, Rampini M, Sbordoni V (1982) Absolute population censuses of cave dwelling crickets: congruence between mark-recapture and plot density estimates. Int J Speleol 12:29-36

Carchini G, Rampini M, Severini C, Sbordoni V (1983) Population size estimates of four species of *Dolichopoda* in

natural and artificial caves of Central Italy (Orth. Rhaph.). Mem Biospeol 10:341-347
Carchini G, Di Russo C, Sbordoni V (to be published) Contrasting age structures in cave cricket populations: patterns and significance. Ecol Entomol
Cavalli Sforza LL, Edwards AW (1967) Phylogenetic analysis: models and estimation procedure. Evolution 21:550-570
Cesaroni D, Matarazzo P, Allegrucci G, Scardi M, Fresi E, Sbordoni V (1989) Multivariate morphometrics and geographic variation of *Dolichopoda* cave cricket populations. 2nd Congr Europ Soc Evol Biol, Roma 25-29 Sept 1989, Abstracts, p 17
Culver DC (1982) Cave life. Harvard University Press, Cambridge Massachusetts London
Di Russo C, Vellei A, Carchini G, Sbordoni V (1987) Life cycle and age structure of *Dolichopoda* populations (Orth. Rhaph.) from natural and artificial cave habitats. Boll Zool 54:337-340
Farris JS (1972) Estimating phylogenetic trees from distance matrices. Am Nat 106:645-668
Felsenstein J (1988) The detection of phylogeny. In: DL Hawksworth (ed) Prospects in systematics, The Systematics Association, special vol.36. Clarendon Press, Oxford, p 112
Fitch WM, Margoliash E (1967) Construction of phylogenetic trees. Science 155:279-284
Hillis DM (1987) Molecular versus morphological approaches to systematics. Ann Rev Ecol Syst 18:23-42
Hinton HE (1981) Biology of insect eggs. Pergamon Press, Oxford
Kimura M, Crow JF (1964) The number of alleles that can be maintained in a finite population. Genetics 49:725-738
Kimura M, Otha T (1971) Protein polymorphism as a phase of molecular evolution. Nature 229:467-469
Kimura M, Otha T (1973) Mutation and evolution at the molecular level. Genetics 73(suppl.):19-35
Kohne DE (1970) Evolution of higher organism DNA. Q Rev Biophys 33:327-375
Mantel N (1967) The detection of disease clustering and a generalized regression approach. Canc Res 27:209-220
Matheron G (1969) Le krigeage universel. Cah Cent Morphol Math 1:1-83
Matheron G (1970) La théorie des variables régionalisées et ses applications. Cah Cent Morphol Math 5:1-212
Mazzini M (1987) An overview of egg structure in Orthopteroid insects. In: Baccetti B (ed) Evolutionary biology of Orthopteroid insects. Horwood Ltd, Chichester, p 358
Nei M (1975) Molecular population genetics and evolution. North-Holland, Amsterdam Oxford New York
Nei M (1978) Estimation of average heterozygosity and genetic distance from a small number of individuals. Genetics 89:583-590
Nei M, Graur D (1984) Extent of protein polymorphism and the neutral mutation theory. Evol Biol 17:73-118
Nei M, Stephens JC, Saitou N (1985) Methods for computing the standard errors of branching points in an evolutionary tree and their application to molecular data from humans and apes. Mol Biol Evol 2:66-85
Powell JR, Caccone A (1989) Intraspecific and interspecific

genetic variation in *Drosophila*. Genome 31:233-238

Powell JR, Caccone A (1990) DNA-DNA hybridization: principles and results. In: GM Hewitt (ed) Molecular techniques in taxonomy, NATO ASI Series (this volume). Springer, Berlin Heidelberg New York

Rampini M, Saltini G (1988) Osservazioni preliminari sulla ultrastruttura delle uova degli Ortotteri Rafidoforidi. LII Congr Naz Un Zool Ital, Camerino 12-16 Sett 1988, Abstracts, p 71

Rogers JS (1972) Measures of genetic similarity and genetic distance. Studies in genetics VII. Univ Texas Publ 7213:145-153

Rohlf FJ, Sokal RR (1981) Comparing numerical taxonomic studies. Syst Zool 30: 459-490

Sansotta A, Pusterla M, Allegrucci G, Cesaroni D, Corsi F, Scardi M, Fresi E, Sbordoni V (1989) The Cerveteri Etruscan necropolis: habitat partitioning and genetic structure of a cave cricket population. 2nd Congr Europ Soc Evol Biol, Roma 25-29 Sept 1989, Abstracts, p 57

Saitou N, Nei M (1987) The neighbor-joining method: a new method for reconstructing phylogenetic trees. Mol Biol Evol 4:406-425

Sbordoni V (1982) Advances in speciation of cave animals. In: Barigozzi C (ed) Mechanisms of speciation. AR Liss Inc, New York, p 219

Sbordoni V, De Matthaeis E, Cobolli Sbordoni M (1976) Phosphoglucomutase polymorphism and natural selection in populations of the cave cricket *Dolichopoda geniculata*. Z Zool Syst Evolut -forsch 14:292-299

Sbordoni V, Allegrucci G, Cesaroni D, Sammuri G (1979) Sulla posizione sistematica e le affinita' di *Dolichopoda etrusca* in base a dati elettroforetici (Orth. Rhaph.). Fragm Entomol 15:67-78

Sbordoni V, Allegrucci G, Caccone A, Cesaroni D, Cobolli Sbordoni M, De Matthaeis E (1981) Genetic variability and divergence in cave populations of *Troglophilus cavicola* and *T.andreinii* (Orth. Rhaph.). Evolution 35:226-233

Sbordoni V, Allegrucci G, Cesaroni D, Cobolli Sbordoni M, De Matthaeis E (1985) Genetic structure of populations and species of *Dolichopoda* cave crickets: evidence of peripatric divergence. In: Sbordoni V (ed) Genetics and ecology in contact zones of populations, Boll Zool 52. Erredici, Padova, p 95

Sbordoni V, Allegrucci G, Caccone A, Carchini G, Cesaroni D (1987) Microevolutionary studies in Dolichopodinae cave crickets. In: Baccetti B (ed) Evolutionary biology of Orthopteroid insects. Horwood Ltd, Chichester, p 514

Sbordoni V, Caccone A, Allegrucci G, Cesaroni D (1990) Molecular island biogeography. In: A Azzaroli (ed) Biogeographical aspects of insularity, Atti dei Convegni Lincei,vol .Accademia Nazionale dei Lincei, Roma, (in press)

Sokal RR, Michener CD (1958) A statistical method for evaluating systematic relationships. Univ Kansas Sci Bull 38:1409-1438

Taylor CE (1976) Genetic variation in heterogeneous environments. Genetics 83:887-894

Taylor CE, Powell JR (1978) Habitat choice in natural populations of *Drosophila*. Oecologia 37:69-75
Thorpe JP (1982) The molecular clock hypothesis: biochemical evolution, genetic differentiation and systematics. Ann Rev Ecol Syst 13:139-168

SPERM and EVOLUTION in *DROSOPHILA* : morphological and molecular aspects

D. Lachaise & D. Joly
Laboratoire de Biologie et Génétique Evolutives
CNRS
91198 Gif-sur-Yvette Cedex
France

There is increasing evidence that the use of sperm structure can be of great help in phylogenetic reconstruction at any taxonomic level [see Baccetti and Afzelius' (1976) and Jamieson's (1987) extensive reviews]. Quoting the latter author, 'opposition to its (sperm structure) use is usually dispelled when its effectiveness for resolving hitherto intractable problems of taxonomic placement and relationship is demonstrated'. The range of sperm structural diversity is so large that it is often possible to ascribe quite reliably any single sperm cell to a hierarchy of taxa (i.e. phylum, order, family, genus and even species). A primitive sperm architecture can be recognized throughout the metazoans in unrelated species which have retained the primitive mode of fertilization (Franzén 1977). As a consequence, animals with a primitive spermatozoon are very unlikely to have arisen from animals with a derived one. Implicit in this, apomorphies and plesiomorphies between related taxa can be clearly determined and hence questionable phylogenetic relationships solved.

The primitive spermatozoon has a middle piece consisting of one or a few mitochondria (Baccetti 1987). In modified spermatozoa the mitochondrial material is enlarged resulting from the confluence of several small mitochondria, and extends along part or the whole of the sperm tail (Fig. 1). Such highly modified spermatozoa are found in *Drosophila*. The enlargement of the mitochondrial mass occurs without any concomitant increased need for oxidative phosphorylation and, therefore, sperm mitochondria are assumed to serve different functions in diverse sperm types (Baccetti & Afzelius 1976). Interestingly, there are on the one hand mitochondrion-less spermatozoa capable of rapid and durable movements (e.g. *Acanthocephala*), and on the other hand giant spermatozoa with prominent mitochondrial derivatives incapable of forward movement, albeit not devoid of peculiar kinetics (e.g. *Drosophila*).

NATO ASI Series, Vol. H 57
Molecular Techniques in Taxonomy
Edited by G. M. Hewitt et al.

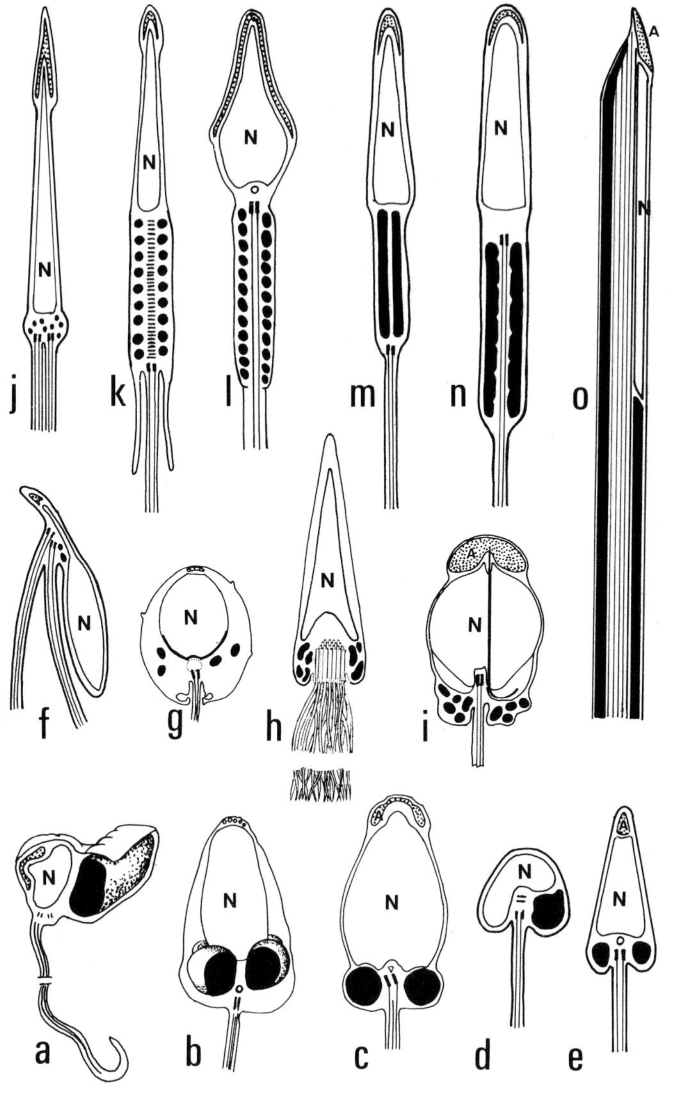

Fig. 1. Variation of sperm structure in Metazoans from a primitive sperm architecture (below) to a modified morphology (above). a. Porifera *Oscarella lobularis* ; b. Cnidarians *Nausithoe* ; c. Polychaeta *Lepidonotus* ; d. Actinopterygians *Periophthalmus* ; e. Echinodermata Sea Urchin ; f. Brachiopterygians *Polypterus* ; g. Arachnida Opiliones *Siro rubens* ; h. Insects *Mastotermes darwiniensis* ; i. Horseshoe crab *Limulus polyphemus* ; j. Dipnoi *Protopterus annectens* ; k. Selachii, Guitar fish ; l. Human spermatozoon ; m. Oligochaeta *Lumbricus terrestris* ; n. Molluscs *Paludina* ; o. Insects *Drosophila bifasciata*. Mitochondria or mitochondrial derivatives in black. N : nucleus ; A : acrosome. Redrawn and simplified after a. Baccetti 1987 ; b. Afzelius & Franzen 1971, J Ultrastruct Res 37, 186 ; c. Jamieson 1987a ; d,f,j,k. Mattei 1969, Thèse Doctorat Univ Montpellier in Favard & André 1970, Comparative Spermatology, Baccetti B ed, Academic Press, NY, p 415 ; e. Afzelius 1955, Z Zellforsch Mikrosk Anat 42,134 ; g. Juberthie & al. 1976, J Mikrosc Biol Cell 25, 137 ; h. Baccetti & Dallai 1978, J Cell Biol 76, 569 ; i. André 1965, Annls Fac Sci Univ Clermont-Ferrand 26, 27 ; l. Fawcett 1958, Int Rev Cyt 7, 195 ; m. Anderson & al. 1967, J Cell Biol 32, 11 ; n. Kaye 1958, J Morph 102, 347 ; o. Takamori & Kurokawa 1986.

Sperm length Diversity and Phylogeny

Worth emphasizing is that the extent of sperm length diversity in the *Drosophila* genus alone is tremendous, being as large (i.e. ranging from 56 to 20,000 µm) as that of the other metazoans altogether including both invertebrates and vertebrates with flagellate sperm (i.e. ranging from 33 to 16,500 µm) (Joly et al. 1990a). Moreover, the number of first spermatocytes per cyst and hence of mitotic divisions varies from one *Drosophila* species to another being either $64(2^6)$, $32(2^5)$, $16(2^4)$ or $8(2^3)$. The two subsequent meiotic divisions lead to the production of 256, 128, 64 and 32 spermatids per cyst respectively (Kurokawa & Hihara 1976). A sperm bundle within a cyst results from the synchronous divisions of cells which remain interconnected by cytoplasmic bridges throughout spermatogenesis. Diffusion of both nuclear and cytoplasmic material through them results in equal sperm length within a cyst. Clearly, there is a general tendency that the spermatozoa increase in length as the first spermatocytes increase in size and decrease in number (Hihara & Kurokawa 1987), even though inconsistencies exist within subgroups (Fig. 2a). Thus, unequal number of mitotic divisions may be one possible, albeit controversial, reason accounting for the markedly different sperm lengths observed in *Drosophila*. When species are ranked according to the mean species specific sperm length the *obscura* and *melanogaster* groups appear almost entirely disjunct (Fig. 2b). Consistently, all members of the former have shorter sperm and show 32 cells per cyst whereas those of the latter have longer sperm and show 16 cells per cyst. These two species groups belong to the subgenus *Sophophora* which overlaps to some extent for sperm length the subgenus *Drosophila*, except that this latter includes in addition all the species with exceedingly giant sperm (e.g. *D. virilis, D. funebris, D. hydei*) all having 8 cells per cyst. At the opposite extreme, the number 64 cells per cyst is found only in the subgenus *Scaptodrosophila* where the first measures of sperm length appear to be at least as short as those in the *obscura* group. Generally, it is widely admitted in insects that the number of spermatozoa per bundle and the overall number of sperm required to fertilize one egg decrease in more advanced forms (Virkki 1969). If this statement is valid, it could be concluded that the subgenus *Drosophila* comprises more advanced forms than the subgenus *Sophophora*, and the *obscura* group is the most primitive group within the Sophophoran radiation (Kurokawa & Hihara 1976).

Have giant sperm tails a role in early embryonic development ?

Mitochondria of *Drosophila* sperm form two long strands extending all along the flagellum over the bewildering length of more than 15 mm in some species

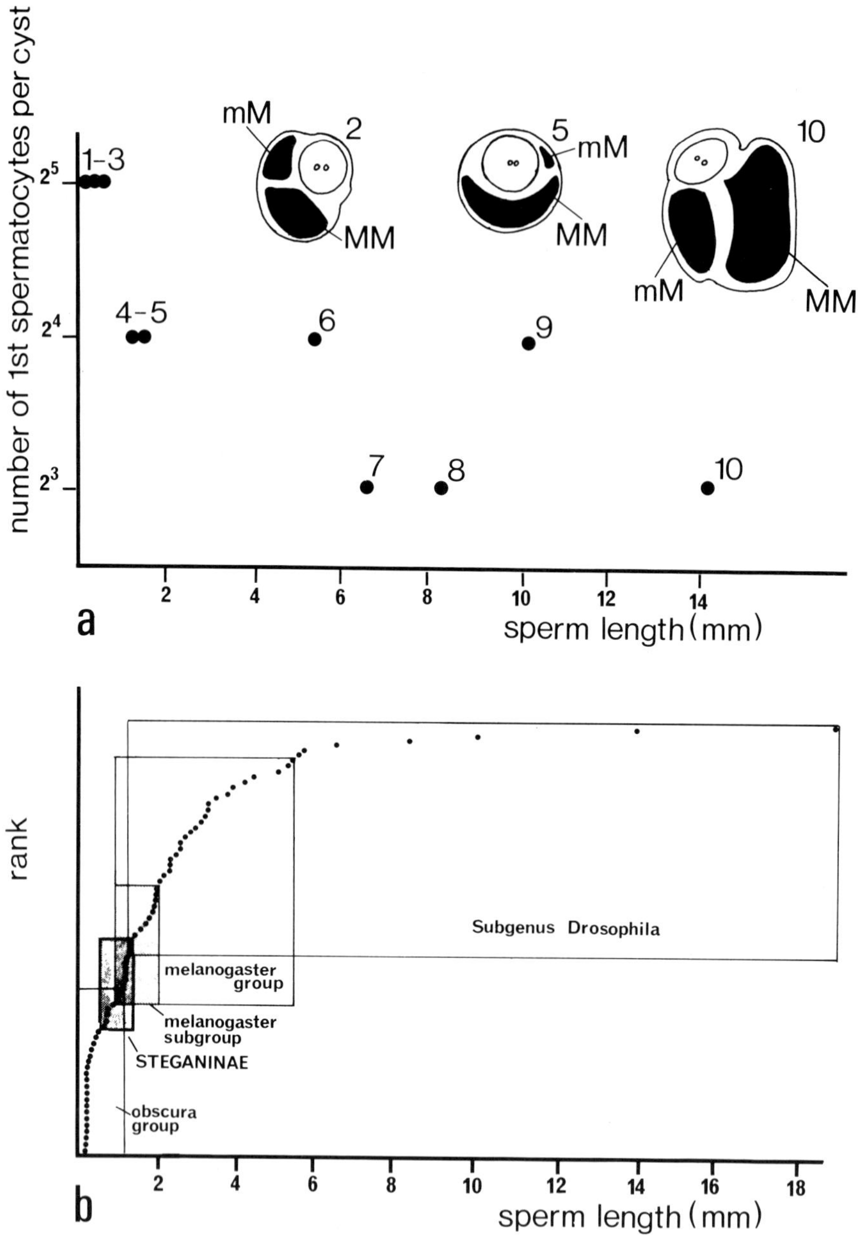

Fig. 2. Variation of sperm length in Drosophilidae. a. Sperm length (from Joly 1989) plotted against the number of first spermatocytes per cyst and hence of mitotic divisions (from Kurokawa & Hihara 1976). 1. *Drosophila pseudoobscura* ; 2. *D. bifasciata* ; 3. *D. obscura* ; 4. *D. simulans* ; 5. *D. melanogaster* ; 6. *Zaprionus tuberculatus* ; 7. *D. virilis* ; 8. *D. funebris* ; 9. *D. repleta* ; 10. *D. hydei*. Transverse sections of mature axonemes show interspecific differences in the size of the two major (MM) and minor (mM) mitochondrial derivatives. Redrawn and simplified (2) from Takamori & Kurokawa 1986 ; (5) from Perotti 1969 ; (10) from N. Kociok in Jamieson 1987a. b. Mean sperm length for 75 species of Drosophilidae ranked in ascending order and irrespective of their relatedness which is superimposed in boxes (from Joly et al., 1990a).

(e.g. *D. littoralis*). The two mitochondrial derivatives of one spermatozoon may be disparate in size as in *D. melanogaster* or more similar as in *D. hydei* (Fig. 2). They contain a proteinaceous crystal which may represent over 50 per cent of the sperm volume and, due to its low cytochrome-C oxidase content, does not seem to contribute to locomotion (Perotti 1973). Indirect immunofluorescence evidence has definitely established that the entire 1800 µm long *D. melanogaster* sperm tail is transferred into the egg during fertilization and persists during early embryonic development (Karr 1988). The incorporation of the sperm tail into the egg is not exactly a new finding ; as early as 40 years ago Demerec (1950) already noticed that as in most insects, the sperm tails of *Drosophila* enter the eggs but, he then claimed, degenerate. Also, Baccetti & Afzelius (1976) mentioned numerous reports from the literature, in sea urchins, annelid worms, and even mammals, where not only the nucleus but also the mitochondria and tail structures are taken up by the egg (only the acrosome is dispersed before sperm entry). This phenomenon seems common in insects (Friedländer 1980).

If the incorporation of the entire sperm tail into the eggs, observed in *D. melanogaster*, could be generalized to those *Drosophila* species with giant sperm, it would provide a mechanism for transmission of a strikingly large amount of paternal mitochondrial material into the egg. Considering that an autonomous synthesis of mitochondrial RNA has been shown to continue until condensation of the paracrystalline material is complete (Curgy & Anderson 1972), this implies that transcription determined by mitochondrial DNA occurs in the paracrystalline derivatives at a stage when nuclear RNA synthesis has stopped. The distribution of sperm proteins into the cytoplasm of the early embryo raises the unorthodox and intriguing possibility of paternal cytoplasmic inheritance (Perotti 1973). It should be borne in mind that mitochondrial DNA has hitherto been considered inherited only as a maternal haploid.

In that respect, it is worth noting that in the sperm of the pentatomid bug *Murgantia histrionica* branched mitochondria fuse gradually to form a cluster of interlocked networks (Pratt 1968), a fusion that was seen as a mechanism to give the mitochondrial DNA a chance to recombine. Thus, an appreciable quantity of recombined paternal mitochondrial DNA is assumed to be transferred to the fertilized egg (Nass et al. 1965 ; Baccetti & Afzelius 1976). However, it is often critical to detect an influence of the paternal mitochondrial DNA in a fertilized egg or embryo (Baccetti & Afzelius 1976). Thus, by making a cross-fertilization between two *Xenopus* toad species, and using molecular hybridization techniques, Dawid (1972) failed to find any influence of the paternal mitochondrial DNA in the hybrid embryo. Similarly, testing the backcross sterility in *Heliothis* interspecific crosses, to study the interaction of nuclear genes, cytoplasm and self-replicating cytoplasmic elements (i.e. mitochondria), LaChance (1984) and Lansman et al. (1983) failed to detect any paternal mitochondrial DNA in the BC lines.

In contrast, in *Drosophila melanogaster* there is increasing evidence that the sperm tail plays a role in early embryonic development. Perotti (1973, 1974) showed that the main derivative undergoes gradual but extensive modification and fragmentation during the period between the first nuclear cleavage of the zygote and the onset of the cellular blastoderm stage, i.e. after the 12th mitotic stage of the embryo. After Perotti, sperm mitochondria separate from the nucleus after entry into the egg and become surrounded by egg mitochondria, a possible way for paternal and maternal mitochondrial DNA to recombine. According to Karr (1988) the sperm tail is confined to the anterior end of the developing embryo and remains intact until shortly after the cellular blastoderm. Using sperm-specific antibodies (the testes tissues being used as antigen) Karr (1988) showed that during that time a number of sperm proteins are delivered to the fertilized egg.

Regarding those *Drosophila* species with giant sperm, a compelling hypothesis is to assume that the greater the length of the sperm tail, the longer it remains intact in the developing embryo. If this is valid one could expect at least a part of the 2 cm-long sperm tail of *D. littoralis* to be still intact at a stage when hardly any remnants of the mitochondrial derivative are visible in the ooplasm of *D. melanogaster*. That the amount of sperm proteins delivered to the egg during fertilization could increase with the sperm length is a provocative possibility that deserves attention.

Within-ejaculate sperm dimorphism

Tiny, giant and dimorphic sperm are the three major Evolutionary Stable Strategies exhibited by *Drosophila*. The coexistence of two sperm length classes in one male (i.e. within-ejaculate) is probably one of the most striking peculiarities in *Drosophila* (Fig. 3). Within-ejaculate sperm dimorphism is a common trait in all the 18 species of the *obscura* group so far investigated (Beatty & Sidhu 1970 ; Beatty & Burgogne 1970 ; Sanger & Miller 1973 ; Joly 1989 and Joly et al. 1989). In outgroups, it has only been found in *D. teissieri* while not in its seven relatives of the *melanogaster* species subgoup (Joly 1989 ; Joly et al. 1990b). Sperm dimorphism has a genetic control (Beatty & Sidhu 1970) and a polygenic system is involved (Joly 1989). Sperm length

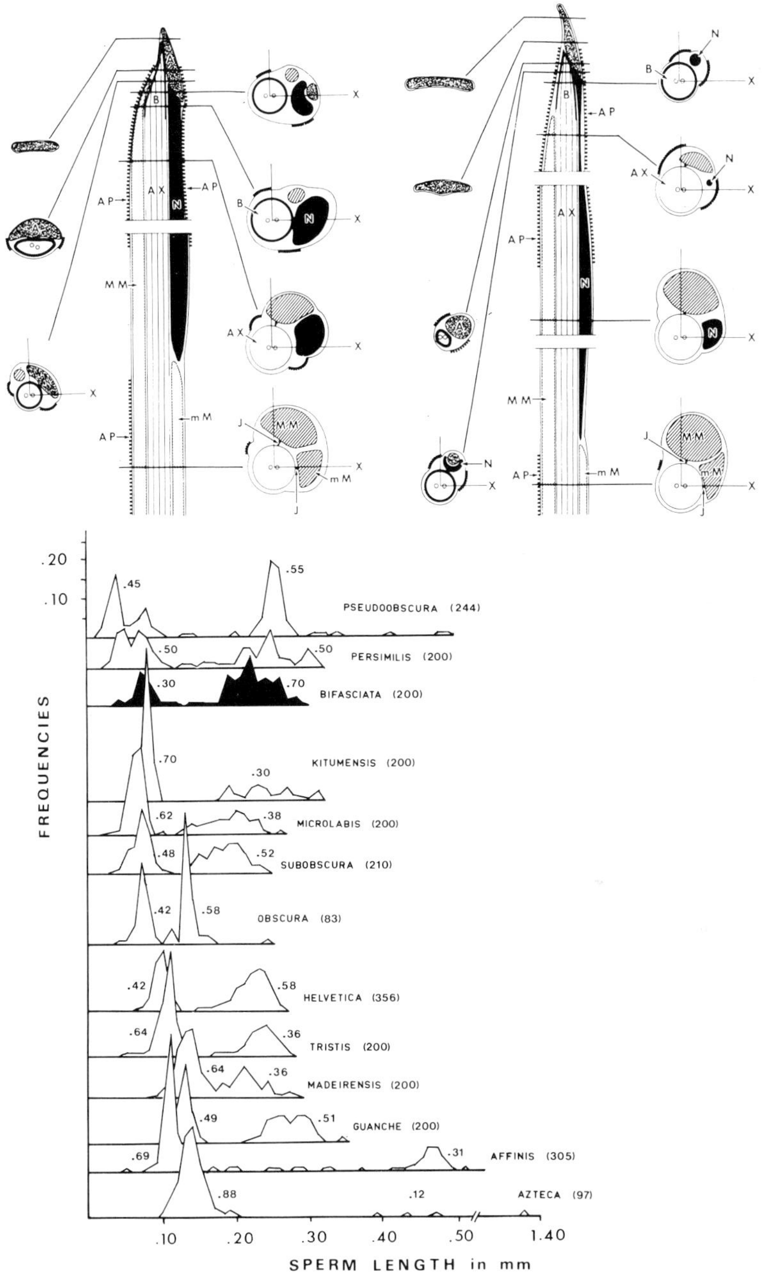

Fig. 3. Within-ejaculate sperm dimorphism in 13 species of the *Drosophila obscura* group. The relative frequencies of the two sperm length classes is given and the number of sperm measured indicated in brackets (from Joly 1989). In the upper part are shown generalized diagrams of short sperm (left-hand) and long sperm (right-hand) in *D. bifasciata* (from Takamori & Kurokawa 1986).

differences are observed between cysts but not within-cyst and therefore no factor of meiotic segregation can be involved. Sperm dimorphism is not specific to *Drosophila*, it also occurs in some tubicid Oligochaeta, Rotifera, Pogonophora, internally prosobranch molluscs, Symphyla, non-homoneurous Lepidoptera and a few rare chilopods and Hymenoptera (Jamieson 1987b). In most cases typical fertilizing sperm (eusperm) can be distinguished from atypical non-fertilizing sperm (parasperm). The former are eupyrene, the latter generally apyrene or oligopyrene. However, *Drosophila* differ from most other sperm dimorphic organisms in that the two sperm morphs are eupyrene and are presumably both involved in karyogamy. The justification for this statement comes from the evidence that the nuclei of both sperm have a similar volume (Takamori & Kurokawa 1986) and undergo a similar arginine-rich nucleoprotein transition (Hauschteck-Jungen & Rutz 1983) as shown in *D. bifasciata* and *D. subobscura* respectively. However, according to the former authors, the irregularly shaped acrosome observed in the head region of the short sperm, different from that in the long sperm, could call into question that point (Fig. 3). The within-ejaculate coexistence of two presumably fertile sperm morphs (short and long) in some *Drosophila* species provides a suitable situation for understanding the adaptive value of sperm morphs to success in fertilization.

Sperm variability and competition

Additionally to sperm dimorphism there may be between-male variability within a population and between-population variability within a species. The difficulty in studying competition between sperm length morphs in a male comes from the lack of genetic markers to identify which of the two sperm morphs has predominantly sired the offspring. A possible bias is to test the outcome of sperm competition in species showing both geographic variability in sperm morph distributions and a repeat mating system. The assumption is then made that one (bimodal or unimodal) sperm length distribution for one population and a second, different, distribution for another genetically distinct population may be considered to act in females, with regard to one another, like short and long sperm originating from a single male. Such a situation is rare but was found in *Drosophila teissieri* which shows not only a clinal geographic variability in mean sperm length but also a shift from sperm length dimorphism in central west-African populations to sperm monomorphism in peripheral east-African populations (Fig. 4). By crossing consecutively two allopatric males, which differ in sperm morphology, with females of one or the other population it was possible to use natural genetic markers (i.e. strain-specific amylase alleles) to determine which male has fathered the offspring (Joly 1989 ; Joly et al. 1990b). Returning to one single dimorphic male, results suggest that the male fitness is enhanced by either one or the other sperm morph according to whether the female mates singly or repeatedly. It was therefore proposed that the stable coexistence of two fertile sperm morphs (short and long) in one ejaculate is a mixed strategy which has evolved via sperm competition in response to a selective alternative, possibly facultative polyandry (Joly et al. 1990b). The basic idea is that, in a given population, there is a direct relationship between the level of sperm heterogeneity in males and the prevailing mating system in females (Bressac et al. 1990). Having sperm which vary in size may be for males a way to provide for all contingencies. It is assumed that as long as the monoandry-polyandry alternative remains, sperm dimorphism can be maintained. Should the alternative disappear and be replaced by a stricter female behaviour (obligatory monoandry or polyandry), sperm dimorphism will become blurred to the advantage of one or the other morph. Another, different, example of the functional significance of sperm dimorphism in fertilization exists in the chalcid wasp *Dahlbominus fuscipennis*. Here dimorphism deals with the spiralization of the sperm head. Sperm with a dextrally coiled head fertilize to give females while those with a sinistrally coiled head do not penetrate the egg membrane and are thus responsible for haploid male offspring. Thus, coiling dimorphism directly affects the sex ratio (Lee & Wilkes 1965).

However, it remains questionable in *Drosophila* whether the process of selection involved is between- or within- ejaculate selection, that is diploid or haploid selection. On the one hand, between-ejaculate selection requires between-male polymorphism of sperm-length distributions (including unimodal and bimodal patterns) in natural populations the evidence of which is still lacking. On the other hand, the prerequisite of haploid selection is post-meiotic gene expression, a condition that was shown to exist in mammalian testis (Distel et al. 1984 ; Hecht 1987) but still not in *Drosophila* testis. A major argument against comes from the occurrence of disjunction mutants, in which chromosomes are divided unequally between daughter cells during meiosis. Within a cyst sperm morphogenesis is complete in all such cells, even in those without chromosomes. The reason for this may be that the products of missing chromosomes are supplied by neighbouring cells via cytoplasmic bridges (Lindsley & Grell 1969). Therefore, although diverse testis-specific .-tubulins are known as major components of the sperm tail axoneme (Kemphues et al. 1979, 1982), the belief is that the differentiation of a sperm uses products of both sets of parental chromosomes. Thus, while morphogenetic changes occur during the spermatid stage, these changes depend upon developmental preparatory activities (presumably in the form of long-lived mRNA) that occur earlier in diploid spermatogonia or primary spermatocytes (Lifschytz & Hareven 1977). However, the question remains to know whether or not short and long sperm originating from different cysts in a male are not only phenotypically but also genotypically distinguishable.

Sperm and speciation

A species seen as a 'field for gene recombination' can be best defined by a shared fertilization system. Positive assortative mating is determined by those adaptations which are involved in the sending, receiving and processing of the signals between mating partners or their sex cells. These adaptations constitute what Paterson (1985) calls the Specific-Mate Recognition System. The fertilization system of an organism in a given habitat is subject to stabilizing selection and is therefore very stable. This raises the question of whether modifications in sperm structure or size can destabilize the coadapted gene complex of the fertilization system (Carson & Templeton, 1984), or be affected by its destabilization. In other words can changes in sperm structure or properties be involved, as cause or effect, in the process of early speciation?

Between-population variability in sperm structure or size, the prerequisite of incipient speciation, is poorly documented. In the polychaete *Platynereis dumerilii* two physiological races with different sperm morphology and fertilization biology are recognized (Franzén 1956). In another Polychaetan species *Tharynx marioni* two populations differ by the spawning season and the shape of the sperm nucleus (Gibbs 1971). Strain differences were reported in characteristics of rabbit and mouse spermatozoa (Beatty & Sharma 1960). We have seen above that *Drosophila teissieri* was unique in *Drosophila* in that it exhibits not only a clinal geographic variability in mean sperm length but also a shift from sperm length dimorphism in central populations (Ivory Coast, Cameroon, Congo) to sperm monomorphism in peripheral populations (Zimbabwe) (Joly 1989 ; Joly et al. 1990b). Interestingly, the variation in sperm morphology parallels that in male genitalia which concerns the arrangement of thick teeth on the male anal plates (Lachaise et al. 1981). Populations from Mt Nimba or the Taï lowland rainforests in the Ivory Coast comprise sperm dimorphic males with short and moderately long sperm and anal plates each bearing one external tooth row. At the opposite end of the cline, the population from Mt Selinda in Zimbabwe comprises sperm monomorphic males with very long sperm and anal plates each bearing two short,

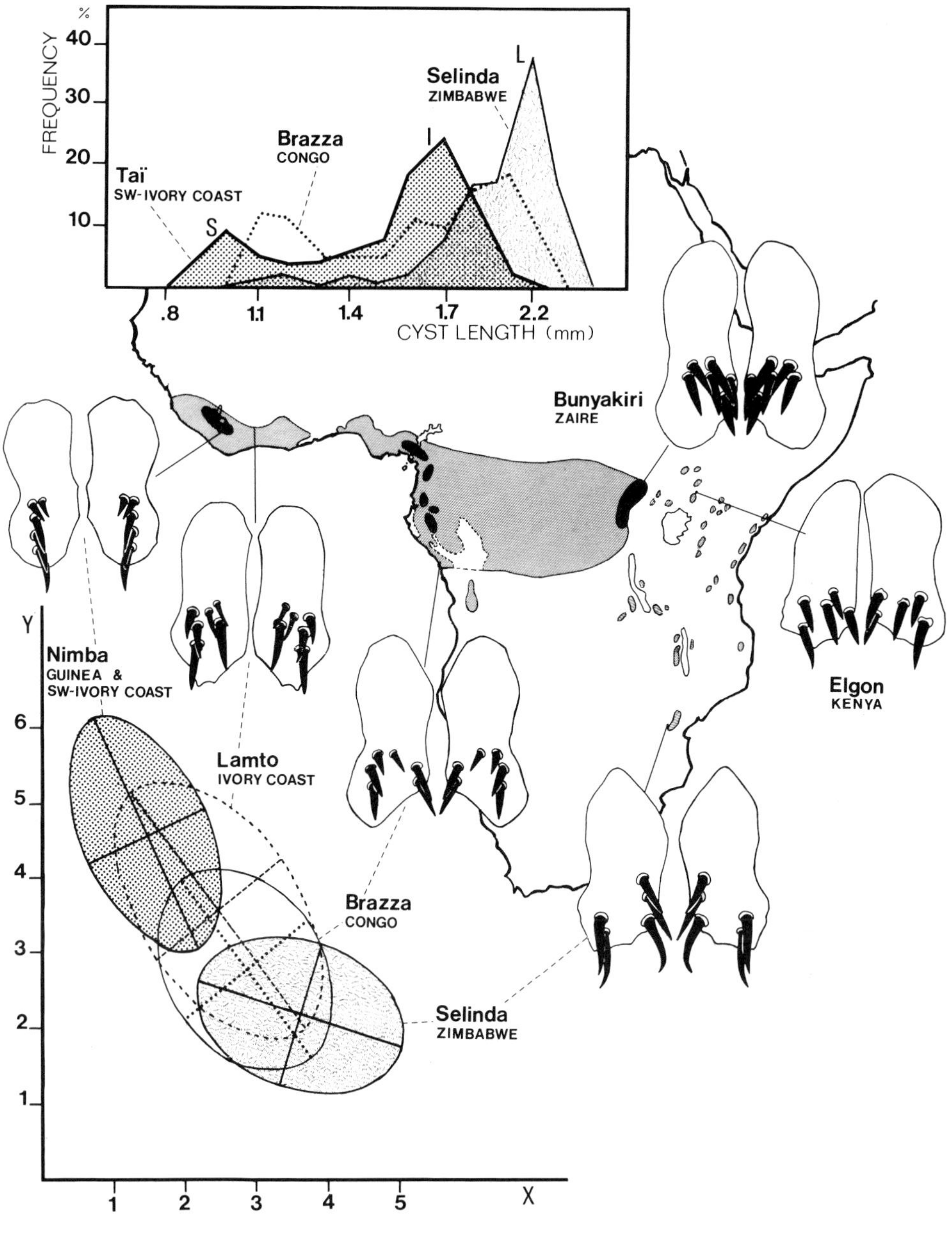

Fig. 4. Concomitant geographic variation of sperm length (curves) and anal plates of male genitalia (ellipses) in *Drosophila teissieri* throughout sub-Saharan Africa. S : short sperm, l : moderately long sperm, L : very long sperm ; X and Y are conventional partitioning of the anal plate surface. (Cyst length from Joly et al. 1990b ; Genitalia from Lachaise et al. 1981). The shaded area represents the present-day forest limits. The black spots indicate the putative lowland forest refuges during the last maximal glaciation (The refuges of Upper Guinea and eastern Zaïre are from Endler 1982 and Mayr & O'Hara 1986 ; see also Maley 1987 for a more complete scheme)

independent (external and internal), tooth rows (Fig. 4). On the basis of the common diagnostic criteria, and without data from intermediate populations, the Mt Nimba (or Taï) population would probably have been made one subspecies (at least) and the Mt Selinda population another (Lachaise et al. 1988). Using these allopatric, fully interfertile, populations for repeat mating experiments where one of the two males was homogamic with the female, it was demonstrated that homogamy was advantageous for males irrespective of their mating order (Joly et al. 1990b). This suggests a better compatibility between sperm and female storage organs within homogamic than within heterogamic pairs. It is worth relating this conclusion to Coulthart and Singh's (1988) statement that proteins from reproductive tissues may have diverged more extensively than proteins from non-reproductive tissues. A shift in *D. teissieri* sperm features including their polypeptide environment could be indicative of a process of incipient speciation.

It is of interest to notice on Fig. 4 that the geographic differentiation observed in *D. teissieri* in sub-Saharan Africa seems to produce a circular overlap surrounding the Congo Basin. Although records from the central Congo Basin are critically lacking, the possibility remains that *D. teissieri* exhibits one of the rare cases of ring speciation in tropical Africa. The repeated fragmentation of the African forest throughout Pleistocene times and the 'forest refuge' effect (Endler 1982 ; Mayr & O'Hara 1986 ; Maley 1987) may have created the conditions favouring differentiation in small subdivided populations. Strictly forest species like *D. teissieri* are particularly prone to limited dispersal and gene flow (Lachaise et al. 1988). The forest refuge environment is probably as much conducive to the formation of isolated founder populations as islands. However, not knowing whether or not the ancestral *D. teissieri* population was extensively outcrossing and extended over the entire Congo Basin area we cannot decide in favour of founder-induced speciation (Carson & Templeton 1984) or more classical allopatric speciation.

Conclusion

The wealth of possibilities offered by the tremendous diversity of spermatozoa in *Drosophila* is doubtlessly a promising field of investigation for reproductive and evolutionary biologists. Most fundamental questions are outstanding questions. Among these are the following ones : Does the evolution of spermatozoa result from sperm competition? Are short and long sperm in dimorphic species genetically distinguishable? What are the underlying mechanisms of gametic selection?

Are sperm tails entirely transferred to the egg irrespective of their length? Do short and long sperm provide an unequal specific protein supply to the developing embryo? Is the hypothesis of paternal cytoplasmic inheritance fallacious or real? Does the shape of the sperm length distribution and hence the frequency of sperm morphs in a local population vary during the yearly cycle? Do the predominant sperm morphs in males change according to the prevailing mating behaviour in females? Is there a close relationship between the evolution of spermatozoa and that of the fertilization system? And, finally, have sperm modifications a role in early speciation?

REFERENCES

Baccetti B (1987) News on phylogenetical and taxonomical spermatology. In : Mohri H (ed) New horizons in sperm cell research. Japan Sci Soc Press, Tokyo/Gordon & Breach Sci Publ, New York, p 333

Baccetti B, Afzelius BA (1976) The biology of the cell, Monographs in Developmental Biology, vol X. Karger, Basel München

Beatty RA, Burgoyne PS (1971) Size classes of the head and flagellum of *Drosophila spermatozoa*. Cytogenetics 10 : 177-189

Beatty RA, Sharma KN (1960) Genetics of gametes III - Strain differences in spermatozoa from eight inbred strains of mice. Proc R Soc Edinb B 68 : 25-53

Beatty RA, Sidhu NS (1970) Polymegaly of spermatozoan length and its genetic control in *Drosophila* species. Proc R Soc Edinb B 71 : 14-28

Bressac C, Joly D, Devaux J, Lachaise D (to be published) Can we predict the mating pattern of *Drosophila* females from the sperm length distribution in males ? Experientia

Carson HL, Templeton AR (1984) Genetic revolutions in relation to speciation phenomena : the founding of new populations. Ann Rev Ecol Syst 15 : 97-131

Coulthart MB, Singh RS (1988) High level of divergence of male-reproductive-tract proteins, between *Drosophila melanogaster* and its sibling species, *D. simulans*. Mol Biol Evol 5 : 181-191

Curgy JJ, Anderson WA (1972) Synthèse d'ARN dans le chondriome au cours de la spermiogenèse chez la drosophile. Z Zellforsch Mikrosk Anat 125 : 31-44

Dawid IB (1972) Cytoplasmic DNA. In : Biggers & Schuetz (eds) Oogenesis. Univ Parak Press, Baltimore, p 215

Demerec M (1950) Biology of *Drosophila*. Wiley, New York/Chapman & Hall, London

Distel RJ, Kleene KC, Hecht NB (1984) Haploid expression of a mouse testis .-tubulin gene. Science 224 : 68-70

Endler JA (1982) Pleistocene forest refuges : fact or fancy ? In : Prance GT (ed) Biological diversification in the tropics. Columbia Univ Press, New York, p 641

Franzén A (1956) On spermiogenesis, morphology of the spermatozoon, and biology of fertilization among invertebrates. Zool Bidr Upps 31 : 355-482

Franzén A (1977) Sperm structure with regard to fertilization biology and phylogenetics. Verhandl deutch zool Gesellsch 1977 : 123-138

Friedländer M (1980) Monospermic fertilization in *Chrysopa carnea* (Neuroptera, Chrysopidae) : Behaviour of the fertilizing spermatozoa prior to syngamy. Int J Insect Morph Embryol 9 : 53-57

Gibbs PE (1971) A comparative study of reproductive cycles in four polychaet species belonging to the family Cirratulidae. J mar biol Ass UK 51 : 745-769

Hauschteck-Jungen E, Rutz G (1983) Arginine-rich nucleoprotein transition occurs in the two size classes of spermatozoa of *Drosophila subobscura* male. Genetica 62 : 25-32

Hecht NB (1987) Haploid gene expression in the mammalian testis. In : Mohri H (ed) New Horizons in sperm cell research. Japan Sci Soc Press, Tokyo/Gordon & Breach Sci Publ, New York p 451

Hihara F, Kurokawa H (1987) The sperm length and the internal reproductive organs of *Drosophila* with special references to phylogenetic relationships. Zool Sci 4 : 167-174

Jamieson BGM (1987a) The ultrastructure and phylogeny of insect spermatozoa. Cambridge Univ Press, Cambridge

Jamieson BGM (1987b) A biological classification of sperm types, with special reference to Annelids and Molluscs, and an example of spermiocladistics. In : Mohri H (ed) New Horizons in sperm cell research. Japan Sci Soc Press, Tokyo/Gordon & Breach Sci Publ, New York, p 311

Joly D (1989) Diversité des spermatozoïdes et compétition spermatique chez les *Drosophila*. Thèse Doctorat Univ Paris Sud, Orsay

Joly D, Bressac C, Devaux J, Lachaise D (1990a) Sperm length diversity in Drosophilidae. DIS 69

Joly D, Cariou ML, Lachaise D (1990b) Can female polyandry explain within-ejaculate sperm dimorphism ? Proc VI Int Congress on Spermatology, Siena

Joly D, Cariou ML, Lachaise D, David JR (1989) Variation of sperm length and heteromorphism in drosophilid species. Genet Sel Evol 21 : 283-293

Karr TL (1988) Persistence of the sperm tail during early embryonic development. Proc. EMBO Workshop on the molecular & developmental biology of *Drosophila*, 28 August-4 Sept 1988, Kolymbari, Crete

Kemphues KJ, Raff RA, Kaufman TC, Raff EC (1979) Mutation in a structural gene for a .-tubulin specific to testis in *Drosophila melanogaster*. Proc Natl Acad Sci USA 76 : 3991-3995

Kemphues KJ, Kaufman TC, Raff RA, Raff EC (1982) The testis-specific .-tubulin subunit in *Drosophila melanogaster* has multiple functions in spermatogenesis. Cell 31 : 655-670

Kurokawa H, Hihara F (1976) Number of first spermatocytes in relation to phylogeny of *Drosophila* (Dipt : Drosophilidae). Int J Insect Morph & Embryol 5 : 51-63.

Lachaise D, Cariou ML, David JR, Lemeunier F, Tsacas L, Ashburner M (1988) Historical Biogeography of the *Drosophila melanogaster* species subgroup. Evolutionary Biology 22 : 159-225

Lachaise D, Lemeunier F, Veuille M (1981) Clinal variation in male genitalia in *Drosophila teissieri* Tsacas. Am Nat 117 : 600-608

LaChance LE (1984) Hybrid sterility : eupyrene sperm production and abnormalities in the backcross generations of interspecific hybrids between *Heliothis subflexa* and *H. virescens* (Lepidoptera : Noctuidae). Ann Entomol Soc Am 77 : 93-101

Lansman RA, Avise JC, Huettel MD (1983) Critical experimental test of the possibility of "paternal leakage" of mitochondrial DNA. Proc Natl Acad Sci USA 80 : 1969-1971

Lee PE, Wilkes A (1965) Polymorphic spermatozoa in the hymenopterous wasp *Dahlbominus*. Science 147 : 1445-1446.

Lifschytz E, Hareven D (1977) Gene expression and the control of spermatid morphogenesis in *Drosophila melanogaster*. Develop Biol 58 : 276-294

Lindsley DL, Grell EH (1969) Spermiogenesis without chromosomes in *Drosophila melanogaster*. Genetics 61 : 46-67

Maley J (1987) Fragmentation de la forêt dense humide africaine et extension des biotopes montagnards au quaternaire récent : nouvelles données polliniques et chronologiques. Implications paleoclimatiques et biogéographiques. Palaeoecology of Africa and the surrounding islands, Vol XVIII. Balkema, Rotterdam Brookfield, p 307.

Mayr E, O'Hara RJ (1986) The biogeographic evidence supporting the Pleistocene forest refuge hypothesis. Evolution 40 : 55-67

Nass MMK, Nass S, Afzelius BA (1965) The general occurrence of mitochondrial DNA. Expl Cell Res 37 : 516-539

Paterson HEH (1985) The recognition concept of species. In : Vrba ES (ed) Species and Speciation. Transvaal Museum Monograph 4, Pretoria, p 8

Perotti ME (1969) Ultrastructure of the mature sperm of *Drosophila melanogaster* Meig. J submicrosc Cytol 1 : 171-196

Perotti ME (1973) The mitochondrial derivative of the spermatozoon of *Drosophila* before and after fertilization. J Ultrastruct Res 44 : 181-198

Perotti ME (1974) Ultrastructural aspects of fertilization in *Drosophila*. In : Afzelius BA (ed) The functional anatomy of the spermatozoon. Pergamon, Oxford, p 57

Pratt SA (1968) An electron microscope study of nebenkern formation and differentiation in spermatids of *Murgantia histrionica* (Hemiptera, Pentatomidae). J Morph 126 : 31-66

Sanger WG, Miller DD (1973) Spermatozoan length in species of the *Drosophila affinis* subgroup. Am Midl Nat 90 : 4859

Takamori H, Kurokawa H (1986) Ultrastructure of the long and short sperm of *Drosophila bifasciata* (Diptera : Drosophilidae). Zool Sci 3 :847-858

Virkki N (1969) Sperm bundles and phylogenesis. Z Zellforsch 101 : 13-27

EPISODIC EVOLUTIONARY CHANGE IN LOCAL POPULATIONS

H. L. Carson
Department of Genetics and Molecular Biology
John A. Burns School of Medicine
University of Hawaii
Honolulu, Hawaii 96822 USA

THEORETICAL BACKGROUND

Active evolutionary change occurs in a sexual deme. This local population is the site where natural selection processes genetic variability. Within the constraints of chance in populations of finite size, selection determines what the genetic composition of the descendent population will be. What kind of deme structure best serves the process of novel evolutionary change? This question continues to be a major area of research in evolutionary genetics. Two hypotheses have recently been discussed. Barton and Charlesworth (1984) and Barton (1989) hold a traditional gradualist view, namely that such change proceeds most actively in a large undivided population that is widely open to gene flow. The incorporation of changes in the gene pool is viewed as a continuous process, with only small increments of change imposed by natural selection in each successive generation.

On the other hand, Carson and Templeton (1984) support a theory holding that novel character change occurs episodically in populations of small size. When deme size is constricted by a population bottleneck, change is thought to occur most actively in the generations that immediately follow. Although rare alleles may be lost from a bottlenecked population by random drift, much genetic variability nevertheless can pass into the daughter population (Nei et al. 1975). In the parental generation, genetic variability for quantitative traits is likely to be tied up in balanced coadaptive polymorphisms and is thus not freely available to selection.

NATO ASI Series, Vol. H 57
Molecular Techniques in Taxonomy
Edited by G. M. Hewitt et al.

Following the founder event, these equilibrated polymorphisms may be destabilized. There is evidence that this process can release novel genetic recombination products or other genetic variance to the action of natural selection (see Carson 1990a for a recent review).

A population bottleneck thus merely acts as a trigger that increases the array of genetic variability available to natural selection. Although the participation of major genes or macromutations is not excluded, new single mutants of this sort are not, in my view, required by the theory (Carson 1989, 1990b). Novel change can result from polygenic restructuring and novel selection of quantitative characters (Lande 1981). Over a series of generations immediately following a bottleneck, directional evolution would be expected to be rapid. Later, this active process slows down, as a new equilibrium is reached within the gene pool. Character change is thought to be accomplished through the establishment by selection of new systems of interacting genes rather than substitution of old alleles by new ones. Accordingly, this theory is less dependent than its alternate on the idea that improbable point mutations having specific phenotypic effects must arise and be individually fixed by selection before character change is realized.

Both the gradual and episodic views are strictly neodarwinian and microevolutionary. Both hold that the local population or deme is the site of active evolutionary change and that natural selection is of all-pervading importance. Both theories invoke natural selection to maintain the genetic *status quo* when the population is running along at equilibrium. Thus, the main point of contention between the opposing theories centers on the demography of the population that is able to actively incorporate novel genetic change. It is the purpose of this paper to discuss those population structures that appear to be conducive to episodic genetic change.

A few years ago, the paleontological theory of "punctuated equilibrium" (Gould and Eldredge 1977) was proposed. The episodic theory just outlined might be superficially verbalized as multi-generation episode of rapid change (=punctuation?) followed by equilibrium. At this stage of scientific advance, however, the similarity between these ideas is largely a matter of analogy, since the two sets of hypotheses are inferences drawn from wholly different data bases. One is from paleontology and the other from population genetics of contemporary populations. Unlike punctuated equilibrium, episodic theory is based on strictly neodarwinian concepts. In population genetics, a "punctuation" surely extends over a number of generations; in one laboratory experiment, for example, the episode of genetic change took 8-10 generations to be accomplished (Carson 1958). With regard to the evolution of novel character sets, the theory depends heavily on natural selection impinging on polygenic systems. Thus I cannot agree with the rejection of Darwinian selection and the invocation of macroevolutionary ideas such as saltation, species selection, group selection, macromutatation or "hopeful monsters" (see Newman et al. 1985 for an interesting neodarwinian model of punctuation). Nevertheless, it seems quite possible to me that the punctuated equilibria evident in the fossil record may ultimately find an explanation in the population genetics of neodarwinian episodes in small populations (see arguments of Hoffman 1989). Since events at the microevolutionary level are open to direct experimental tests, using demes of contemporary species, this possibility continues to pose an exciting challenge in evolutionary biology.

Both the data and the theory of population genetics provide ample evidence that most of the time natural selection maintains a population in a dynamic equilibrium with environmental selective forces. Thus, directional selection quickly exhausts the genetic variance in the gene pool that is immediately available to selection. Accepting this, it is tempting to look to a demographic episode, such as a founder event or local hybrid swarm, as a means whereby the inertia of the equilibrated state may be broken, leading, in some instances at least, to a new directional advance.

Much of the discussion of the continuous and episodic views of evolutionary change have centered on events that appear to accompany the origin of genetically-based reproductive isolation between populations. In my view, this preoccupation has obfuscated the central problem of the evolutionary process, namely, the manner of origin of novel sets of characters of all sorts, not just those that isolate. Thus, for example, Nei et al. (1983), Barton (1989) and Coyne and Orr (1989), to name only a few, equate the "origin of species" with the "origin of intrinsic isolating mechanisms". My view has been that "isolating mechanisms", such as behavioral incompatabilities or sterilities, are merely incidental byproducts or pleiotropic effects of the differing adaptive changes that emerge within separately evolving allopatric populations. To me, the origin of species is nothing more than the origin of a unique reproductive community (or gene pool). Originally produced by past evolutionary events, it is a population that currently maintains itself as an integrated and unique set of characters. Selection holds the species gene pool at equilibrium and keeps it stable toward introgression of foreign genes.

DEMES ON A GROWING SHIELD VOLCANO

I turn now to some examples of geographic and environmental forces that require both repeated extinction and repeated founding of populations. In earlier discussions of the founder event, the tendency has been to think of a colonizing founder event as a highly unusual episode in the life of a species. Examples that come to mind are the successful chance dispersal of the propagules of a species from one continent or remote island to another, resulting in the colonization of an area that was previously unoccupied by that species. In such cases, a single rare event appears to sometimes be accompanied by a profound break in genetic condition between donor and founder populations. It has been theorized that such a founder event may indeed have far-reaching evolutionary consequences with regard to the formation of new species (see Carson 1990b). Here, however, I discuss a quite different population situation, one that requires local populations in a given geographical area to be exterminated and re-founded at relatively frequent intervals. If a species is made up of many small local populations, some these may be vulnerable to extinction by environmental adversity or demographic accident. If recolonization maintains an approximate equilibrium with extinction, then the larger species population of such a species could be described as a "shifting mosaic". The result has been called a "metapopulation" (Levins 1970; Harrison et al. 1988). Various examples and a review of attempts to model metapopulations has recently appeared (Oliveri et al. 1990).

One of the problems of dealing with transient populations is the difficulty of determining whether complete extinction of a local population has indeed occurred prior to an apparent recolonization. I discuss here a case where this uncertainty does not exist, namely, local populations that have been incinerated by volcanic action prior to recolonization of the site by propagules from adjacent intact areas. Such conditions exist on two new actively-growing tropical shield volcanoes in Hawaii.

The Hawaiian islands are volcanic in origin and each island is believed to have formed separately by the emergence above sea level of lava flows emanating from a single fixed melting source in the Earth's mantle. This is located under the central area of the Pacific Ocean. The northwestward movement of the Pacific tectonic plate over this "hotspot" is considered to have resulted in a succession of volcanoes, with the youngest at any one time being always located at the southeastern end of the archipelago (Clague and Dalrymple 1989).

Mauna Loa and Kilauea, located on the island of Hawaii are currently the two youngest terrestrial volcanoes in the archipelago; they are eruptively among the most active on the planet. Mauna Loa is massive, rising to 4100 m above sea level; Kilauea is a younger volcano that lies as a satellite on the southeast flank of Mauna Loa at 1200 m. Extensive radiocarbon dates of the age of lava flows on both mountains have recently become available (Lockwood and Lipman 1987; Holcomb 1987). These data, combined with historical records of lava flows, show that the surfaces of both volcanoes are being buried at rates of about 40% (for Mauna Loa) and 90% (for Kilauea) per 1000 years. This exceedingly rapid surface coverage means that there are almost no forests on either volcano older than 8000 to 10,000 years. Ancient refugia are lacking and the concept of a climax forest cannot apply.

Although these volcanoes have summit calderas that occasionally erupt, most of the surface of these mountains is made up of flows of lava that emerge along rifts on their flanks. The new lava finds its way by gravity down towards the ocean either in natural lava tubes or on the surface of flows that are often no more that 200-500 years old. Accordingly, the geological substrate consists at any one time of a complex mosaic of lava flows of different ages. Even more significant is the fact that, although apparently irregular, this complex mosaic of substrates is continually shifting as new flows cover the older ones.

Each new lava flow becomes quickly colonized with life so that even in as little as 200 years a substantial and complex ecosystem has occupied the lava. This is aided by the fact that in the presence of an equable, mesic tropical climate the

Hawaiian biota have evolved exuberant colonizing ability. Under heavy rainfall in a tropical climate, even a seemingly impermeable and forbidding pahoehoe surface (Figure 1a) yields a complex ecosystem in a surprisingly short time (Figure 1b); (see also Mueller-Dombois et al. 1981).

Figure 1: a) a 1990 pahoehoe lava flow on Kilauea volcano
b) a naturally-vegetated pahoehoe flow about 350 years old.

Descriptions of the forests on the flanks of these growing shield volcanoes have stressed the formation of "kipukas", that is, areas of slightly older forest that become surrounded and cut off from one another by the newer molten flows. Kipukas vary continuously from very small to very large. Being slightly lower in altitude than the newer flows around them ("kipuka" in Hawaiian means "a small hole"), the relentless flows from the rifts will sooner or later find and fill these depressions and destroy the biota in them. Although they may be temporarily spared, kipukas, like the rest of the surface features, are very short-lived.

Isolation of populations within kipukas has sometimes been evoked as a demographic factor that might be conducive to genetic differentiation of populations within these small areas. The rapid rate of turnover indicated by the radiocarbon data, however, suggest that these areas on a growing shield volcano are too ephemeral to provide such an oppportunity. The metapopulation concept, moreover, emphasizes that the kipukas are serving as a source of propagules for continual founding of populations on the newer flows. The demographic situation is, therefore, best viewed as promoting an enforced series of founder events rather than isolation as the source of possible genetic differentiation.

In summary, if the organisms living on these volcanoes are to remain as viable elements of the ecosystem, they are required to be continually "on the move", making necessary a repeated series of founder events. In areas where the source of propagules is adjacent to a newer flow, a large number of founders may be involved in the colonization. Thus the genetic effects of the founding events would be expected to be minimal. On the other hand, as the complexity of the new ecosystem builds up, the founders of certain secondary members of the ecosystem in particular may arrive from a distance, even possibly from a neighboring volcano in which active volcanism has been in decline. This may provide the opportunity for a more extreme founder effect, similar to those that have been inferred between the volcanoes of an older island and those of a newer one across an abysmally deep channel (see Carson 1990b). In such cases, genetic change in the generations following founding might lead towards speciation.

POPULATIONS OF *Drosophila silvestris* NEAR THE CALDERA OF KILAUEA VOLCANO

The new radiocarbon dating of lava flows has led to the above theoretical metapopulation interpretation of the species on Mauna Loa and Kilauea. Nevertheless, virtually no organized gathering of empirical data have yet been obtained correlating the age of lava flows and the genetic composition of the species found within the newer forests. Nevertheless, I mention now some incidental observations on a few local populations of the species *Drosophila silvestris*; these suggest that further study would be profitable.

Drosophila silvestris is endemic to Hawaii, the newest of the Hawaiian Islands. The species occurs on all five major volcanoes on the island, including Mauna Loa and Kilauea. Following up on an earlier study of the island-wide populations of this species (Carson 1982), Craddock and Carson (1989) have conducted a rigorous reeaamination of the data on the distribution of genetic variation within the species over the island. Data include gene frequencies of 22 polymorphic isozyme loci out of 25 studied, 11 polymorphic chromosomal inversions and quantitative genetic variation in a secondary sexual character of males, namely, the number of cilia on the tibia of the foreleg. The *silvestris* inversions are not found in any of the presumptive ancestral species on the older islands. The oldest volcano on the island of Hawaii is only about 400,000 years old, so it can be inferred that *silvestris* is a recently-evolved species.

Craddock and Carson (1989) have divided The total population of the species (T) into two regions (R) and then further into 17 sampled localities or demes (D). The two major regions comprise wholly allopatric populations that manifest qualitatively and quantitatively different leg-bristle morphotypes in males. This character, which is genetically variable both within and between demes in both regions, appears to play a role in sexual selection. The differentiation of populations over the island allows the recognition of four chronological groups of inversions and permits inference of their probable geographical sites of origin.

These data form the basis of a quantification of the genetic population structure of the species by hierarchical F statistics (Wright 1965). An extraordinary level of genetic differentiation between the demes of this species, F^{DT}, is revealed by these calculations. Thus, about 84% of the total chromosomal diversity in the species is observed within demes. The dispersion of variability in this species is thus extraordinarily great, being far greater than that found in older continental species of Drosophila that have very wide geographical distributions.

Three polymorphic inversions (3m, $4k^2$ and 4t) are common to all 17 demes sampled, that is, they occur within both of the leg-bristle regions and may thus be judged as the oldest of the inversions. Two of these (3m and $4k^2$) show very high negative correlations with altitude, forming striking altitudinal clines in a number of places on the island; the frequencies of these two variants are closely correlated.

Carson, Lockwood and Craddock (1990) have scrutinized the genetic properties of five local populations of *silvestris* that are distributed over an altitudinal range from about 1200 to 1700 m above sea level in the immediate vicinity of Kilauea Crater. The situation is especially interesting since an extensive area that is currently occupied by dense forest having *silvestris* populations is known to have been recently incinerated twice by explosive volcanic events, once about 100 BC and again in AD 1790. Both events consisted of massive, ground-hugging pyroclastic hurricanes that produced total devastation, leaving horizons of carbonized organic material.

Despite these destructive events, the areas have been subsequently colonized by exuberant cloud-forest vegetation along with *Drosophila silvestris*. Figure 2 shows the frequencies of two chromosomal variants in the five localities over the 500 m altitudinal sector.

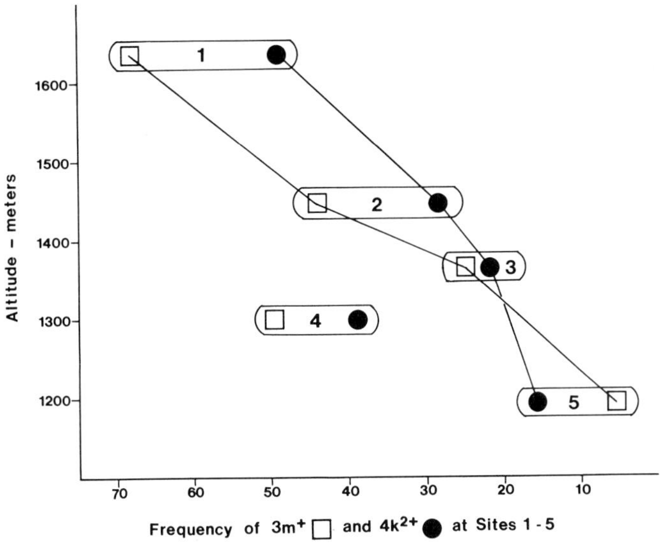

Figure 2. Frequency of two chromosomal variants in five population samples taken over an altitudinal transect on Mauna Loa, Hawaii, near and above the vent of Kilauea volcano. Samples 3, 4 and 5 and a portion of 2 were taken in areas where the forest had been recently destroyed by volcanic activity. Reproduced by permission from Carson et al. 1990.

A plot of ash distribution indicates that the forest at lower-altitude sites 3, 4 and 5 and a portion of 2 were destroyed in the older, larger explosion; the areas at 4 and 5 were destroyed again in the 1790 event. In localities 2, 3 and 5, natural selection appears to have restored the cline of frequencies of the two relevant inversions. The correlation coefficients of the two chromosomal variants with altitude at sites 1, 2, 3 and 5 are very high ($r=0.992$, $P<0.01$), in close agreement with values that would be predicted from their island-wide altitudinal relationships. Locality 4, however, is anomalous in that the frequencies of both inversions found there are discordant with altitude. This implies that altitudinal adjustment is still not complete after an estimated 75 generations since the 1790 event.

These events occurring near the summit of Kilauea may represent in microcosm a paradigm of the type of population destruction and replacement that appears to be the norm for large areas on the flanks of a growing shield volcano. Accordingly, in the larger sense, *D. silvestris* appears to exist as a metapopulation. The current wide availability of molecular genetic markers offers a powerful technique, that, when correlated with specific dating, should provide a unique kind of historical population data of extraordinary interest in future research on evolutionary events in populations.

Why are there so many species and so much dispersion of population genetic variability in Hawaii? Carson et al. (1990) suggest that the opportunity for episodic evolutionary change in demes is particularly active during the growth phase of each successive Hawaiian volcano. As each volcano becomes dormant, opportunity for such episodes declines. Volcanic activity has always been greatest at the youngest, southeastern island or volcano of the archipelago. It is suggested that metapopulation structure on the youngest island or volcano at any one time elicits substantial evolutionary activity. As a volcano becomes dormant, the evolutionary products of the active phase tend to remain relatively stable, serving as historical remnants of the former exuberant evolutionary activity.

SUMMARY

Both theory and data are reviewed suggesting that important evolutionary character change is often episodic and may be iniatated in small local populations, or demes. Under outcrossed sexual reproduction, the requirement of continual founding of new populations from a limited number of individuals may have a destabilizing influence on the balanced genetic structure that underlies certain quantitative character states. Thus, novel recombinational genetic variance of quantitative characters may be released episodically to the action of natural selection. Losses of rare alleles by random drift or aquisition of novel point mutations are likely to be of minor importance compared with recombination.

Examples are adduced from the biota of Hawaii, the newest island of the Hawaiian chain. This island comprises five shield volcanoes, none of which is older than 0.4 million years bp. An extensive study of radiocarbon dating of individual lava flows on Mauna Loa volcano is now nearing completion. These geological data have important implications for population structure and genetic change in species inhabiting the mountain. Dating of Holocene flows (less than 10,000 years bp), indicates that surface burial on Mauna Loa is about 40% per 1000 years. On Kilauea, a new satellite of Mauna Loa, the burial rate of older surface is even higher (90% per 1000 years). This suggests that during the growth phase, the aerial surface of such a mountain (i.e. above sea level) may have been replaced many times. No climax forest or stable, ancient biological refugia can exist. Far from being rendered bare by the continual destructive force of the volcanism, however, these volcanoes support "patchworks" of tropical forests of different young ages, containing a wealth of species. The latter must survive by virtue of their ability to be continually "on the move", undergoing repeated colonizations. They thus form *metapopulations*. Such populations are characterized by a process of local population extinction that is balanced by the continual founding of colonies in sites that are entirely new.

Data on the genetic structure, altitudinal clines and demography of *Drosophila silvestris* populations that are found on Mauna Loa near Kilauea crater are reviewed and discussed. It is suggested that evolutionary activity is greatest on active volcanoes; when dormancy comes, the mountain is clothed with a mosaic of historical vestiges of the active evolutionary phase.

LITERATURE CITED

Barton NH (1989) Founder effect speciation. In Otte D, Endler JA (eds) Speciation and its consequences. Sinauer, Sunderland Massachusetts, p 229

Barton NH, Charlesworth B (1984) Genetic revolutions, founder effects, and speciation. Annu Rev Ecol Syst 15:133-164

Carson HL (1958) Increase in fitness in experimental populations resulting from heterosis. Proc Nat Acad Sci USA 44:1136-1141

Carson HL (1982) Evolution of Drosophila on the newer Hawaiian volcanoes. Heredity 48:3-25

Carson HL (1989) Genetic Imbalance, Realigned Selection, and the Origin of Species. In Giddings LV, Kaneshiro KY, Anderson WW (eds) Genetics, Speciation and the Founder Principle. New York Oxford, p 365

Carson HL (1990a) Increased Genetic Variance after a Population Bottleneck. Trends Ecol Evol 5:228-230

Carson HL (1990b) Evolutionary process as studied in population genetics: clues from phylogeny. Oxford Surv Evol

Biol 7:(in press)

Carson HL, Lockwood JP, Craddock EM (1990) Extinction and recolonization of local populations on a growing shield volcano. Proc Nat Acad Sci USA 87: (in press)

Carson HL, Templeton AR (1984) Genetic revolutions in relation to speciation phenomena: the founding of new populations. Annu Rev Ecol Syst 15:97-131

Clague DA, Dalrymple GB (1989) Tectonics, geochronology, and the origin of the Hawaiian-Emperor Chain. In Winterer EL, Hussong DM, Decker RW (eds) The Geology of North America. Geol Soc Amer, Boulder, Colorado. Vol N: p 187

Coyne JA, Orr HA (1989) Two rules of speciation. In Otte D, Endler JA (eds) Speciation and its consequences. Sinauer, Sunderland Massachusetts, p 180

Craddock EM, Carson, HL (1989) Chromosomal inversion patterning and population differentiation in a young insular species, *Drosophila silvestris*. Proc Nat Acad Sci USA 86:4798-4802

Gould SJ, Eldredge N (1977) Punctuated equilibria: the tempo and mode of evolution reconsidered. Paleontology 3:115-151

Harrison S, Murphy DD, Ehrlich PR (1988) Distribution of the bay checkerspot butterfly, *Euphydryas editha bayensis*: evidence for a metapopulation model. Amer Nat.132:360-382

Holcomb RT (1987) Eruptive history and long-term behavior of Kilauea Volcano. In Decker RW, Wright TL, Stauffer, PH (eds) Volcanism in Hawaii. US Geological Survey Professional Paper 1350, Washington p 261

Hoffman A (1989) Arguments on Evolution: A Paleontologist's Perspective. Oxford University Press

Lande, R (1981) Models of speciation by sexual selection on polygenic traits. Proc Nat Acad Sci USA 78:3721-3725

Levins R (1970) Extinction. In Gerstenhaber M (ed) Lectures on mathematics in the life sciences. American Mathematical Society, Providence, Rhode Island Vol 2, p 77

Lockwood JP, Lipman PW (1987) Holocene eruptive history of Mauna Loa Volcano. In Decker RW, Wright, TL, Stauffer, PH (eds) Volcanism in Hawaii. US Geological Survey

Professional Paper 1350, Washington p 509
Nei M, Maruyama T, Chakraborty, R (1975) The bottleneck effect and genetic variability in populations. Evolution 29:1-10
Nei M, Maruyama T, Wu C-I (1983) Modes of evolution of reproductive isolation. Genetics 103:557-579
Newman CM, Cohen JE, Kipnis C (1985) Neo-darwinian evolution implies punctuated equilibria. Nature 315:400-401
Olivieri I, Couvet D, Gouyon P-H (1990) The Genetics of Transient Populations: Research at the Metapopulation Level. Trends Ecol Evol 5:207-210
Wright, S (1965) The interpretation of population structure by F-statistics with special regard to systems of mating. Evolution 19:395-420

SEARCHING FOR SPECIATION GENES IN THE SPECIES PAIR *Drosophila mojavensis* AND *D. arizonae*

E. Zouros
Department of Biology
University of Crete
Iraklion, Crete, Greece and
Department of Biology
Dalhousie University
Halifax, N. S., Canada B3H 4J1

Life is continuous in time but discontinuous in form. Molecular biology has revolutionized the description of discontinuity in the living world (i.e. phylogeny), but has so far shed little light on the mechanisms responsible for it (i.e. speciation). An important reason for this is that present attributes of extant species are generally assumed to contain useful information for the construction of phylogeny, but it is not clear what information, if any, they can provide about the speciation process. One way of overcoming this difficulty is to view speciation as a process that differentiates the probabilities that genes from different contemporaneous individuals (of generation *t*) will co-occur in the same individual of some future generation *t+n*. If it is assumed that the genes that make it impossible for two populations at generation *t+n* to exchange genes are the same ones that affected the probabilities of gene exchange in the preceding *n* generations, then there can be an experimental basis for the study of the genetics of speciation. This assumption cannot be tested in principle, but appears the more reasonable the smaller is *n*, i.e. the time since the divergence of the two species.

My colleagues and I have been studying the sibling species pair *Drosophila mojavensis* and *D. arizonae* (formerly *arizonensis*) with the aim of discovering something about the genetic nature of barriers to gene exchange that appear to exist

NATO ASI Series, Vol. H 57
Molecular Techniques in Taxonomy
Edited by G. M. Hewitt et al.

in these species. The basic advantage to using this pair of species is that it produces hybrids with relative ease. All F_1 hybrids are fertile, except the sons of arizonae females to mojavensis males. Backcrossing fertile F_1 males to one or the other species allows for the production of flies that carry non-recombinant chromosomes from the two species in all possible combinations. Thus the effect of an entire chromosome substitution on some attribute related to reproductive isolation can be studied in the progeny of backcrosses, provided that the specific origin of homologous chromosomes can be recognized. As a recognition system we have used species-diagnostic allozymes. This strategy is most appropriate when the objective is to estimate the relative effects of chromosome substitutions on the character under study. This appears to be a reasonable first step in the genetic analysis of speciation. After chromosomes with major effects have been recognized, further progress in the genetic dissection of the character may be obtained by the identification of small chromosomal regions (or even genes) that are responsible for most (or even all) of the chromosome's effect. This necessitates using female hybrids, so that progressively smaller segments of one's species genome are filtered into the genome of the other species through interspecific recombination. The high fertility of female hybrids makes this process relatively easy in this pair of species.

Genetics of sexual isolation

The degree of sexual isolation as measured in mate-choice experiments vary according to the geographical origin of the strains of the two species. This distribution of interspecific mating discrimination is not random, but rather depends on whether the strains originate from areas where the two species do or do not coexist. This was demonstrated by Wasserman and Koepfer (1977), who provided one of the most convincing cases of character displacement for sexual isolation. Strains of D.

mojavensis from areas in which the two species coexist exhibit larger indeces of sexual isolation from D. arizonae than strains of D. mojavensis from areas of allopatry. Interestingly enough, the allopatric/sympatric status does not seem to affect the mating choice of D. arizonae. A similar pattern of asymmetrical discrimination exists among females of D. mojavensis from the mainland (Sonora, Mexico) and the peninsula (Baja California, Mexico) parts of the species' distribution when presented with different males of their own species (Zouros and d'Entremont, 1980). When given the choice, females of D. mojavensis from Sonora (where the two species coexist) would mate more often with males from Sonora than males from Baja California, but females from Baja California (where D. arizonae does not occur) do not prefer one male over the other. Both of these observations support the classical view (Dobhzansky, 1970) that sexual discrimination develops under selection against loss of fitness through hybridization. The findings of Wasserman and Koepfer could be explained as the result of selection in the area of sympatry against the relaxation of interspecific sexual discrimination. Likewise, the preference of D. mojavensis females from Sonora for males from their own area can be explained as a by-product of their high level of vigilance against matings with D. arizonae, a vigilance that is reduced in females from Baja California. If male courtship pattern and female recognition have tightly co-evolved in D. mojavensis populations of Sonora under the pressure of coexistence with D. arizonae, it is quite possible that the courtship pattern of conspecific but allopatric males may quite often fall outside the range of acceptance by the female. Thus, selection for interspecific sexual isolation may be responsible for incipient sexual isolation among allopatric conspecific populations. Koepfer's selection experiments (1987a, 1987b) have demonstrated the presence of genetic variability for mating preference in D. mojavensis populations, which is a necessary prerequisite for these explanations of interspecific and intraspecific variation in sexual behavior.

The technique of whole chromosome substitution described

above was applied to the analysis of sexual isolation of the two species (Zouros,1981a). The effect of each chromosome on male mating behavior was studied by using the sons of the F_1 male (from the cross female mojavensis to male arizonae) to mojavensis female. These backcross males (symbolized as B_1 for backcross # 1) are of the type Ya/XmAm/(m or a), i.e they have the Y chromosome of arizonae, the X chromosome of mojavensis and for each autosome pair they can be either homospecific for mojavensis (m/m) or heterospecific (a/m). Because there are four autosomes (the fifth small dot-like chromosome was not studied), there are sixteen different types of B_1 males. A B_1 male was placed in competition with an arizonae male and offered an arizonae female. The B_1 male was characterized as successful if the mating was to B_1 male and as unsuccessful if the mating was to arizonae male. After so characterized, the B_1 male was scored electrophoretically for four enzymes, each marking one of the autosomes. The results showed a strong effect for the Y and the fourth chromosome, but all other chromosomes had significant even though less pronounced effects. The chromosome effects on the female's sexual discrimination was studied in flies of the type Xa/XmAa/(m or a). These females were put in a cage with equal number of mojavensis males for 24 hours. Females were then removed and their sperm receptacles were examined for the presence of sperm. Each female was, thus, characterized as "inseminated" or "not inseminated" and was then put to electrophoresis for the identification of its chromosomal complement. All chromosomes showed an effect on the female's ability to recognize and reject the foreign male. More over these effects appear to be additive rather than epistatic: the more the chromosomes of one species the female possessed the more it would discriminate against the male of the other species. The main conclusion from this study is that all chromosomes carry at least one, most likely many loci that affect the male's courtship behavior and the female's ability to discriminate among these behaviors. If sexual isolation is as polygenic as it appears to be from this analysis and from similar ones in other Drosophila groups (Tan 1946, Ehrman 1961), it would be reasonable to expect that natural populations will always contain some amount of genetic variability for sexual isolation on which natural selection could act. Evidence for the existence of such variation comes from Hawaiian Drosophilids (Carson and Lande 1984, Carson 1985) and from D. mojavensis (Koepfer 1987a and 1987b).

Genetics of male hybrid sterility

The only detectable post-mating reproductive isolation between D. mojavensis and D. arizonae is the sterility of the F_1 male hybrid from the cross female arizonae to male mojavensis. There appears to be no reduction of fertility in female hybrids or of viability in hybrids of either sex (Zouros, 1981b). The degree of post-mating isolation between these two species is so mild that it is highly unlikely that it alone could prevent the two species from merging into one. At the present time, the prevention is achieved, most likely, through a combination of ethological and ecological barriers to gene flow. The emphasis of our analysis of male hybrid sterility is on genetic events and processes that may have triggered the process of speciation rather on those that have sustained it.

A complete chromosome analysis of male sterility requires the examination of each of the four autosomes in all three possible combinations (m/m, a/m, a/a) and under all possible backgrounds for the other chromosomes. This clearly demands an inordinate amount of effort. As an alternative, we have examined a number of cases in which the two sex chromosomes occurred in different combinations and appeared, from the outset, to be of special interest (Zouros et al., 1988; Zouros, 1989). Males of the type Xm/Ym are mostly fertile (more accurately, they have motile sperm; sperm motility is the phenotype that we score in these experiments), except when they carry an arizonae third chromosome (i.e. when they are Xm/YmIIIa/m), in which case many have immotile sperm. Males of the type Xa/Ya can be fertile or sterile depending on whether they carry mojavensis chromosomes at the third or fifth chromosome pair, i.e. males of the type Xa/YaIIIa/m or XaYaVa/m are mostly sterile. It is obvious that in this pair of species, homospecificity of the two sex chromosomes is not enough to secure male fertility. Some autosomal states are largely incompatible with fertility in a homospesific sex-chromosome background (as is the state IIIa/m), whereas other autosomal states (like Va/m) are causing sterility in Xa/Ya but not in Xm/Ya males. Males of the type Xa/Ym are almost always sterile. This might be due to incompatibilities among the two sex chromosomes or to interactions between autosomes and sex chromosomes. The experiments that can differentiate between these alternatives cannot be performed because the males that have to be used as parents are themselves sterile.

Males of the type Ya/Xm are the most interesting in terms of the genetics of hybrid sterility and have been the subject of most of our research. Because the F_1 males from the cross female mojavensis to male arizonae are fertile, repeated backcrossing of these males to mojavensis produces Ya/Xm males with the mojavensis cytoplasm and a decreasing number of arizonae chromosomes. In a typical experiment of this type we used as mojavensis line the strain moj br, which is homozygous for a recessive eye mutant on the fourth chromosome, and as arizonae line the strain a875. After repeated backcrossing (now more than 50 generations) we eventually obtain two types of males in equal numbers: sterile ones that are mojavensis for all their chromosomes except the Y, and fertile that in addition to Ya are heterospecific for the fourth chromosome (and phenotypically wild-eyed). This established a clear and unconditional interaction between the Y and the fourth chromosomes: they both must be of arizonae origin for males to be fertile (in more precise terms: in Ya/Xm males there is an incompatibility between the Y chromosome and the m/m status of the fourth chromosome). We took this genetic analysis one step further. Through a scheme of crosses we are generating females that are heterospecific for the fourth chromosome but mojavensis in any other respect. These are repeatedly backcrossed to mojavensis. Through recombiantion, this produces fourth chromosomes that carry less and less arizonae material. After several generations of backcrossing, females are crossed to Ya/XmIVa/m, i.e males with an entire arizonae fourth chromosome. br males from this cross are in their majority sterile, but few are completely fertile. When crossed to moj br, they produce fertile and sterile sons in a 1:1 ratio. These males must then carry the arizonae fourth chromosome material that is required for fertility in Ya/Xm males. The repetition of this scheme must eventually result in the extraction from the original arizonae fourth chromosome only the genetic material that is necessary for fertility in Ya/Xm males. These males are referred to as mojYaSMFa (Pantazidis and Zouros, 1988), where SMF stands for the putative one-locus factor that restores sperm motility in mojavensis males carrying the arizonae Y chromosome. Figure 1 (parts a, b and c) presents graphically the sperm motility or motility caused by substituting arizonae genetic material into mojavensis males.

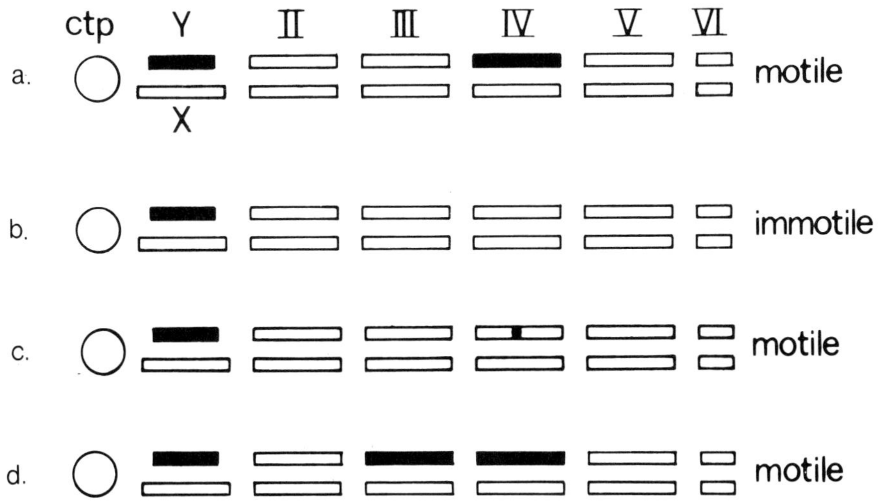

Figure 1. Four types of males carrying D. mojavensis (empty) and D. arizonae (filled) genetic material. Ctp: cytoplasm of the egg that produced the male; Y, X, II, III, IV, V, VI: chromosomes (Y and X: sex chromosomes, II to VI: autosomes).

All D. mojavensis material comes from stock moj br. In parts a, b and c the D. arizonensis material comes from stock a875; in part c the filled part of chromosome IV stands for the sperm motility factor (SMF). In part d the D. arizonensis material comes from stock a806; both the IV and the III arizonensis chromosomes are needed for sperm motility in these males.

Morphological differences in spermatogenesis between sterile and fertile mojavensis males carrying arizonae genetic material

We have completed a comparison of the major phases of spermatogenesis in five types of males: mojavensis (moj), arizonae (ari), mojavensis with the Y and one fourth chromosome of arizonae (mojYaIVa), mojavensis with the Y and the SMF of arizonae (mojYaSMFa), and mojavensis with only the Y of arizonae (mojYa) (Panatzidis et al., unpublished). Of these only the last cannot produce motile sperm.

Under a phase-contrast microscope no differences are obvious in the stages of gonial proliferation or spermatocyte growth between mojYa and mojYaSMFa. The first obvious differences appear in the pre-elongation stage immediately after completion of meiosis. In mojYaSMFa all 64 spermatids of the cyst can be seen, each with a nucleus and a nebenkern (mitochondrial derivative) structure. In mojYa cysts also contain 64 nebenkerns, but the number of nuclei are less than 64 and spermatids can be found without any or with more than one nuclei and without any or with more than one nebenkerns. In the early elongation stage the differences become more pronounced. In mojYaSMFa spermatids align themselves normally to form tight bundles, while in mojYa the spermatids become disoriented and diffused in the cyst. These differences are accentuated in the following stage of elongation when in mojYa spermatids fail to align against each other so that they form loose and abnormal bundles. The subsequent stages of individualization and coiling do not exist in mojYa. The electron microscopy comparison of spermiogenesis in moj (normal mojavensis) and mojYa revealed differences in all three major structural components of the sperm: the nucleus, the mitochondrial derivatives and the axoneme. In the early post-elongation stage chromatin condenses in the normal sperm forming a crest around the nuclear membrane and microtubules are seen to assemble in a regular formation outside the nuclear membrane. In mojYa the chromatin remains undiffused and microtubules are irregularly arranged. At post-elongation the mitochondrial derivatives in these males appear to contain large "empty" cytoplasmic areas and some of them may have two paracrystaline bodies instead of one. Sections at the end of the tail of the spermatid reveals absence of one or both of the central microtubules. There are no differences between the four types of males that produce normal sperm (ari, moj, mojYaIVa, mojYaSMFa) in any of characteristics that differentiate the spermatogenesis of fertile genotypes from the sterile mojYa. It can be, therefore, said that these abnormalities are solely caused by the substitution of the Ya for the Ym in mojavensis males. By the same token, it can be said that the restoration of all normal features of spermatogenesis in mojYaSMFa is due to a single genetic factor. In combination, the two observations suggest that the abnormalities in mojYa are due to an incompatibility between the Y chromosome of arizonae and the mojavensis SMF allele (SMFm).

There is, however, one difference among genotypes producing motile sperm and this refers to the number of individualized spermatids per cyst. D. mojavensis or arizonae males appear to have always 64 chromatids per cyst. In mojYaSMFa this number is often less. In a preliminary experiment for the quantification of this difference, the number of spermatids per cyst varied within and among individuals from 64 to 60. In mojYaIVa the corresponding numbers were 59 and 45. Further experiments are needed (including varying the X chromosome) before these observations can be interpreted. It is, however, clear that there must be several other genes on the fourth chromosome, in addition to SMF, that affect sperm production provided that normal sperm can be produced. These background genes appear to function in an additive fashion. In males with a compatible Y/SMF combination, the more of the background genes are of the same specific origin the smaller the loss of spermatids. In mojYaSMFa nearly all these genes are of mojavensis origin, thus the number or spermatids per cyst is close to the maximum of 64. In mojYaIVa an entire fourth chromosome is of the foreign species, thus the larger reduction in the average number of spermatids. Genes with a similar effect may exist in the other autosomes. A proper analysis of this variation may provide a tangible example of the commonly postulated genetic system that consists of one gene with a major effect and of many with minor effects.

Polymorphism for SMFa

The results outlined in the preceding section were obtained by using the stocks mojavensis br and arizonae a875. When arizonae a806 was used instead, we found that substitution of an arizonae fourth chromosome in mojavensis males carrying the arizonae Y chromosome was necessary but not sufficient for sperm motility (Zouros et al., unpublished). For these males to be fertile it is required that both the fourth and the third chromosome be heterospecific (as shown in Fig. 1d). This difference between a875 and a806 is independent from the mojavensis stock used in interspecific hybridizations. By creating special arizonae lines that carried the Y of a875 and the autosomes of a806 and vice versa we have shown that the difference does not stem from the Y chromosome or the third chromosome, but from the fourth chromosome. The fourth chromosome of a875 is necessary and sufficient to restore sperm motility in mojavensis males with an arizonae Y. The fourth chromosome of a806 is necessary but not sufficient, as the simultaneous presence of a third arizonae chromosome is required. No other arizonae chromosome could replace the third in this role. Since our previous studies suggest that the fourth chromosome's effect on sperm motility is due to a single locus, SMF, a875 and a806 must differ at this locus. Thus, SMFa appears to occur in two forms: the strong, SMFa(s), that suffice for restoration of fertility and the weak, SMFa(w),that requires contribution from the third chromosome. A survey of a limited collection of arizonae stocks has shown that the strong and weak states are widely distributed among D. arizonae populations, so that we can speak about an SMFa polymorphism. Interestingly, the geographical distribution of the polymorphism appears to obey a pattern. Two stocks that originated from areas far away from the area of sympatry with mojavensis (Guatemala and Jalisco, Mexico) contain only the weak allele. All stocks that are fixed for the strong allele came from the area of sympatry (Sonora, Mexico). Three stocks segregate for both alleles and two of these came from the region where the areas of sympatry and allopatry meet (Southern Sonora and Northern Sinaloa, Mexico).

Discussion

The ability to assess the effects of whole chromosome substitutions on pre-mating and post-mating reproductive isolation has been the main reason for using D. mojavensis and D. arizonae for the study of the genetics of speciation. Of all types of isolation, male sterility is the one most amenable to genetic analysis. Scored as sperm immotility, this character is shown to be under strict genetic control with no developmental or environmental noise. Many changes have occurred since the divergence of the two species (estimated at 2.5 Myr; Mills et al., 1986) at loci involved in spermatogenesis. Within a specific genetic background many of these substitutions are compatible with sperm formation in the heterospecific state and others are not. The latter can be, thus, singled out and studied in considerable detail. The most intriguing aspect of these genetic changes is that they are compatible with spermatogenesis when they co-occur with conspecific substitutions at other loci. The discovery of such co-adapted and cross-incompatible genetic complexes are of special value in the genetics of speciation. What we must be interested in is not simply the genetic changes that may lower the fitness in the hybrid, but also the subsequent changes at other loci that may restore the fitness and pave the way to divergence. Most explicit genetic models of speciation are indeed about the co-evolution of two or more loci (e.g. Nei et al., 1983; Zouros,1986; Charlesworth et al., 1987).

The Y/SMF is one example of a coevolved two-locus interaction that we have studied in some detail. Provided that the egg-cytoplasm, the X and a complete haploid set of autosomes in a male are of mojavensis origin, this male will be sterile if it is Ya/SMFm, but it will be fertile if it is Ya/SMFa. There is strong evidence that SMF represents a single locus (of one or more tightly linked genetic elements; Pantazidis and Zouros, 1988). The replacement of Ym by Ya in an otherwise mojavensis male causes severe and multiple abnormalities in spermatogenesis that become apparent in the early stages of spermiogenesis. These abnormalities occur both at the level of the spermatid and at the level of the cyst as a whole, with the result that spermatogenesis fails to complete its course. These adverse effects of the foreign Y chromosome can be completely eliminated if the Y substitution is coupled with a similar substitution at the SMF locus. SMF must, therefore, be a major locus in spermatogenesis. It becomes of major interest to find out what its function is under normal circumstances, why it has changed in the course of divergence of these two species and if it has also changed in other pairs of closely related species known to produce sterile male hybrids.

The discovery of a polymorphism at the arizonae homologue of SMF and its non-random distribution within the geographical range of the species adds another layer of interest for this locus. An interpretation of the polymorphism as the result of selection for enhanced post-mating isolation of the two species, while tempting, cannot be easily supported. It is certainly difficult to see how this mechanism may maintain the polymorphism at present, and it is also not clear how it may have acted in the past. In the areas of sympatry, arizonae populations have the "strong" allele, i.e. sons from fertile hybrids need only be heterospecific for the fourth chromosome to be fertile. Allopatric populations have the "weak" allele.

If hybrids were ever produced in these areas, their sons would be fertile if they were heterospecific for both the fourth and the third chromosome. Thus, everything else being equal, the post-mating barriers to gene flow are more severe in the areas of allopatry than sympatry. Since the only phenotypic distinction between SMFa(s) and SMFa(w) occurs when these alleles are carried by *mojavensis* males with an *arizonae* *Y* chromosome, one may assume that within *arizonae* the polymorphism is neutral. This might be an unwarranted assumption, and in fact the polymorphism may owe its existence to functional differences expressed in *arizonae*. Further understanding of the possible role of the Y/SMF interaction in the *mojavensis/arizonae* speciation and of the role of the SMFa polymorphism could come only through the extension of the analysis at the molecular and developmental levels.

Acknowledgments

Support for this research has been provided by NSERC (Canada) and the Institute of Molecular Biology and Biotechnology (Crete, Greece). I thank A. Pantazidis for permission to refer to his unpublished work.

Literature cited

Carson HL (1985) Genetic variation in a courtship related male character in *Drosophila silvestris* from a Hawaiian locality. Evolution 39:678-686

Carson HL, Lande R (1984) Inheritance of a secondary sexual character in *Drosophila silvestris*. Proc. Natl. Acad. Sci. U.S.A. 81:6904-6907

Charlesworth B, Coyne JA, Barton NH (1987) The relative rates of evolution of sex chromosomes and autosomes. Am. Nat. 130:113-146

Dobzhansky Th (1970) Genetics of the evolutionary process. Columbia University Press New York

Ehrman L (1961) The genetics of sexual isolation in Drosophila paulistorum. Proc. Natl. Acad. Sci. U.S.A. 49:155-158

Koepfer HR (1987a) Selection for sexual isolation between geographic forms of Drosophila mojavensis. I. Interactions between the selected forms. Evolution 41:37-48

Koepfer HR (1987b) Selection for sexual isolation between geographic forms of Drosophila mojavensis. II. Effects of selection of mating preference and propensity. Evolution 41:1409-1413

Mills LE, Batterham P, Alegre J, Starmer WT, Sullivan DT (1986) Molecular genetic characterization of a locus that contains duplicate Adh genes in Drosophila mojavensis and related species. Genetics 112:295-310

Nei M, Maruyama T, Wu C-I (1983) Models of evolution of reproductive isolation. Genetics 103:557-579

Pantazidis AC, Zouros E (1988) Location of an autosomal factor causing sterility in Drosophila mojavensis males carrying the Drosophila arizonensis Y chromosome. Heredity 60:299-304

Tan CC (1946) Genetics of sexual isolation between Drosophila pseudoobscura and Drosophila persimilis. Genetics 1:558-573

Wasserman M, Koepfer HR (1977) Character displacement for sexual isolation between Drososphila mojavensis and Drosophila arizonensis. Evolution 31:812-823

Zouros E (1981a) The chromosomal basis of sexual isolation in two sibling species of Drosophila, D. arizonensis and D. mojavensis. Genetics 97:703-718

Zouros E (1981b) The chromosomal basis of viability in interspecific hybrids between Drosophila arizonensis and Drosophila mojavensis. Can. J. Genet. Cytol. 23:65-72

Zouros E (1986) A model for the evolution of asymmetrical male hybrid sterility and its implications for speciation. Evolution 40:1171-1184

Zouros E (1989) Advances in the genetics of reproductive isolation in Drosophila. Genome 31:211-220

Zouros E, d'Entremont CJ (1980) Sexual isolation among populations of *Drosophila mojavensis*: response to pressure from a related species. Evolution 34:421-430

Zouros E, Lofdahl K, Martin P (1988) Male hybrid sterility in *Drosophila*: interactions between autosomes and sex chromosomes of *D. mojavensis* and *D. arizonensis*. Evolution 42:1332-1341

COLONIZING SPECIES OF DROSOPHILA

Antonio Fontdevila
Departament de Genètica i de Microbiologia
Universitat Autònoma de Barcelona
08193 Bellaterra, Spain.

THE THEORETICAL FRAME OF COLONIZATION

Species tend to colonize new territories by dispersal of their individuals. Regardless of the way of dispersal, quite often colonization involves the establishment of a few founder individuals in new territories. The genetic constitution of one such finite founder population may not represent that of the original population, displacing the population to an unstable state in which drift plays an important role (bottleneck effect).

Drift effects must be assesed by themselves and in respect to the other forces of the population dynamics of colonization. The bottleneck effect in colonization is shown by the reduction in gene diversity (expected heterozygosity). However, heterozygosity may be reduced only slightly unless the number of founders is very small. In fact, the greatest reduction in genetic variation is attained if the population remains small for many generations after the foundation. On the other hand, bottleneck size has a large effect in the reduction of the average number of alleles by the elimination of many low frequency alleles (Nei et al., 1975).

Another way to asses bottleneck effects in colonization is by studying genetic differentiation in colonized and original populations. Bottlenecks will reduce within-population variability, but increase among-population variability. Population differentiation in colonization can be studied by computing the among-population variance, but since gene frequency changes occur during colonization, the use of fixation indices will be of great interest to unveil how far the process of differentiation towards fixation has gone (Wright 1978). However, the process of differentiation by drift may be counteracted by migration. In fact, a slow amount of migration (i.e. one individual per generation) is sufficient to prevent local differentiation by drift.

Colonization is a long process that usually starts by a founder event and is followed by a series of population expansions and new foundations. It would be unrealistic to assert that natural selection

NATO ASI Series, Vol. H 57
Molecular Techniques in Taxonomy
Edited by G. M. Hewitt et al.

is not present in this process. The opinions range from those that assert that selection is relaxed in the expansion phase of populations (Carson 1973) to those that believe that natural selection is most active in population flushes (Templeton 1981). The discrimination between selection and drift in colonization is rather elusive, but there are ways to detect selection. The rationale is that drift effects will operate in the same way in all genetic loci under study, but selection will not. Thus, a difference in population structure among loci would suggest the operation of selection. Unfortunately, statistics devised to test this hypothesis cannot be applied to study cases in which there is uncertainty about the independence of the foundation of colonizer populations (Lewontin and Krakauer, 1973).

The study of bottleneck effects in colonization has evolutionary implications because such effects have been advocated as of primary importance in the founder-induced speciation models (Mayr 1963). The current debate is centered between those postulating that reproductive isolation can be achieved by shifting from one "adaptive peak" to another overcoming a selective barrier (the Wrightian view) and those that view isolation as the result of a steady accumulation of favorable alleles (the Fisherian view). In both models selection and drift play a role, although with different intensity. Most of current controversy (see for example Carson and Templeton, 1984; Barton 1989) derives from the dearth of study cases of founder events that have been historically reconstructed and experimentally studied. Here we present some of these cases.

THE STUDY CASE OF *DROSOPHILA BUZZATII*

The biogeographical frame

The increase of comercial exchanges in modern times has allowed transportation of many animals and plants providing historically documented cases of colonization. Among them, the Cactaceae, endemics of the American continent, have experimented a worldwide spread. These colonizer events are relevant to our studies because they involved, presumably, the colonization of many cactus-feeding insects and among them *Drosophila buzzatii*, a species of the repleta group, endemic from South America where it is associated with different *Opuntia* species.

The historical processes establish clear differences between the *D. buzzatii* colonizations of the Mediterranean basin and Australia. We know that the *D. buzzatii* host plants in Argentina (*O. quimilo*,

O. sulphurea, O. vulgaris) are not the main host plants in the colonized areas, being *O. stricta* and *O. ficus-indica* the most common in Australia and the Old World, respectively. These species are not of South American origin and their wide utilization by *D. buzzatii* talks about the adaptability of this species to new food resources. In fact, *D. buzzatii* has been also reported to emerge from columnar cacti (e.g. *Cereus validus, Trichocereus pasacana*) in South America (Pereira et al., 1983; Hasson, 1988).

This differential host shift may not be the only thing that distinguishes both colonizations. We know that in 1920 a large area of Australia was infested by *Opuntia* and a massive introduction of the phycitid moth *Cactoblastis cactorum*, a cactus-feeding insect, was initiated to control this agricultural pest. The operation involved extensive shipments of *Opuntia* rotting cladodes (rots) from Argentina between 1931 and 1936 and it is taken as responsible of the introduction of *D. buzzatii* to Australia (Barker and Mulley, 1976; Barker et al., 1985). We do not know how large was the bottleneck experimented by founder populations of *D. buzzatii*, it may not have been so large due to the massive import of rots and also because founder individuals encountered an extensive and open area of *Opuntia* stands to be occupied readily. Certainly, this is what happened with *C. cactorum*. In only six years, this species invaded most of the thousands of square kilometers of the infested area (Murray 1982).

Details about the Old World colonization by *D. buzzatii* are more speculative. Fontdevila et al., (1981, 1982) have argued that it was not until the direct trade between Spain and Argentina was open in the middle of the eighteenth century that agricultural imports from Argentina allowed for the introduction of *D. buzzatii* to Spain. By this time *Opuntia* plantations, mainly of *Opuntia* from tropical America, were common in the Mediterranean area, providing a suitable niche to shift and occupy. Yet, there is no reference reporting that the *Opuntia* area was extensive and invasive, rather it would have been scattered in patchy plantations close to human habitation as it is today. However, the relative size of these plantations may have been larger than they are today due to the flourishing of the exploitation of *Opuntia* products such as cochineal dyes. This historical biogeographical scenario has two main consequences for founder populations of *D. buzzatii*. First, the widespread of founder populations may have taken place by a process of stepping stone, mediated by succesive founder events. This entails a moderate rate of population increase and a leading role of founder episodes in secondary colonizations. Second, most probably the introduction of *D. buzzatii*

Table 1. Mean chromosomal diversity (H) and among population differentiation (F_{DT}) of original and colonizer populations of *Drosophila buzzatii*. Standard errors are given in parenthesis.

		Second chromosome		Fourth chromosome	
POPULATIONS	N^a	H	F_{DT}	H	F_{DT}
ORIGINAL[1]	18	0.4789 (0.0315)	0.151	0.1036 (0.0435)	0.332
COLONIZER					
OLD WORLD[1]	27	0.5483 (0.0230)	0.089	0.2901 (0.0314)	0.080
AUSTRALIA[2]	17	0.4587 (0.0175)	0.049	0.0	0.0

[a]Number of analyzed populations; [1]Data from Fontdevila et al., 1981, 1982; Hasson 1988 and unpublished records; [2]Data from Knibb et al., (1987)

in the Old World is the result of several primary colonizations mediated by succesive importations of *Opuntia* material (cladodes and fruits).

The chromosomal polymorphism

According to the theory of homoselection (Carson 1959), the loss of chromosomal diversity in marginal areas is an adaptive response to the advantage of high levels of recombination in these areas. Carson (1965) advanced the hypothesis that colonizing species originate from these marginal isolates and carry with them low levels of chromosomal polymorphism. Based on some pioneering works with a limited number of populations (Wasserman, 1954, 1962; Carson and Wasserman, 1965; Mather 1957), *D. buzzatii* was taken as an example of colonization mediated by dispersal from homoselected populations from peripheral isolates. Since then, more than 60 populations have been analyzed all over the world and a new picture has emerged from these studies (Fontdevila et al., 1981, 1982; Barker et al., 1985; Knibb et al., 1987; Fontdevila 1989 and unpublished results).

D. buzzatii is polymorphic for the second and the fourth chromosomes, with eight and two inversions each, respectively (Ruiz et al., 1984). The colonization of the Old World did not decrease the overall mean chromosomal diversity (Table 1), but significant changes have occurred in number and frequency of inversions. Concerning the second chromosome polymorphism, all but one (jq^7) of the five low frequency arrangements of the original region (jq^7, y^3, jc^9, r^9, js^9) have been lost. Inversion jq^7 is present in only one original population (Arroyo Escobar: p=0.014), but occur in 13 of the colonizer populations (p= 0.14-0.04). The high frequency inversions (st, j) are present in most of original and colonizer populations in highly variable frequencies. Arrangement jz^3 is present at widely different frequencies (p= 0.43-0.00) in most of the original populations of Argentina and is absent in the populations of Bolivia. This arrangement is present in areas of primary colonization and is lost in areas of secondary colonization. Changes of fourth chromosomal polymorphism in Old World colonization are even more striking. One inversion (s) is present in 13 localities of Argentina, often in low frequencies ($p<0.1$), but ocassionally reaches high values in some populations of the NW of Argentina (p= 0.4-0.5). This inversion has not been detected in Bolivia and Brazil. On the other hand, the great majority of Old World populations show moderate to high frequencies (p= 0.15-0.50) of this inversion and only the populations of the Eastern Mediterranean do not present it. The other inversion (st) is present in all the populations and is usually fixed in the original areas. This pattern gives a high average diversity of the fourth chromosome in Old World colonizing populations (Table 1).

From these data on the Old World colonization three main conclusions can be inferred. First, original chromosomal polymorphism shows a high degree of geographical variability that is evidenced by high values of the fixation (F_{DT}) indices (Table 1). Second, there is no overall loss of polymorphism in colonization as evidenced by the high chromosomal diversity in colonizer populations. Third, there is suggestive evidence of founder effects due to significant changes in inversion frequencies and the loss of low frequency arrangements. These effects are more pronounced in secondary colonizer populations (Fontdevila et al., 1981). These conclusions do not conform with the view that this widespread species originated from marginal isolates subjected to homoselection (Carson 1965) and new explanations have to be searched.

Nonetheless, the colonization of Australia by *D. buzzatii* has produced different results. As in the Old World colonization, rare arrangements have been lost, but here some other arrangements are not present either. Namely, arrangements $2jq^7$ and 4s have never been detected. Thus, the second chromosome polymorphism has been reduced to the two most frequent and ubiquitous arrangements (2St and j), since $2jz^3$ arrangement is only present in some populations at low frequencies, and the fourth chromosome has become monomorphic for the 4St arrangement. In this case there has been a reduction of polymorphism and a certain kind of homoselection might be invoked. Fortunately, we have another episode of colonization in the Old World to compare and at least two explanations can be put forth without resorting to the homoselection theory. First, since host plants are so different in both colonized areas, selection is responsible for the differences in polymorphism. Second, founder effect has changed the initial frequencies and some arrangements have not even been present in founder populations. I am inclined to the second explanation since many Old World populations (e.g. in secondary colonizations) have lost several arrangements because of founder effects, regardless of the host plant present. This does not mean strictly that cactus species is not a selective factor, but that selection may be more effective in the original areas where a high diversity of cactus exist.

The allozyme polymorphism

The allozyme polymorphism has been analyzed for a total of 62 *D. buzzatii* populations in several electrophoretic studies. More than 35 Australian populations have been studied for a total of 36 loci (Barker and Mulley, 1976; Barker 1981); five original populations from South America (Argentina and Bolivia) and 13 from the Old World have also been studied in my laboratory for 22 loci (Sánchez, 1986 unpublished) and a limited number of isofemale lines from several Argentinian and Brazilian localities were also electrophoresed for 31 presumptive loci (Barker et al., 1985). I have made an attempt to compare these studies and found that 20 loci were common among them. Table 2 gives a summary of relevant parameters of this polymorphism using these common loci. The average gene diversities are larger than those reported in other studies with more loci (Barker and Mulley, 1976) unveiling a biased tendency of researchers to use polymorphic loci preferentially (Singh 1990). Clearly, number of alleles per locus, percentage of polymorphic loci and gene diversity are smaller in colonizer populations due to founder effects. Yet, Australian and Old World colonizations differ in the magnitude of these effects. Although gene diversities (H), averaged across populations are not significantly different between the two sets of independently studied original populations ($F = 0.199$), Australian colonizer

Table 2. Compared allozyme polymorphism between original and colonized populations of *Drosophila buzzatii*.

POPULATIONS	Alleles per locus	Polymorphism (%)	Gene diversity (H)
ORIGINAL			
Argentina & Bolivia[1]	1.60	32.7	0.098
Argentina & Brazil[2]	1.61	33.3	0.102
COLONIZER			
Mediterranean basin & Canary Islands[1]	1.23	17.1	0.067
Australia[3]	1.42	17.6	0.085

([1]) Sánchez, 1986. ([2]) Barker et al., 1985. ([3]) Barker and Mulley 1976; Barker, 1981.

populations show an H mean value higher than that of Old World populations (0.085 vs 0.067, $F = 33.422$, $P < 0.001$). This difference is consistent with the hypothesis that colonization in Australia has been followed by a fast population increase that prevented much loss of variability.

POPULATION DIFFERENTIATION IN *D. BUZZATII* COLONIZATION

Table 3 gives an account of the intra- and inter-population fixation for allozyme loci in original and Old World colonizer *D. buzzatii* populations. On average, colonization of the Old World has decreased the individual intrapopulation diferentiation (F_{IS}), but has not increased differentiation among populations (F_{ST}).

The allozyme differentiation is most illustrative when different colonizing events are compared for the same species. Table 4 shows

Table 3. Mean values of fixation indices of individuals relative to populations (F_{IS}) and of populations (F_{ST}) for allozymes in original and Old World colonizer populations of *D. buzzatii*. *D* stands for Nei's genetic distance (standard error in parentesis).

	F_{IS}	F_{IT}	F_{ST}	D
ORIGINAL	0.372	0.452	0.128	0.0199 (0.0041)
COLONIZER	0.248	0.345	0.129	0.0116 (0.0010)

Data from Sánchez (1986).

the fixation indices (F_{DT}) in the original and the colonizer populations of *D. buzzatii.* Two main conclusions can be inferred. First, there is a general, strong tendency to fixation due to founder events. Second, taking in account only the polymorphic loci, population differentiation is reduced in the Australian colonization only. There are three main possible explanations of this difference. First, migration among populations may be much higher in Australia than in the Old World. Second, a fast population expansion after Australian foundation led to no differentiation among populations. Third, the recent Australian colonization has not allowed time to population differentiation. In view of the evidence of low founder effect in the Australian colonization (Table 2), I become inclined towards the second explanation. One cannot discard the effects of gene flow among colonizing populations, as it has been demonstrated by Barker et al., (1989) but there is no reason a priori to believe that there is more gene flow among Australian than among Old World populations. The third explanation has been also advocated to understand the low geographic differentiation of *D. simulans* compared to that of *D. melanogaster* (Singh 1990) and it should be considered in our case.

Nonetheless, fast population expansion after foundation cannot explain all the population differentiation for individual loci. There is spatial heterogeneity of F_{DT} across loci detected by the Lewontin-Krakauer test in original and colonizer areas ($P < 0.001$), yet natural selection cannot be inferred because a hierarchical series of foundation events can also generate heterogeneity among neutral loci. In Australian colonizer populations, Sokal et al., (1987) using directional spatial autocorrelation techniques found suggestive evidence that selection operates on different spatial scales in each allozyme.

Table 4. Allozyme differentiation among original and colonizer populations of *Drosophila buzzatii* measured by F_{DT}

LOCUS	NEW WORLD[a]	OLD WORLD[a]	AUSTRALIA[b]	CHR[c]
Xdh	0.047***	0.000***	0/0	2
Ao	0.001**	0.117***	0.028***	2
Est-1	0.057***	0.060***	0.052***	2
Est-2	0.198***	0.186***	0.024***	2
Em-1	0.001**	0.000 ns	0/0	2
Pgm	0.006 ns	0.001 ns	0.010***	4
Adh-1	0.000 ns	0.055***	0.045***	3
Adh-2	0.000 ns	0/0	0.000**	3
6Pgdh	0.573***	0.532***	0/0	X
Pep-2	0.044*	0.037***	0/0	2
Pep-1	0.000 ns	0/0	0/0	5
Lap-1	0.062***	0/0	0/0	2
α Gpdh	0.000 ns	0/0	0/0	3
Mdh	0.008 ns	0/0	0/0	3
Fum	0.004*	0/0	0/0	X
Aph	0.012**	0/0	0/0	4/5
Acph	0.032*	0/0	0/0	4/5
Hk	0.000 ns	0/0	0/0	X

Heterogeneity χ^2 among populations is indicated by *:
* significant at $P < 0.05$; ** at $P < 0.01$; *** at $P < 0.001$; ns, non significant.
a Data from Sánchez (1986); *b* Data from Barker and Mulley (1975);
c CHR means chromosome where the locus is situated.

These results do not discard the founder effects unveiled by the comparative study between both colonizations. Rather, they demonstrate the joint operation of selection and drift in shaping population differentiation.

Chromosomal differentiation among populations after colonization has been also computed (Table 1). There is a significant decrease in F_{DT} values in both colonizations. This low differentiation in colonizing populations is correlated with a high chromosomal polymorphism in the Old World colonization. The same is true for the second chromosome differentiation in the Australian populations, but not for the fourth chromosome. In the latter case founder effects are most probably responsible for the loss of the 4s rearrangement and the absence of among population differentiation. Hierarchical analyses of fixation indices show that most of the Old World population differentiation can be accounted for by demes inside small regions ($F_{DR} = 0.064$) and very little by regions inside geographical areas (subdivisions) ($F_{RS} = 0.015$). Fontdevila et al., (1981) working on chromosomal polymorphism suggested that macroenvironmental factors must be of low importance to explain population differentiation in Old World populations and adaptation should be regarded at the microenvironment level inside populations. The work of Santos et al., (1988) with inversion polymorphism agrees in that population structure cannot be explained completely by random differentiation in a Spanish population and some kind of selection must be invoked. This is in contrast with differentiation in original South American areas, where differentiation among ecological regions ($F_{RS} = 0.261$) inside geographical areas (subdivisions) account for most of population differentiation (Figure 1). The easiest interpretation is that differentiation in the original areas is dependent on ecological gradients inside the arid zones, whereas Old World populations are mainly differentiated according to historical events.

These results show that chromosomal polymorphism changes in colonization are different from those of allozyme polymorphism. One is tempted to atribute this difference to selection operating in a different way on each kind of polymorphism. Yet, often it is difficult to disentangle the operation of selection in both systems due to the linkage relationships between allozymes and rearrangements. However, in some privileged cases of no linkage it may be possible to gain some insight. In *D. buzzatii*, Adh-1 is located in the third chromosome that is free of inversions. This allozyme locus has increased its gene diversity in colonization and shows low non-significant intrapopulational differentiation in most colonizing populations (data not shown). This is just the reverse trend observed by the remaining allozymes. Moreover, interpopulational differentiation of Adh has increased in both cases of colonization by a similar amount (Table 4). This suggests that the among population differentiation is independent of the dynamics of population colonization. In summary, the Adh population structure is difficult to be explained by random non-selective forces operating in the colonizing process.

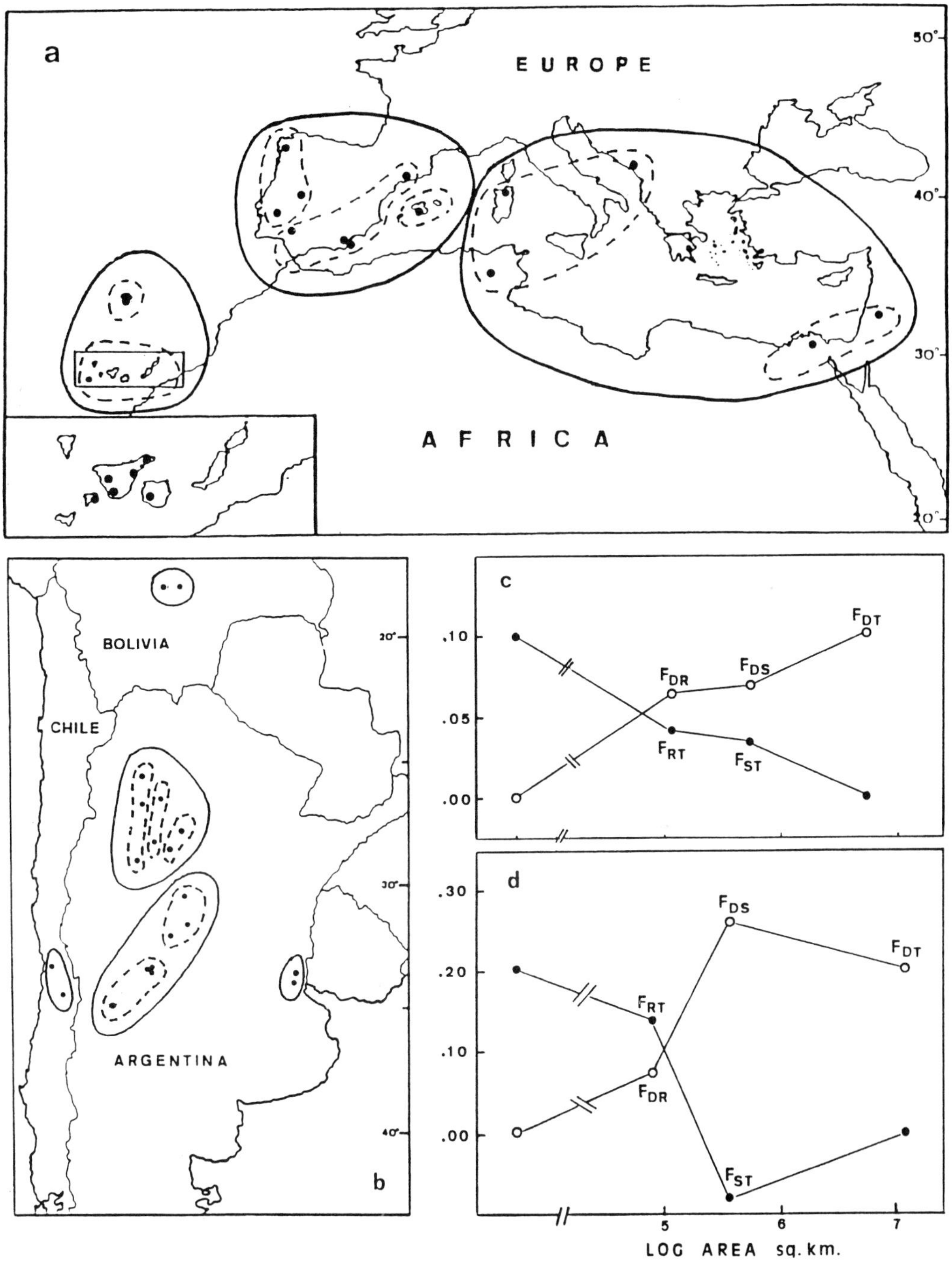

Figure 1. a,b: Geographic distribution of *D. buzzatii* demes analyzed in South America and the Old World for the chromosomal polymorphism. Dotted lines group demes within regions (R), solid lines group regions within subdivision (S). c,d: Hierarchical F statistics combined across chromosomes plotted against log area for the approximate mean size of a region (R), subdivision (S) and total area (T): (c) Old World; (d) South America.

COMPARATIVE STUDIES OF POPULATION DIFFERENTIATION IN WIDESPREAD DROSOPHILA SPECIES

Table 5 shows a summary of distributions of fixation indices (F_{DT}) among populations of different species of *Drosophila* in order of increasing population allozyme differentiation. Three classes of species can be defined.

Class I. Species that show low differentiation. Species of the willistoni group (*D. willistoni* and *D. equinoxialis)* and of the obscura group (*D. subobscura*, *D. obscura* and *D. pseudoobscura*) show a low differentiation. In this case high migration rates have been documented for the members of the obscura group (Powell et al. 1976; Coyne et al., 1982), but not for those of the willistoni group (Burla et al., 1950). However, gene flow depends not only on dispersal but also on the feeding and breeding structure. *D. willistoni* is a fruit breeder with an abundance of resources in the tropical forest, allowing a lot of movement among sites that may produce much gene flow. Nevertheless, it would be most advisable to measure *D. willistoni* dispersal rate anew with wild flies.

D. subobscura, a Paleartic species, has invaded South and North America in the past decade. This invasion has been accomplished in a few years by means of a fast population expansion that covers latitudes ranging from 30 to 50 degrees in both hemispheres (Chile and Western United States). The most important outcome of these studies on the recent colonization of *D. subobscura* is the establishment of chromosomal latitudinal clines that mimic those already found in the Old World (Prevosti et al., 1988). These results lend support to the adaptive value of this polymorphism. This differentiation process can be studied by computing fixation indices. Table 6 shows that, in spite of the clinal differentiation, the process of colonization has decreased the degree of differentiation among colonizing populations. This genetic homogeneity can be accounted for by the population explosion following the founder event. Additional evidences to test the theoretical frame of colonization come from allozyme differentiation studies. Average allozyme diversity show no significant differences between original ($H = 0.244$) and colonizing populations ($H = 0.242$), but the number of alleles per locus has decreased from 4.4 to 1.7 (Balañà 1989). These results conform quite well to the theoretical expections (Nei et al., 1975) under rapid population expansion after a founder event. Yet, changes in allozyme population differentiation during colonization are more difficult to interpret. I

Table 5. Mean and distribution of fixation indices (F_{DT}) of allozyme loci from several widespread species of *Drosophila*

Species	loci	F_{DT}	0/0	0-0.04	0.05--0.14	0.15--0.24	0.25--0.49	0.5--1.0
D. subobscura								
Original (OW)[1]	15	0.004	0.25	0.75				
Colonizer (NW)[2]	15	0.013	0.60	0.40				
D. willistoni[3]	25	0.022	0	0.88	0.12			
D. pseudoobscura (US)[3]	24	0.028	0.46	0.46	0.04	0.04		
D. obscura[3]	30	0.067	0.47	0.43	0.07	0.03		
D. buzzatii								
Original (NW)[4]	22	0.113	0.20	0.55	0.15	0.05	0	0.05
Colonizer (OW)[4]	22	0.116	0.60	0.20	0.10	0.05	0	0.05
Colonizer (Aust.)[5]	22	0.034	0.70	0.25	0.05			
D. melanogaster[3]	10	0.241	0.10	0.10	0	0.50	0.30	
D. paulistorum[3]	17	0.255	0.06	0.41	0.18	0.18	0.06	0.12
D. pseudoobscura[3]	24	0.268	0.42	0.25	0.21	0	0.08	0.04

Data from: ([1]) Loukas et al., (1979); ([2]) Prevosti et al., (1989); ([3]) Wright (1978); ([4]) Sánchez (1986); ([5]) Barker and Mulley (1975)

have computed fixation indices for a set of original and colonizing populations where there is a good mobility correspondence among alleles. Table 6 shows that no changes in population differentiation have been observed, although the low F_{DT} values detected in the original populations are quite surprising and need future confirmation.

Class II. Species that show moderate differentiation. *D. buzzatii* is representative of this class. We have already discussed the *D. buzzatii* differentiation and attributed the differences between both colonizing events to the rate of population expansion after colonization. This evidences the importance of the dynamics of colonization to understand the present population structure of any colonizing species. It is interesting to notice that both original and Old World colonizer populations show the same degree of mean population differentiation, yet fixation index distribution across loci is rather different. Fixation of loci due to founder effects has displaced the distribution leftward

Table 6. Genetic differentiation among original and colonizing populations of *Drosophila subobscura* measured by the fixation index (F_{DT})

Populations	Chromosomal polymorphism			Allozyme polymorphism		
	N^a	F_{DT}	Ref.	N^a	F_{DT}	Ref.
Original	4	0.127	(1)	4	0.004	(3)
Colonizer	16	0.013	(2)	5	0.013	(4)

Data from: [a] Number of analyzed populations; (1) Brncic et al., (1981); (2) Prevosti et al., (1989); (3) Loukas et al., (1979); (4) Balañà (1989).

towards low F_{DT} values in the Old World colonizer populations. This effect is most dramatic in Australian populations.

Barker et al., (1986) have found low spatial variation, but high temporal variation in Australian populations for allozyme polymorphism. They found heterogeneity for temporal Fst among loci suggesting that selection is operating at the microspatial level inside populations.

Under the infinite-island model the estimated Nm values are greater than 1 for most allozyme loci in original and colonizer populations. This may suggest that migration is important in opposing differentiation. Direct estimates of dispersal obtained by the author and his associates (unpublished data) and also by Barker and East (1989) corroborate that gene flow may be important among *D. buzzatii* populations. Yet, indirect estimates may be biased because the infinite-island model is unrealistic. Using the finite-island model in which two island populations are not independent and can receive migrants from the same population, under certain restrictive assumptions (Weir 1990), Nm values can be estimated from F_{IS} and shown to be less than 1 for the majority of loci in original and Old World colonizing populations (data not shown).

These considerations and the comparative study among both documented colonizations suggest that population differentiation in *D. buzzatii* allozyme loci may be affected by gene flow but historical processes such bottlenecks and founder events produce significant differences in population structure. Not all loci are affected equally by such random processes and some similarities between colonizations suggest the action of selection in some loci (e.g. in Adh locus).

Class III. Species that show high differentiation. Some of these species have high F_{DT} values due to their subdivision in taxa that show incipient speciation. The case of *D. paulistorum* is most illustrative. This species ranges from Central America to Southern Brasil and its populations can be divided in six taxa named semispecies or incipient species (Dobzhansky and Powell 1975) that show several degrees of partial reproductive isolation among them. Then, the high average fixation index can be explained by this process of incipient speciation. When a hierarchical study is performed, the differential fixation among localities within semispecies ($F_{DS} = 0.099$ in the leading loci) is less than half that among semispecies ($F_{ST} = 0.204$ in the leading loci). Wright (1978) thinks that this result does not support the hypothesis that the formation of semispecies and the local differentiation within them are the outcome of the same conditions. Rather, he argues that random differentiation is more important than any directional force. A similar interpretation is given to the high value of fixation index in *D. pseudoobscura* when the population of Bogota is included ($F_{DT} = 0.268$).

On the other hand, the high among-population differentiation of *D. melanogaster* cannot be interpreted in relation to incipient speciation. No evidence of reproductive isolation exists among *D. melanogaster* populations of different geographical origins. This species is highly cosmopolitan and its world colonization is reasonably well understood. David and Capy (1988) divide the world populations in three categories: ancestral, ancient and new. Ancestral populations are found in tropical Africa, where the species originated about 2-3 million years ago. After the last glaciation, 10-15 thousand years ago, *D. melanogaster* colonized Eurasia without the participation of man. Ancient populations are defined as those ranging from Western Europe to the Far East, in a rather continuous occupation. Finally, populations found in the American and Australian continents are new, having been introduced by man a few centuries ago.

It is a well established fact that *D. melanogaster* exhibits a high degree of diferentiation not only among geographically separated regions but also among local populations separated only by a few hundred meters (Lemeunier et al., 1986; Voudibio et al., 1989). In recently introduced populations such as those of North America this differentiation ($F_{DT} = 0.241$) could be explained because there has not been enough time for genetic homogenization through gene flow. However, this consideration breaks down when one sees the extreme differentiation among long established ancestral Afrotropical populations. This differentiation requires a strong restriction to gene flow or high levels of local adaptation. The intuitive assumption that *D. melanogaster*, being a domestic species, has always been transported along with human activities and the whole species would thus function as a panmictic unit is untenable due to its population structure. Some early measures of dispersal with mutants gave extremely low estimates of gene flow (Wallace 1970), and recent experiments with rare alleles have given estimates that may explain the majority of population differentiation (Singh and Rhomberg 1987 a,b). Yet, at least some polymorphic loci (about 15 %) show a pattern of differentiation that should be accounted for by local adaptation. The situation in a small Spanish area of wine production may illustrate this point. David and his collaborators have shown that fixation indices are higher in field populations than in cellar populations for Adh, Gpdh and Est-6 loci (David et al., 1989). The effects are much larger for Adh than for the other loci and suggests that the field population structure may be the outcome of yearly recolonization from panmictic cellar populations and the combined efects of gene flow and local Adh selection.

CONCLUDING REMARKS

Widespread Drosophila species show all kinds of spatial genetic (allozyme) differentiation, regardless whether they have dispersed following a defined habitat or shifting to a new one. Recent colonizations are illustrative and show that the dynamics of the population expansion after foundation is crucial to the ensuing population differentiation. After a founder event, allozyme genetic diversity is reduced but not as much as number of alleles. However, fast population growth and high invasiveness after colonization maintains large amounts of the original variability and prevents local geographic differentiation. This is in accordance with the theoretical expections of changes in genetic variability after a bottleneck.

Chromosomal and allozyme polymorphisms do not behave equally. Large scale colonizations do not diminish the overall expected heterozygosity in chromosomal polymorphisms, only when colonization is mediated by a severe bottleneck some inversions are lost. This is similar to allozyme polymorphisms. However, documented colonizations produce an important reduction in chromosomal population differentiation, as measured by the Wright fixation index. In *Drosophila buzzatii*, colonizing populations do not show any correlation between macroenvironmental factors and population differentiation, but in *D. subobscura* they do. In original populations, though, both species show a population differentiation that is dependent of some macroenvironmental factors. Differences between *D. subobscura* and *D. buzzatii* may be related to the new colonizing environment experienced by them. *D. subobscura* colonizing populations have found in the American continent an ecological latitudinal gradient very similar to that in the Paleartic zone. On the other hand, the ecological gradients of the arid zones in South America may be very different from the more uniform arid areas in the Old World where *D. buzzatii* has been naturalized. In these areas chromosomal rearrangements are mainly subjected to microenvironmental conditions that maintain their polymorphism.

The high population differentiation shown by some widespread species is not easy to interpret. In some cases, such as *D. paulistorum* and *D. pseudoobscura*, high fixation indices are accounted for by incipient speciation. This apparent correlation between among population divergence and speciation, as measured by incipient reproductive isolation, may seem to favor the adaptive divergence hypothesis in speciation. Yet, the negative correlation across loci between heterozygosity and interspecific genetic distance in closely related species (Choudhary et al., 1988) sheds some doubt on the adaptive hypothesis. In fact, population differentiation does not guarantee the presence of incipient speciation. *D. melanogaster* shows an F_{DT} value as high as that of *D. paulistorum* and *D. pseudoobscura*, yet no case of local incipient speciation has ever been described in the former species.

The study of *D. buzzatii* colonizations is very illustrative in that it sheds light on the relative importance of historical, ecological and genetic constraints in population differentiation. Dispersal and habitat uniformity has not been sufficient to impede genetic differentiation among colonizing populations in the Old World. Since we have an Australian parallel colonizing experiment in which low population differentiation exists, historical (founder) events may be mainly responsible for among population differentiation. This may not apply to chromosomal polymorphisms, demonstrating the operation of genetic constraints in colonization. Chromosomal rearrangements are more sensitive to ecological scenarios and, although afected by founder events, chromosomal differentiation in colonizing populations is kept low because of habitat uniformity.

In summary, widespread species of Drosophila show that population diversification after colonization is highly dependent of the dynamics of founder events. In very rare ocassions founder effects are conducive to incipient speciation, and even in these cases it is unknown what kind of genetic changes have been operated in colonization. Certainly, allozyme polymorphism is unable to distinguish a speciation process from simple population differentiation. However, the fact that many monomorphic loci are clearly diagnostic among species substantiates the idea that random and/or mutational processes may participate in species formation.

ACKNOWLEDGMENTS

Research with *D. buzzatii* by the author and his associates has been supported by CICYT and DGICYT grants n°s. 2920/76; 0910/81; 2825/83 and PB-85/0071. My associates Alfredo Ruiz, Armand Sánchez, Esteban Hasson, Horacio Naveira and Mauro Santos are greatly acknowledged by their collaboration in obtaining a great deal of the original information. However, the ideas and statements advanced in this paper are the sole responsability of the author. I am indebted to Dr. J. Balañà for allowing me to use some of his *D. subobscura* data. Mr. Antonio Barbadilla and Ms. Julia Provecho typed and prepared the camera ready version of this paper and their cooperation was of utmost value for its completion.

BIBLIOGRAPHY

Balañà J (1989) Estudi de l'associació entre els polimorfismes cromosòmic i enzimatic en poblacions nord-americanes de *Drosophila subobscura*. PhD thesis. University of Barcelona, Spain

Barker JSF (1981) Selection at allozyme loci in cactophilic Drosophila. In: Gibson JB, Oakeshott JG (eds) Genetic studies of Drosophila populations. The Australian National University, Canberra

Barker JSF and Mulley JC (1976) Isozyme variation in natural populations of *Drosophila buzzatii*. Evolution 30: 213-233

Barker JSF, East PD (1989) Estimation of migration from a perturbation experiment in natural populations of *Drosophila buzzatii* Patterson & Wheeler. Biol. J. Linn. Soc. 37: 311-334

Barker JSF, Sene FM, East PD, Pereira MAQR (1985) Allozyme and chromosomal polymorphism of *Drosophila buzzatii* in Brazil and Argentina. Genetica 67: 161-170

Barker JSF, East PD, Weir BS (1986) Temporal and microgeographic variation in allozyme frequencies in a natural population of *Drosophila buzzatii*. Genetics 112: 577-611

Barton NH (1989) Founder effect speciation. In: Otte D, Endler, JA (eds) Speciation and its consequences. Sinauer Massachusetts

Brncic D, Prevosti A, Budnik M, Monclús M, Ocaña J (1981) Colonization of *Drosophila subobscura* in Chile I. First population and cytogenetic studies. Genetica 56: 3-9

Burla H, da Cunha AB, Cavalcanti AG, Dobzhansky Th, Pavan C (1950) Population density and dispersal rates in Brazilian *Drosophila willistoni*. Ecology 31: 393-404

Carson HL (1959) Genetic conditions which promote or retard the formation of species. Cold Spr Harb Symp Quant Biol 24: 87-105

Carson HL (1965) Chromosomal morphism in geographically widespread

Carson HL (1973) Reorganization of the gene pool during speciation. In: Morton NE (ed) Genetic structure of populations. Univ Press Hawaii

Carson HL, Templeton AR (1984) Genetic revolutions in relation to speciation phenomena: The founding of new populations. Annu. Rev. Ecol. System. 15: 97-131

Carson HL, Wasserman M (1965) A widespread chromosomal polymorphism in a widespread species, *Drosophila buzzatii*. Amer Natur 99:111-115

Choudhary M, Coulthart MB, Singh RS (1988) Genetic divergence and models of species formation in Drosophila. Science

Coyne JA, Boussy IA, Prout T, Bryant JH, Jones JS, Moore JA (1982) Long-distance migration of Drosophila. American Naturalist 119: 589-595

David JR, Capy P (1988) Genetic variation of *Drosophila melanogaster* natural populations. Trends in Genetics 4: 106-111.

David J, Alonso-Moraga A, Capy P, Muñoz-Serrano A, Vouidibio, J (1989) Short range genetic variations and alcoholic resources in *Drosophila melanogaster*

Dobzhansky Th, Powell JR (1975) The willistoni group of sibling species of Drosophila. In: King RC (ed) Handbook of Genetics, vol 3, Invertebrates of Genetic Interest. Plenum Press, New York, London, pp 589-622

Font-Quer P (1973) Plantas medicinales: El Dioscórides renovado. Barcelona.

Fontdevila A (1989) Founder effects in colonizing populations: The case of *Drosophila buzzatii*. In: Fontdevila A (ed) Evolutionary Biology of Transient Unstable Populations. Springer-Verlag Berlin Heidelberg

Fontdevila A, Ruiz A, Alonso G, Ocaña, J (1981) Evolutionary history of *Drosophila buzzatii*. I. Natural chromosomal polymorphism in colonized populations of the Old World. Evolution 35: 148-157

Fontdevila A, Ruiz A, Ocaña J, Alonso G (1982) The evolutionary history of *Drosophila buzzatii* II. How much has chromosomal polymorphism changed in colonization? Evolution 36: 843-851

Hasson ER (1988) Ecogenética evolutiva de *D. buzzatii* y *D. koepferae* (complejo mulleri; grupo repleta; Drosophilidae; Diptera) en las zonas áridas y semiáridas de la Argentina. Ph. D. Dissertation, Universidad de Buenos Aires, Argentina

Knibb WR, East PD, Barker JSF (1987) Polymorphic Inversion and Esterase loci complex on chromosome 2 of *Drosophila buzzatii* I. Linkage disequilibria. Aust J Biol Sci 40: 257-269

Lemeunier F, David JR, Tsacas L, Ashburner M (1986) The melanogaster species group. In: Ashburner M, Carson HL, Thompson JN (eds) The genetics and biology of Drosophila, Acad Press London, vol 3e, pp 147-256

Lewontin RC, Krakauer J (1973) Distribution of gene frequency as a test of the theory of the selective neutrality of polymorphisms. Genetics 74: 175-195

Loukas M, Krimbas CB, Vergini Y (1979) The genetics of *Drosophila subobscura* populations IX. Studies on linkage disequilibrium in four natural populations. Genetics 93: 497-523

Mayr E (1963) Animal species and evolution. Cambridge Mass. Harvard Univ Press

Mather WB (1957) Genetic relationships of four *Drosophila* species from Australia. Texas Univ Publ. 5721: 221-225

Murray ND 1982 Ecology and Evolution of the *Opuntia- Cactoblastis* Ecosystem in Australia. In: Barker JSF, Starmer WT (eds) Ecological Genetics and Evolution Academic Press, Sydney, New York

Nei M, Maruyama T, Chakraborty R (1975) The bottleneck effect and genetic variability in populations. Evolution 29: 1-10

Pereira MAQR, Vilela CR, Sene FM (1983) Notes on breeding and feeding sites of some species of the repleta group of the genus *Drosophila* (Diptera, Drosophilidae). Ci. e Cult., Sao Paulo 35: 1313-1319

Powell JR, Dobzhansky Th, Hook JE, Wistrand HE (1976) Genetics of natural populations. XLIII. Further studies on rates of dispersal of *Drosophila pseudoobscura*. Genetics 92: 613-622

Prevosti A, Ribó G, Serra L, Aguadé M, Balañá, J. Monclús M, Mestres F (1988) Colonization of America by *Drosophila subobscura*: Experiment in natural populations that supports the adaptive role of chromosomal-inversion polymorphism. Proc Natl Acad Sci USA 85: 5597-5600

Ruiz A, Naveira H, Fontdevila A (1984) La historia evolutiva de *Drosophila buzzatii*. IV. Aspectos citogenéticos de su polimorfismo cromosómico. Genét. Ibér. 36: 13-35

Sánchez A (1986) Relaciones filogenéticas en los clusters buzzatii y martensis (grupo repleta) de Drosophila. PhD dissertation, Universitat Autònoma de Barcelona, Spain.

Singh RS (1990) Patterns of species divergence and genetic theories of speciation. In: Wöhrmann K, Jain SK (eds) Population Biology. Springer-Verlag Berlin Heidelberg, pp 231-65

Singh RS, Rhomberg LR (1987a) A comprehensive study of genic variation in natural populations of *Drosophila melanogaster*. I. Estimates of gene flow from rare alleles. Genetics 115: 313-322

Singh RS, Rhomberg LR (1987b) A comprehensive study of genic variation in natural populations of *Drosophila melanogaster*. II. Estimates of heterozygosity and patterns of geographic differentiation. Genetics 117: 255-271

Sokal RR, Oden NL, Barker JSF (1987) Spatial structure in *Drosophila buzzatii* populations: simple and directional spatial autocorrelation. Am. Nat. 129: 122-142

Templeton A (1981) Mechanisms of speciation: A population genetic approach. Ann Rev Ecol Syst 12: 23-48

Vouidibio J, Capy P, Defaye D, Pla E, Sandrin J, Csink A, David J (1989) Short-range genetic structure of *Drosophila melanogaster* populations in an Afrotropical urban area and its significance. Proc. Natl. Acad. Sci. USA 86: 8442-8446

Wallace B (1970) Observations on the microdispersion of *Drosophila melanogaster*. In: Hecht M, Steere W.S. (eds) Appleton-Century-Crofts New York

Wasserman M (1954) Cytological studies of the repleta group. Texas Univ Publ 5422: 130-152

Wasserman M (1962) Cytological studies of the repleta group of the genus *Drosophila*. V. The *mulleri* subgroup. Texas Univ Publ 6205: 85-117

Weir BS (1990) Intraspecific differentiation. In: Hillis DM, Moritz C (eds) Molecular Systematics. Sinauer Sunderland, Massachusetts, pp 373-410

Wright S (1931) Evolution in Mendelian populations. Genetics 16: 97-159

Wright S (1978) Evolution and the genetics of populations Vol 4. The University of Chicago Press, Chicago London

MOLECULAR TAXONOMY IN THE CONTROL OF WEST AFRICAN ONCHOCERCIASIS

R. J. Post[1]
K. A. Murray[1]
P. Flook[1]
A. L. Millest
Department of Biological Sciences
University of Salford
Salford M5 4WT
United Kingdom

Traditionally taxonomy is divided into three interrelated disciplines: descriptive taxonomy, relationships between taxa and identification. Descriptive taxonomy is the original recognition and description of new taxa which were not previously known to exist. The relationships between taxa are often studied by phylogenetic reconstruction. Identification is the process by which individual specimens are assigned to previously known taxa, and does not necessarily involve the same characters used for the original recognition of those taxa. A great many of the taxonomic problems faced by field biologists are problems of identification. For example, a species may have been first described on the basis of adult characters, but the applied biologist may also need to identify the juvenile stages. It is this sort of problem that we shall consider in the control of onchocerciasis.

Onchocerciasis is a blinding disease caused by infection with the filarial nematode parasite, *Onchocerca volvulus*. This disease is endemic in many parts of the tropics, particularly

---

1. Previously at: Department of Medical Entomology, Liverpool School of Tropical Medicine, Pembroke Place, Liverpool L3 5QA, United Kingdom.

NATO ASI Series, Vol. H 57
Molecular Techniques in Taxonomy
Edited by G. M. Hewitt et al.

West Africa where the worst afflicted villages show rates of infectivity to almost 100% and blindness in 15% of the total community (Duke, 1990). Adult worms are found in the subcutaneous tissues, with females producing around 1000 microfilariae per day (Schulz-Key, 1990). The viviparous microfilariae migrate through the skin and eyes causing pathological lesions which can result in blindness. Parasites are transmitted from person to person by blood-sucking flies of the *Simulium damnosum* complex. These flies have riverine, filter-feeding larvae, but adult females require a blood meal for maturation of the eggs. Microfilariae picked up with a human-blood meal develop to the infective L3 stage and can be passed on to another person at a future blood meal. Control of transmission of the disease has normally involved the use of insecticides against the aquatic larvae of the vector (Philippon et al., 1990) with the possible integration of chemotherapy of patients using the drug ivermectin (Habbema et al., 1990). These control strategies have undoubtedly been made more difficult by multiple taxonomic problems in both the vector and the parasite.

The vector, *S. damnosum* s.l., is a complex of at least eight sibling species, in West Africa, originally established on the basis of larval polytene chromosome analysis. There is geographic variation in vectoral significance, and to some extent this relates to taxonomic variation. However, the differences in the abilities of the different sibling species to act as vectors is not clear, and this has made it more difficult to develop an effective control strategy that discriminates against only the important vectors. Hence there is a clear requirement for a technique of adult fly identification for transmission studies.

The running of a control programme is normally assessed by the Annual Transmission Potential, which is an index related to the level of infectivity in the fly population. However, *S. damnosum* s.l. will also bite cattle and can pick up microfilariae of *Onchocerca ochengi*, which do not infect man, but develop to an infective L3 stage indistinguishable from *O. volvulus* (Omar et al., 1979). Hence, the success of a control programme, particularly a programme based on chemotherapy, can be masked by the continued presence of cattle *Onchocerca*. There is a clear requirement for a technique for the identification of larval parasites within the vector fly.

The use of DNA sequences for identification offers the potential for specificity and sensitivity using a variable character that is highly unlikely to show life-stage mediated variation. However, for the field biologist the technology requirement should be reduced to a minimum so that large numbers of specimens can be identified quickly, and sometimes in relatively unsophisticated laboratories. One of the molecular techniques that most nearly fits this requirement is the use of species-specific DNA probes with dot-blots, or related blotting methods such as squash-blot (Post and Crampton, 1987). Probably the most suitable sorts of DNA sequences for use in such a system are repetitive sequences because of the greater degree of sensitivity they confer, which is particularly important for small organisms such as parasites. It is not clear whether different classes of repetitive DNA (John and Miklos, 1988) are more likely to show species-specific variation, and hence applied biologists, such as parasitologists, have widely adopted an empirical approach (ICOPA, 1990). This approach, as detailed by Post and Crampton (1988), generally involves the construction of genomic libraries from DNA of known species, and then the libraries are screened comparatively using labelled genomic DNA from each of the species in turn. Recombinant colonies which have cloned a repetitive DNA sequence which differs in copy number between species will be revealed by relative variation in the intensity of hybridisation. Such colonies can be isolated, and characterised if necessary, before use as a potentially diagnostic probe to DNA from unknown specimens on dot-blot.

We shall consider probes isolated in this way and currently being developed for the identification of the sibling species of the *S. damnosum* complex, and also the separation of *O. volvulus* from *O. ochengi*. We shall also compare the nature of the interspecific sequence variation, and how this variation might also be used in a probe-less identification system.

DNA PROBES FOR THE *SIMULIUM DAMNOSUM* COMPLEX

The *S. damnosum* complex can be divided cytotaxonomically into three subcomplexes; firstly the *S. damnosum* subcomplex which consists of *S. damnosum* s.s., *S. sirbanum* and *S. dieguerense*; secondly the *S. squamosum* subcomplex which consists of *S. squamosum* and *S. yahense*; and thirdly the *S. sanctipauli* subcomplex which consists of *S. sanctipauli*, *S. soubrense* and *S. soubrense* 'B'. Three DNA probes (pSO3, pSO11 and pSQ1) have been developed for the identification of individual specimens to subcomplex (Post and Crampton, 1988). The sequences represented by these clones are present in all species, but the amount of hybridisation revealed by dot-blot varies between species. Hence specimens can be identified to subcomplex according to the relative amount of hybridisation to each of the three probes. Hybridisation with pSO3 is greater than pSO11 to members of the *S. sanctipauli* subcomplex, but this pattern is reversed for the *S. damnosum* subcomplex and the *S. squamosum* subcomplex. These can then be separated according to the amount of hybridisation to pSQ1 (Figure 1).

Such differences in the amount of hybridisation could result from two different classes of molecular variation. Firstly, differences in copy number between species, and secondly sequence divergence between species such that, even with equal copy number, there would be less homology and hence less hybridisation. These possibilities can be distinguished by melting curve analysis. Dot-blots of known amounts of genomic DNA from known species were hybridised to the radiolabelled probe, and the amount of hybridisation measured by Cherenkov counting through a series of washes of increasing temperature (see Figure 2 for example). For each probe clear differences were found between species for the amount of hybridisation, but neither the mean melting temperatures (TM_{50}) nor the slope of the melting curves were different when tested by parallel curve analyses using maximum likelihood to fit a generalised logistic curve. This suggests very strongly that the differences were due to copy number.

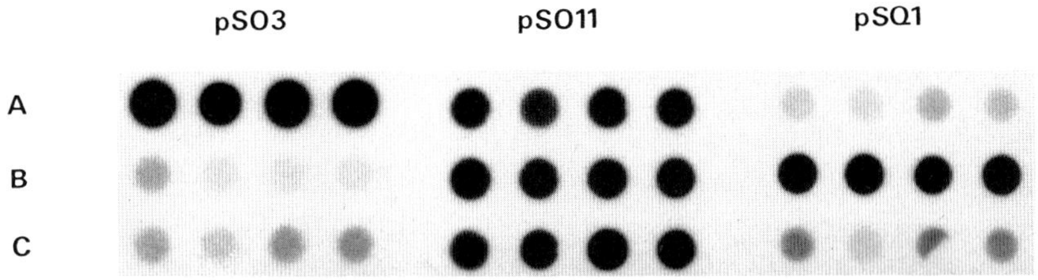

Figure 1: Replica dot-blots of genomic DNA extracted from four adult female S. soubrense 'B' (row A), S. yahense (Row B) and S. sirbanum (row C), and probed with pSO3, pSO11 and pSQ1.

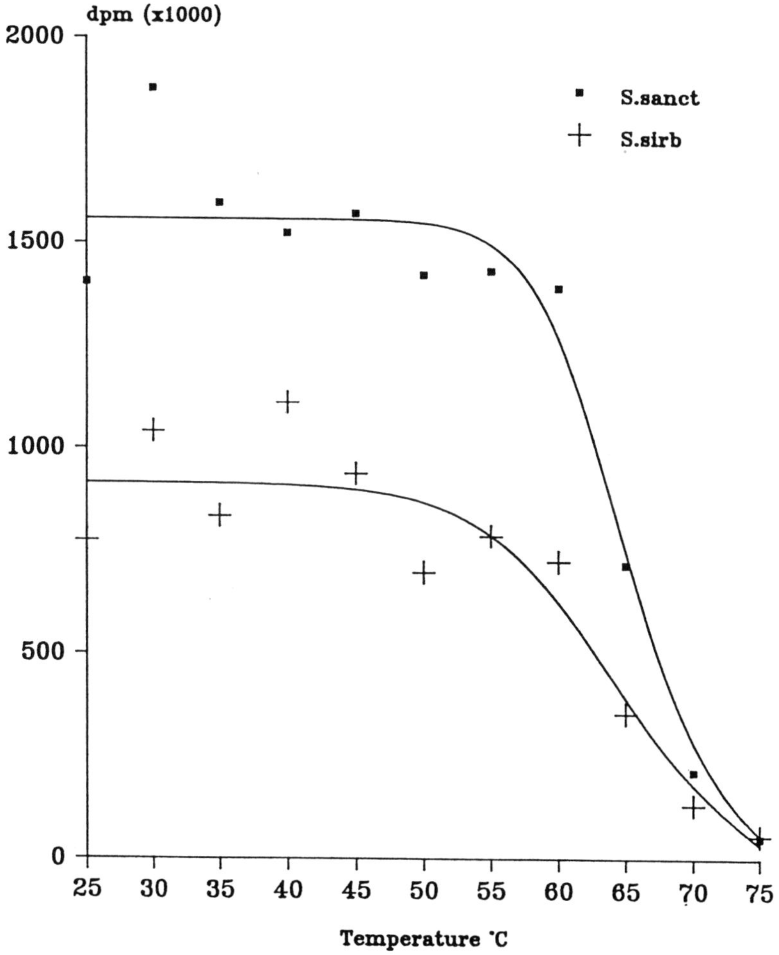

Figure 2: Melting curves of dot blots of genomic DNA of S. sanctipauli and S. sirbanum hybridised to ^{32}P-labelled pSO3 and washed at increasing temperatures.

DNA PROBES FOR *ONCHOCERCA* SPECIES

A range of sequences have been isolated from a number of *Onchocerca* species, and these sequences show various levels of taxonomic specificity. For example pOG17 (Post and Crampton, 1988), which was originally isolated from *O. gutturosa* from cattle, hybridises to all species of *Onchocerca* tested, but not to the potential vectors or hosts. A related sequence pOG5 (Post, unpublished) which was isolated from the same genomic library also hybridises widely. However, at 42°C hybridisation to 200ng *O. volvulus* is about 0.25 times the hybridisation to the same amount of *O. gutturosa* DNA. When the washing temperature was increased to 65°C the *O. volvulus* hybridisation was reduced to about 0.06 times *O. gutturosa* (Figure 3). In other words the TM_{50} is lower for *O. volvulus* that for *O. gutturosa*, which suggests that the sequence represented by p065 is present in both species, but it has diverged between the species.

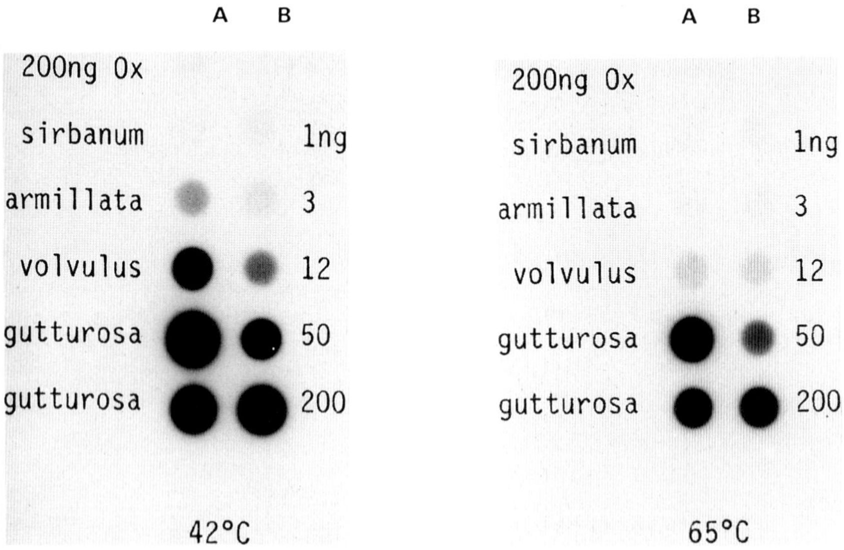

Figure 3: Dot-blot of 200ng of genomic DNA of Ox, *S. sirbanum*, *O. armillata*, *O. volvulus*, *O. gutturosa* from Britain and *O. gutturosa* from Mali (column A). Also a titration of genomic DNA of *O. gutturosa* from Mali (column B). The dot-blot was hybridised to pOG5 and washed at 42°C and 65°C.

In contrast the probe pOA1, isolated from *O. armillata* from cattle, does not cross-hybridise to other species, and seems to represent a family of repeats unique to *O. armillata* (Murray et al., 1988)

The probe pOV2 (Murray, 1990) was isolated from *O. volvulus*. It represents a family of repeats which do not cross-hybridise with pOG17 or pOA1. At low stringency it will cross-hybridise to several species of *Onchocerca*, but at high stringency it is specific to *O. volvulus* (Figure 4). This probe represents a family of repeated sequences (pFS1 family) which have formed the basis for taxonomic probes isolated in a number of laboratories including puOvs3 (Shah et al., 1987), pFS1 (Erttmann et al., 1987), CIAI-2 (Harnett et al., 1989), pOvs134 (Meredith et al., 1989), and pSS1 (Erttmann et al., 1990). The specificity of these clones is largely dependent upon the stringency at which they are used, and is evidently the result of interspecific sequence divergence.

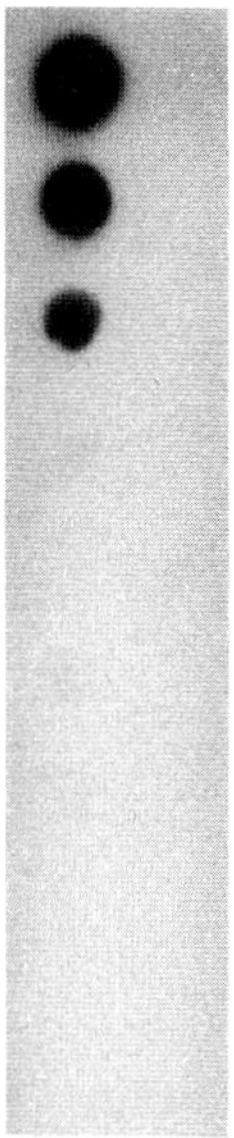

Figure 4: Dot-blot showing specificity of pOV2 at high stringency.

Some members of the pFS1 family have been sequenced (Figure 5), and show between 7-21% sequence divergence. Of these published probe sequences, all of them were isolated from <u>O. volvulus</u> and all of them will hybridise to that species. However, at more or less equivalent stringency puOvs3 will also hybridise to <u>O. ochengi</u>, and hence that probe appears to show less divergence from <u>O. ochengi</u> than the other probes. A number of restriction site differences were consistent between puOvs3 and the other clones, and work has begun to test these as a possible basis for a probe-less identification system.

```
                                                  NruI
pOvs134    TTAATCAATT TTGCAAACTG CGTTTTT-CG CCGGAAAAAT CGCCGTGTAA  50
puOvs3-1   .....T.... .......... ....G..-.. .....T.T.. ...GA.A...
puOvs3-2   .....T.... .......... ....G..-..         .. ...GA.A...
pFS1       .....T............... ..C.G..-.. T......... ...G.CA...
C1A1       A......... .......A.. .......-.. ...A...... ..........

                                                      AvaII
pOvs 134   ATGTGGAAAT TCACCAAAAT ATAGTCGAAT ATTTTTCTTA GGACCCAATT 100
puOvs3-1   .....A.... .......... .......... ......T... ....T.....
puOvs3-2   ..T..-.... .......... ........GC ......T... .C..T.....
pFS1       ..C..C.... .......... .......... ......T... .C..T.....
C1A1       ......-... ...AA..... .......... .......... .......G..

                 SnaBI
pOvs134    TGAAGGTACG TACCCGTTTT TTGAAATTAG AGGTCATAGG TCATCAGTTA  150
puOvs3-1   ....A.C... .....C...C G.....C... G.......T. ...C......
puOvs3-2   ...G....T. .........C ....T..... ..A.....T. GT.C......
pFS1       ........T. .........C .......... G.A....... .........C
C1A1       .......... .......... .........T .T........ .....T....
```

Figure 5: Published report sequences from probes based on the pFS1 family, except for the pOvs134 sequence which is a concensus of those sequences published for that probe (for references see text). Probes pOvs134 and C1A1 are <u>O. volvulus</u> species specific. Probe pFS1 is <u>0 volvulus</u> forest strain specific. Probe puOvs3 hybridises to both <u>O. volvulus</u> and <u>O. ochengi</u>.

Twenty-base pair oligonucleotides were synthesised homologous to each end of the repeat unit, and the sequence amplified by polymerase chain reaction from genomic DNA of O. volvulus and O. ochengi. The amplified DNA produced could be visualised as a smear by agarose gel electrophoresis stained with ethidium bromide. The PCR product was diluted in restriction enzyme buffer and incubated with AvaII or SnaBI. Whilst AvaII did not cut, SnaBI completely digested the O. volvulus smear, as expected, but not the O. ochengi smear (Figure 6).

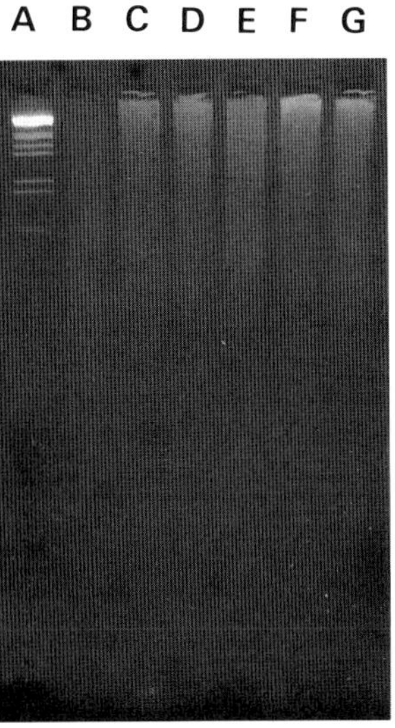

Figure 6: Ethidium bromide stained agarose gel of PCR amplified pFS1 family repeat sequence from O. volvulus (lanes B, C and D) and O. ochengi (lanes E, F and G). Lanes D and G undigested. Lanes C and F incubated with AvaII. Lanes B and E incubated with SnaBI. Lane A lambda phage double digested with EcoRI and HindIII.

CONCLUSIONS AND SUMMARY

Repetitive DNA probes selected by comparative screening of genomic libraries, and investigated because of their taxonomic potential do not show uniform characteristics. Sequences represented by probes such as pOA1 show extreme interspecific variation in copy number such that the sequence is effectively absent from some species and at very high copy number in others. The interspecific difference in copy number is less extreme for sequences represented by clones such as pSO3, pSO11 and pSQ1. In contrast the basis for taxonomic variation for probes based on the pFS1 family of repeats is largely due to sequence divergence.

The usage of probes is affected by the nature of the variation, such that pOA1-type probes are most easily used by "all or nothing" hybridisation to dot-blots. Probes based on the pFS1 family can also be used in an "all or nothing" way, but the stringency of hybridisation is of vital importance. There is a balance between lowered specificity at low stringency, and lowered sensitivity at high stringency. However, it might be possible to use sequence variation in restriction sites, in these probes, to develop probe-less identification systems based on PCR.

ACKNOWLEDGEMENTS

This work was supported by the Medical Research Council, the Wolfson Foundation and the World Health Organisation. We would also like to thank Drs J. M. Crampton, A. J. Trees and P. J. McCall for advice and discussion.

REFERENCES

Duke BOL (1990) Human onchocerciasis - an overview of the disease. Acta Leidensia 59: 9-24

Erttmann KD, Unnasch TR, Greene BM, Albiez EA, Boateng J, Denke AM, Ferraroni JJ, Karam M, Schulz-Key H, Williams PN (1987) A DNA sequence specific for forest form Onchocerca volvulus. Nature 327: 415-417

Erttmann KD, Meredith SEO, Greene BM, Unnasch TR (1990) Isolation and characterisation of form specific DNA sequences of O. volvulus. Acta Leidensia 59: 253-260

Habbema JDF, Plaisier AP, Van Oortmarssen GJ, Remme J (1990) Prospective evaluation of onchocerciasis control strategies. Acta Leidensia 59: 387-398

Harnett W, Chambers AE, Renz A, Parkhouse RME (1989) An oligonucleotide probe specific for Onchocerca volvulus. Molecular and Biochemical Parasitology 35: 119-126

ICOPA (1990) Bulletin de la Societe Francaise de Parasitology 8: Supplements 1 & 2

John B, Miklos GLG (1988) The eukaryote genome in development and evolution. Allen and Unwin London

Meredith SEO, Unnasch TR, Karam M, Piessens WF, Wirth DF (1989) Cloning and characterisation of an Onchocerca volvulus specific DNA sequence. Molecular and Biochemical Parasitology 36: 1-10

Murray KA (1990) Characterisation of repetitive DNA sequences of Onchocerca species. Ph.D thesis University of Liverpool

Murray KA, Post RJ, Crampton JM, McCall PJ, Kouyote B (1988) Cloning and characterisation of a species-specific DNA sequence from Onchocerca armillata. Molecular and Biochemical Parasitology 30: 209-216

Omar MS, Denke AM, Raybould JN (1979) The development of Onchocerca ochengi (Nematoda: Filarioidea) to the infective stage in Simulium damnosum s.l. with a note on histochemical staining of the parasite. Tropenmedizin und Parasitologie 30: 157-162

Philippon B, Remme JH, Walsh JF, Guillet P, Zerbo DG (1990) Entomological results of vector control in the onchocerciasis control programme. Acta Leidensia 59: 79-94

Post RJ, Crampton JM (1987) Probing the unknown. Parasitology Today 3: 380-383

Post RJ, Crampton JM (1988) The taxonomic use of variation in repetitive DNA sequences in the Simulium damnosum complex. In: Service MW (ed) Biosystematics of Haematophagous Insects, Systematics Association special volume 37. Clarendon Press Oxford

Schulz-Key H (1990) Observations on the reproductive biology of Onchocerca volvulus. Acta Leidensia 59: 27-43

Shah JS, Karam M, Piessens WF, Wirth DF (1987) Characterisation of an Onchocerca-specific DNA clone from Onchocerca volvulus. American Journal of Tropical Medicine and Hygiene 37: 376-384

DNA PROTOCOLS FOR PLANTS

Prof Jeffrey Doyle

Bailey Hortorium
462 Mann Library, Cornell University,
Ithaca, New York

CTAB Total DNA Isolation

This procedure has been used with success on a wide variety of plant groups and even some animals. The method is used to isolate total genomic DNA (nuclear, chloroplast, and mitochondrial). It is a rapid, inexpensive method that is suitabie for use in conjunction with other protocois, such as isolation of DNA enriched for cpDNA. it is also easy to scale down for use in population sampling, using 0.01g or less of fresh tissue. Other applications include isolation of DNA from herbarium specimens (Doyle & Dickson, 1987. *Taxon* 36:715-722), and isolation of RNA. A brief word on the history of the protocol is in order. This procedure was modified by us (Doyle and Doyle, 1987. *Phytochemical Bulletin* 19:11-15) for use with fresh piant tissue from a method of Saghai-Maroof et al. (1984, *PNAS* USA 81:8014-8019) who used lyophilized tissue. They in turn had developed their procedure from earlier protocols. We were recently asked to publish a slightly modified version of our procedure (Doyle and Doyle, 1990 *Focus* 12:13-15). We recently learned from Brian Taylor (Texas A&M University, USA) that he had published a virtually identical procedure for fresh tissue, also in *Focus,* in 1982 (Taylor & Powell, *Focus* 4:4-6) of which we (and apparently the editors of *Focus!)* were entirely unaware. It is indeed a useful procedure, thus independently confirmed.

Procedure:

1. Preheat 5-7.5 ml of CTAB isoiation buffer (2% CTAB [Sigma H-5882], 1.4 M NaCI, 0.2% 2-mercaptoethanol, 20 mM EDTA, 100 mM Tris-HCl, pH 8.0) in a 30 ml glass centrifuge tube to 60°C in a water bath.

2. Powder 0.5-1.0g fresh leaf tissue in liquid nitrogen in a chilled mortar and pestle. Scrape powder directly into preheated buffer and swirl gently to mix.

    If preferred, fresh tissue may be ground in 60°C CTAB isolation buffer in a preheated mortar.

3. Incubate sample at 60°C for 30 (15-60) minutes with optional occasional gentle swirling.

4. Extract once with chloroform-isoamyl alcohol (24:1), mixing gently but thoroughly.

5. Spin in clinical centrifuge (swinging bucket rotor) at room temperature to concentrate phases. We use setting 7 on our IEC clinical (around 6,000 x g) for 10 min.

    Generally the aqueous phase will be clear, though often colored, following centrifugation, but this is not always the case.

6. Remove aqueous phase with wide bore pipet, transfer to clean glass centrifuge tube, add 2/3 volumes cold isopropanol, and mix gently to precipitate nucleic acids.

NATO ASI Series, Vol. H 57
Molecular Techniques in Taxonomy
Edited by G. M. Hewitt et al.

In some cases this stage yields large strands of nucleic acids that can be spooled out with a glass hook for subsequent preparation. In most cases, this is not the case, however, and the sample is either flocculent, merely cloudy-looking, or, in some instances, clear. If no evidençe of precipitation is observed at this stage, the sample may be left at room temperature for several hours to overnight. This is one convenient stopping place, in fact, when many samples are to be prepped. In neariy all cases, there is evidence of precipitation after the sampie has been allowed to settle out in this manner.

7. If possible, spool out nucleic acids with a glass hook and transfer to 10-20 ml of wash buffer (76% EtOH, 10 mM ammonium acetate).

a. Preferred alternative: Spin in clinical centrifuge (e.g. setting 3 on IEC) for 1-2 min. Gently pour off as much of the supernatant as possible without losing the precipitate, which will be a diffuse and very loose pellet. Add wash buffer directly to pellet and swirl gently to resuspend nucleic acids.

b. Last resort: Longer spins at higher speeds may be unavoidable if no precipitate is seen at all. This will result, generally, In a hard pellet (or, with small amounts, a film on the bottom of the tube) that does not wash well and may contain more impurities. Such pellets are difficult to wash, and in some cases we tear them with a glass rod to promote washing at which point they often appear flaky.

Nucleic acids generally become much whiter when washed, though some color may still remain.

8. Spin down (or spool out) nucleic acids (setting 7 IEC, 10 min) after a minimum of 20 min of washing. The wash step is another convenient stopping point, as samples can be left at room temperature in wash buffer for at least two days without noticeable problems.

9. Pour off supernatant carefully (some pellets are still loose even after this longer spin) and allow to air dry briefly at room temperature.

10. Resuspend nucleic acid pellet in 1 ml TE (10 mM Tris-Ha, 1 mM EDTA, pH 7.4).

Although we commonly continue through additional purification steps, DNA obtained at this point is generally suitable for restriction digestion. If DNA is to be used at this stage, pellets should be more thoroughly dried than indicated above.

Gel electrophoresis of nucleic acids at this step often reveals the presence of visible bands of ribosomal RNAs as well as high molecular weight DNA.

11. Add RNAase A to a final concentration of 10 ug/ml and incubate 30 min at 37°C.

12. Dilute sample with 2 volumes of distilled water or TE, add ammonium acetate (7.5 M stock, pH 7.7) to a final concentration of 2.5 M, mix, add 2.5 volumes of cold EtOH, and gently mix to precipitate DNA.

DNA at this stage usually appears cleaner than in the previous precipitation. Dilution with water or TE is helpful, as we have found that precipitation from 1 ml total volume often produces a gelatinous precipitate that is difficult to spin down and dry adequately.

13. Spin down DNA at high speed (10,000 x g for 10 min in refrigerated centrifuge, or setting 7 in clinical for 10 min).

14. Air dry sample and resuspend in appropriate amount of TE.

Comments: DNA isolated in this manner is generally of high molecular weight, though often some breakdown is observed. The liquid nitrogen alternative procedure consistently yields DNA of higher average molecular weight than does the method in which fresh tissue is ground directly in buffer. However, for most applications, such as screening large numbers of individuals, we routinely use fresh-ground samples.

Yields using these methods often approach 1 mg/g fresh tissue, though this is strongly dependent both on the age and quality of the tissue and on the species used. A_{260} quantification generally gives unreliable results, presumably due to interference of residual CTAB in samples.

Modifications to the basic procedure may improve quality or yield of DNA in some groups. In plants with high concentrations of phenolic compounds, such as oaks and walnuts, 1% (w/v) polyvinylpyrollidine (PVP-40) has been added to the isolation buffer with successful results. For plants containing high polysaccharide levels and/or glutinous sap, which often yield very viscous grindates (e.g. Onagraceae, bromeliads), successful isolations have been achieved by simply increasing the CTAB percentage to 3% or higher.

In some cases, DNA obtained by this procedure is further purified by one or two CsCl ultracentrifugation steps. This modification combines the advantages of the high yields routinely achieved by the CTAB isolation method with the presumably greater purity of CsCl methods.

Rapid Chloroplast Enriched DNA Prep

This procedure is basically that of Bookjans et al. (1984; *Anal. Bichem.* 141:244-247) which enriched for chloroplasts and then isolated the DNA uslng a sephacryl column. We replaced this final step with our standard CTAB DNA isolation procedure. There have recently been two published protocols that are very similar to this modification (Milligan, 1989, *Plant Molecular Biology Reporter* 7:144-149; Sandbrink et al., 1989, *Biochem. Syst. Eiol.* 17:45-49). The Bookjans method was developed for isolating cpDNA from a variety of peas having very delicate chloroplast membranes, which made standard sucrose gradient isolations difficult as they depend on isolating intact chloroplasts. This method depends on using a high ionic strength buffer to prevent DNA from lysed nuclei from associating with the broken chloroplasts that are pelleted In the early stages of the isolation; cpDNA, physically bound to the broken chloroplast membranes, is retained and isolated in the CTAB stage. The procedure works with a variety of plants, but by no means with all taxa we have tried ("not working" in this case means that although high molecular weight DNA is isolated, no visible enrichment for cpDNA is noted). One problem we have noted with some DNAs isolated in this manner is that over periods of storage they become more difficult to digest with various restriction enzymes.

1. Homogenize (10)15-30 g leaf tissue in waring blender with ± 500 ml of Bookjans buffer; use 3-5 5 second bursts. Bookjans buffer: 50 mM Tris, 25 mM EDTA, 10 mM 2-mercaptoethanol, 0.1% bovine serum albumin, 1.25 M NaCl, pH 8.0. Bring buffer, centrifuge tubes, and blender to 4°C and work on ice.

2. Filter homogenate through three layers cheese cloth, 1 layer miracloth into two 250 ml centrifuge bottles.

3. Centrifuge in refrigerated centrifuge at 1500 x g for 12 minutes at 4°C.

4. Pour off supernatant and gently resuspend chloroplast pellet with a soft paint brush in 100-200 ml of Bookjans buffer. Spin at 1500 x g for 12 minutes.

5. Resuspend chloroplast pellet in ± 5 ml of 2x CTAB buffer warmed to 60°C using a soft paint brush. Transfer suspension from the two centrifuge tubes to one Corex tube; and proceed as in Doyle & Doyle (1987) CTAB procedure.

Sucrose Gradient cpDNA Isolation

There are a variety of methods available for isolation of chloroplasts using gradients of several sorts, the most common of which are the sucrose gradient methods. The method given here is a modification of that of Calie & Hughes (1987, *Plant Molecular Biology Reporter* 4:206-212), in which we have replaced the CsCl gradient isolation of DNA from lysed, isolated chloroplasts with a CTAB isolation from the chloroplast gradient band. The Calie & Hughes method differs from more standard sucrose gradient procedures (such as Palmer, 1982, *Nucleic Acids Res.* 10:1593-1605) mainly in substituting dithiothreitol for 2-mercaptoethanol (DTT is a better reducing agent) and including polyamines (spermine and spermidine) in the buffers. These were found to be necessary by Calie and Hughes for obtaining good cpDNA from mosses; we have used this method successfully on a variety of plants (such as walnuts) when the standard buffers did not work.

Grinding Buffer

0.35 M Sorbitol
50 mM Tris
5 mM DTT
0.1% BSA
5 mM Spermine
5 mM Spermidine
10 mM EDTA
pH 9.0

Wash Buffer

0.35 M Sorbitol
50 mM Tris
5 mM Spermine
5 mM Spermidine
25 mM EDTA
pH 9.0

Sucrose step gradients are made in Wash Buffer (15 ml pad of 52% Sucrose overlain with 7 ml 30% sucrose)

1. Grind leaf material in waring blender in 10x (wt:vol; e.g., 30g = 300 ml) Grinding Buffer with three bursts of 5 sec each. Bring buffer, centrifuge tubes, and blender to 4°C and work on ice.

2. Filter grindate through 3 layers cheesecloth and 1 layer mira cloth into 250 ml centrifuge bottles.

3. Spin filtrate 1000 x g for 15 min at 4°C.

4. Discard supernatant. Resuspend pellets in each bottle in 5 ml Wash Buffer, taking care to be gentle (use a soft paint brush).

5. Layer resuspended pellets onto sucrose gradients. Never load more than around 7 ml of resuspended pellets per gradient. (Note: it is best to have a slightly

disrupted interface between gradient layers, to prevent nuclei from aggregating at interface. This can be achieved either by pouring gradients the night before and allowing them to diffuse somewhat, or by pouring them less gently immediately before use, or by pouring them an hour or so before use, capping with parafilm, and gently laying them on their sides. Make sure, though, that gradients are at 4°C when used).

6. Spin 75,000 x 9, 45 min, 4°C in a swinging bucket rotor (We use Beckman SW 28 rotor at 20,000 rpm).

7. During the run, nuclei pellet to bottom, while chloroplasts form a green band at the 30%-52% interface.

   a. Collect chloroplast fraction, pool into 250 ml centrifuge bottle; ignore the potentially large green layer of debris that accumulates at the interface of the aqueous (from resuspended pellet) and 30% sucrose layers which consists of broken cells and organelles. Proceed with step 8.

   b. Pour off remaining sucrose to expose nuclear pellet; proceed with step 10.

8. Dilute with 100-150 ml Wash Buffer, adding slowly and gently. (this step is included to dilute out the sucrose, permitting pelleting of organelles).

9. Spin at 1000 x g 15 mln at 4°C in refrigerated centrifuge.

10. Discard supernatant. Separately resuspend chloroplast and nuclear pellets In 5-7 ml 60°C 2x CTAB buffer and proceed as in Doyle & Doyle (1987) CTAB procedure to isolate cp and nuclear DNAs.

Pouring Gels

The percentage of agarose used in gels is chosen depending on the size of the fragments one wishes to separate. Thus, for a range of fragments typical for chloroplast DNA digests using a restriction enzyme with an intermediate number of recognition sites in the ca. 150 kb genome (e.g., X*ba*I with around 20-30 sites), a 0.7% gel is used. For enzymes such as *EcoRI* or D*ra*I, which cut much more frequently and thus produce smaller fragments (many smaller than 1 kb), a 1% or 1.5% gel is used; other enzymes, such as S*ma*I, typically produce a smaller number of much larger fragments (>20kb), and a lower percentage gel (0.5 or 0.6%) is selected.

Two basic gel running buffer recipes are standard, Tris-acetate (TAE) and Tris-borate (TBE); we use the latter, and this is included in the recipes given below.

1. Example: To make up a 0.7% mini-gels (50 ml gels):

   0.35 9 agarose
   50 ml 1x TBE (running buffer)
   2.5 microliters ethidium bromide (EtBr)

   Bring agarose and buffer solution to boil in microwave. Make sure all the agarose has gone into solution (flakes will be present if not) before adding EtBr (use gloves as it is a mutagen), to final concentration of 0.5 ug/ml.

2. Set up the gel rig, being sure to put in the removable plastic walls. The comb should be ca. 1 mm above the floor of the tray (this protects from tearing the wells when the comb is removed). Connect leads to gel rig (but don't turn on) prior to pouring the gel (this avoids having to jostle the gel after it is loaded).

3. Partially cool the gel solution (to around 65°C) prior to pouring to prevent warping the plexiglas gel rig. Pipette a small amount of the partially cooled solution to seal both ends of the rig and allow to solidify before pouring the remainder of the gel.

4. Once the gel is completely set (cool to the touch), pour 1 x TBE buffer to cover the gel and then carefully remove the walls and comb.

5. Load the wells as follows:

   1st lane: size standard: usually 0.5 ug lamba DNA restricted w/*Hin*dlll. Heat briefly in a small beaker of hot (60°C) water (opens up the cohesive ends to produce the full complement of restriction fragments for standards).

   Load the other lanes with mixed samples (1 ul 10x urea loading dye per 10 ul sample DNA). Warm the dye in a small beaker of water if it is crystallized, before mixing with your samples.

6. Turn on power. TBE gels are run at 5 volts/cm and do not need buffer recirculation. The DNA runs to the red (+) pole. It will take 2-several hours for the gel to run. Progress is tracked with the two dyes including in the loading buffer; bromophenol blue will run at about 300 base pairs in a 0.7% gel, while the xylene cyanol runs with fragments of around 3 kb. [note: TAE gels are run for longer periods of time]

7. After run is complete, turn off power, wear gloves, remove gel and photograph under UV light (ethidium bromide, an intercalating agent, fluoresces in UV).

10 x Urea Loading Dye

10 M Urea
0.1% Bromophenol Blue
0.1% Xylene cyanol

10 x TBE Gel Buffer

0.89 M Tris
0.89 M Boric Acid
0.002 M EDTA
pH 8.0

Southern Blotting

Transfer of DNA fragments to nylon (or nitrocellulose) membranes is a critical stage in all RFLP studies. A great diversity of methods exist (includlng "active" methods such as electo-blotting). Choice of methods depends on many things, but particularly on the type of membrane used. Given here is our modificatlon of the original E. M. Southern (1975; J. Mol. Biol. 98:503) capillary transfer method, using high salt buffer for transfer; we use MSI Nylon 66. Other methods may use either low salt, neutral pH buffers or alkaline solutions. Modifications include double-sided transfer and rapid blot methods that do not use wicks and depend only on the buffer already present in the gel. A number of different methods also exist for binding transferred DNA to membranes, Including UV-crosslinking and vacuum baking. In general it is best to follow the directions provided with the brand of nylon used; this fairly standard method, however, works well with most nylons we have tried.

Prepare gels for transfer:

1. Depurinate DNA (optional step done to break up large fragments for more efficient transfer; omit for gels of 1% or higher, for which smaller fragments are expected):

   Soak gel in just enough 0.2 N HCl to cover.

Shake for 7 min.
Pour off and rinse in H_20.

[Note: Watch gel for color change in light blue band. If it turns yellow remove gel at once!]

2. Denature DNA in NaCl/NaOH:

Soak gel with shaking for 15 min, pour off
Repeat wash
Pour off and rlnse in H_20.

3. Neutralize in Tris/NaCl:

Soak gel with shaking for 10-15 min.
Pour off and rinse in H_20.

Assemble blot boxes while the gel is being treated. Our blot box design consists of a shallow plexiglas buffer reservoir with gel support platform bridge. A cheap alternative is two staining dishes with a plexiglas bridge):

1. Assemble boxes
2. Measure Support and cut Whatman 3 mm paper for wicks
3. Fill box with 10 x SSC
4. Measure gel (from just above wells to bottom)
5. Cut nylon and a piece of 3mm paper to fit

Set up Southern:

1. Place wick over gel support, wet with 10x SSC. Work out bubbles
2. Place gel on wick, work out bubbles
3. Mark filter with gel information, wet in 10x SSC
4. Place filter on gel, work out bubbles
5. Place 3mm paper (wetted in 10x SSC) on filter on gel, work out bubbles
6. Stack on paper towels ($\pm$ 7.5 cm thick)
7. Cover with plexiglas sheet and weigh down (do not use too much weight; <500g) 8. Transfer overnight. Strike set-up and bake filter 60-65°C. 2 hrs in standard oven.

Solutions:

HCl:	30 ml concentrated HCl/1 l H_20
NaCl/NaOH:	0.5 M NaOH; 1.5 M NaCl
Tris/NaCl:	0.5 M Tris; 3.0 M NaCl pH 7.4
20X SSC:	3.0 M NaCl 0.3 M Na-Citrate

Preparation and Transformation of Competent Cells

A. *Preparation of Cells.* In order to facilitate the uptake of plasmids into bacteria, cultures are treated to make them more permeable, or "competent". Competent cells may be purchased commercially and may be preferable for critical cloning steps as they are prepared using methods that generally give better uptake than the one given here. The method provided here (modified from Sambrook et al. 1989) is cheap and simple, does not use obnoxious chemicals such as dimethylsulfoxide (DMSO), and works with a wide variety of bacterial strains. We use it for routine maintenance of our plasmid stocks and for general subcloning.

1. Inoculate 50 ml of sterile L Broth (5 g yeast extract, 8 g bacto-tryptone, 5 g NaCl; to 1 l with H_20) with single colony of desired bacterial strain.
2. Incubate overnight at 37°C with shaking.
3. Use 50 ul of overnight culture to inoculate 50 ml of sterile L Broth.
4. Incubate at 37°C with shaking for 1.5 hrs (to pre-log phase: we assess this by observation, looking for partly cloudy cultures; a more accurate measure can be obtained from OD_{600} measurement)
5. Harvest bacteria by centrifuging at 4000 x g for 5 min. at 4°C; discard supernatant.
6. Resuspend cells in 20 ml of cold 0.1 M $CaC1_2$.
7. Incubate on ice for 20 min.
8. Repeat step 5.
9. Resuspend in 1 ml of cold 0.1 M $CaC1_2$.
10. Store on ice (use within 24 hrs).

B. Transformation of Competent Cells

1. To 100 ul of freshly made competent cells add 30 ng of plasmid DNA.
2. Incubate on ice 20 min.
3. Incubate 75 sec. at 42°C.
4. Incubate 2 min. on ice.
5. Add 1 ml sterile L Broth.
6. Incubate 45 min. at 37°C.
7. Spread 300 ul on 1.1% agar L Broth plate containing proper antibiotic.
8. Incubate overnight at 37°C.

Plasmid DNA Mini-Isolation

This procedure, which is modified from Maniatis et al. (1982) and is one of many available, avoids the use of large scale culturing and time consuming DNA isolations (e.g., CsCl gradients) and provides relatively purified plasmid DNA in quantities sufficient for labelling. Additional purification steps may be required for restriction digestion.

1. Inoculate 1 ml sterile L Broth (containing desired antibiotic) from a single colony bearing the desired plasmid. This is done with a toothpick that has been surface sterilised (in alcohol, flamed prior to use), touched lightly to the desired colony on a plate. A saturated liquid stock may also be used.

2. Incubate overnight to saturation at 37°C with shaking. Pellet bacterial cells by centrifuging 2 min. in microcentrifuge. Discard supernatant.

3. Resuspend cells in 200 ul resuspension buffer by vortexing.

4. Add 20 ul freshly made 10 mg/ml lysozyme (in distilled H_20), mix well.

5. Incubate in boiling H_20 bath for 45 sec. Suspension will become quite viscous.

6. Centrifuge 10 min. in microcentrifuge. Pour supernatant into `clean tube.

7. Add 20 ul 2 M NaCl.

8. Add 200 ul cold isopropanol, mix well. Incubate on ice 10 min.

9. Centrifuge as in step 6. Discard supernatant.

10. Air dry pellet briefly and resuspend in 50 ul TE buffer (TE: 10 mM Tris, 1 mM EDTA pH7.4)

11 . Run 5 ul on 0.7% gel to check size or use 5 ul for restriction.

Note: A cleaner prep can be achieved by phenol/chloroform extraction performed after step 6. Do an additional chloroform/isoamyl (24:1) extraction before continuing with step 7.

Resuspension Buffer

8% Sucrose
5% Triton X-100
50 mM EDTA
50 mM Tris/HCl pH 8.0

L Broth

5 g Yeast Extract
8 g Bacto tryptone
5 g NaCl
H_20 to 1 liter

DNA Fragment isolation using Whatman DE-81 Paper

There are a variety of methods available for isolating restriction fragments from gels, ranging from electroelution to low melt agarose. We often use this method (taken from *Anal. Biochem.* 1982, 117:295), which is easy and rapid and provides DNA that can be used in restrictions and ligations.

A. Preparation of paper

1. Cut paper into strips and soak several hrs in 2.5 M NaCl
2. Wash 1-2 hours in dH_20 (change every 1/2 hr.) to remove salt
3. Store in 1 mM EDTA (pH 7.5) at 4°C

B. Recovery of fragment

1. Run restriction digested DNA on standard TBE agarose gel
2. Locate desired fragment on UV-illuminated gel
3. Cut gel just below fragment
4. Fill cut with treated DE-81 paper (Do not leave any air trapped)
5. Run gel for 10-15 min. check periodically so as to collect only the desired fragment
6. Remove paper and place in Eppendorf tube (removing any big chunks of agarose)
7. Add 300-700 ul elution buffer (20 mM Tris pH 7.5;1 mM EDTA 1.5 M NaCl)
8. Incubate 1 hour at 37°C after vortexing to shred paper
9. Spin through glass wool to collect eluate
10. Butanol extract 3x; phenol/chloroform extract 1x; chloroform/isoamyl (24:1) extract 1 x
11. Precipitate with 2 vol 95% EtOH, spin 20 min. in microfuge, wash with 70% EtOH, spin again for 10 min, and dry inverted at room temperature
12. Resuspend in 50 ul TE
13. Run small amount on gel to assess recovery

Labelling Probes

There are a variety of methods available for making labelled probes for detecting DNA restriction fragments or sequencing products. Two of the most common are given here and are used by us with radioisotope. Non-radioactive methods often employ the same procedures, substituting various reagents for isotope. The nick translation method has been replaced in many labs by the newer random priming method; we most frequently use the latter.

A. *Nick Translation* (modified from Sambrook et al., 1989)

1. Mix together in micofuge tube (pre chilled and on ice):

    2 ul 10x NT Buffer
    0.5 ug DNA to be labelled
    2 ul each of 1-4 different -^{32}P dNTPs
    Unlabeled dNTPs (whichever were not included as labelled dNTP) to final concentration of 20 mM Add 1 ul diluted DNase 1 (10 ng/ml = 1: 10,000 dilution of 1 mg/ml stock)
    Add 1 ul DNA *E. coli* Polymerase 1 holoenzyme (10 units)
    H_20 to final vol of 20 ul

2. Incubate 2 hrs at 15°C

3. Add 15 ul Stop Buffer

4. Separate from unincorporated dNTPs on Sephadex G100-120 column (track light blue Blue Dextran fraction, which migrates with high molecular weight DNA). [alternative methods, such as spun columns, are often used here]

Stop Buffer

0.2 M EDTA
1%SDS
40 mg/ml Blue Dextran
0.1 mg/ml Bromophenol Blue

10x NT Buffer

0.5 M Tris/HCl pH 7.5
0.1 M $MgS0_4$
1 mM DTT (dithiothreitol)
500 ug/ml BSA

B. *Random Hexanucleotide Priming.* (various modifications of: Feinberg & Vogelstein 1983, Anal. Biochem. 132:6-13; Feinberg & Vogelstein 1984, Anal. Biochem. 137:266).

1. Denature double-stranded probe DNA by immersing in boiling water for 3 min. Place on ice. All remaining steps are done on ice.

2. Remove 80 ng DNA, add to clean micofuge tube

3. To tube add:

    20 ul LS
    4 ul dCTP (40 uCi) [we use specific activity >3000 Ci/mM, 10 uCi/ul]
    1 ul Klenow fragment of DNA Polymerase I (5 units)
    H_20 to final vol. of 40 ul

4. Incubation is fairly non-specific; we allow reaction to proceed 2 hr to overnight at room temp or at 37°C, with no apparent differences in incorporation.

5. Add 15 ul Stop Buffer and separate from unincorporated dNTPs on a Sephadex G100-120 column.

LS Buffer is a mixture of 1 M Hepes (pH 6.6): DTM: OL [25: 25: 7]; store at -70°C in 20 ul aliquots

DTM: 1 mM dATP, dTTP & dGTP in:

250 mM Tris/HCL pH 8.0
25 mM Mg Cl_2
50 mM 2-mercaptoethanol

OL: 90 units/ml random hexadeoxynucleotides (Pharmacia) in TE

PREPARATION AND VISUALIZATION OF MITOCHONDRIAL DNA FOR RFLP ANALYSIS

Michel Solignac

Laboratoire de Biologie et Génétique évolutives
CNRS 91190 Gif-sur-Yvette France

INTRODUCTION: Purification of mitochondria and/or of mtDNA

The principal problem in preparation of mitochondrial DNA is to avoid contamination by nuclear DNA. One strategy is to prepare a highly purified fraction of mitochondria, using cellular fractionation (differential centrifugation, sucrose gradients). Another is to prepare a total extract of DNA (or an extract solely enriched in mtDNA) and to separate nuclear and mitochondrial DNA, taking advantage of their differential physical properties (physical conformation, density). The last possibility is to work with the mixture of DNAs from the total extract and to use a probe, complementary of the mtDNA.

1) PURIFICATION OF MITOCHONDRIA

Work always on ice in presence of EDTA. Gloves are not necessary. A way to obtain a purified fraction of mitochondria with few chromatine contamination is to use cells or tissues rich in mitochondria. Ovaries, oocytes and virgin eggs are peculiarly suitable for this purpose since the large cytoplasm of female gametes contains a great number of mitochondria. Occasionally, platelets have also been used as a source of mitochondria (Giles et al. 1980).

Often purification of mitochondria is performed with somatic tissues for which the ratio mtDNA/nuclear DNA is about 1/100. When working with somatic tissues, it is necessary to control homogenization in order to disrupt the cells but not the nuclei: try several homogenizer (clearance of the pestle, teflon or glass pestle, hand or motor and in this last case, speed and number of strokes). The results are always better with fresh tissues (instead of frozen tissues) because the mitochondrial membranes are intact. For large animals, try several tissues; generally, only soft tissues are used for vertebrates (liver, kidney, brain).

Separation of mitochondria is performed in isotonic sucrose (or rarely in mannitol or a mixture of both). The sucrose solutions cannot be sterilized but sucrose is dissolved in sterile TE (Tris-EDTA). Their concentrations can be controled with a refractometer. Concentration of sucrose is given either in moles per liter or in

NATO ASI Series, Vol. H 57
Molecular Techniques in Taxonomy
Edited by G. M. Hewitt et al.

percentage: **weight** of sugar for the final **weight** (w/w). Solutions can be kept at +4°C for a week or can be frozen. A strong shaking is necessary after thawing before use.

Cellular fractionation is an obligate step but the protocols are more or less complicated. It is performed through differential centrifugations: very low speed centrifugation pellets intact cells, nuclei and organic debris; low speed centrifugation pellets large membrane fragments and pigments. In some protocols only very low speed centrifugations are used and they are repeated until disappearance of the pellet. The mitochondrial fraction is pelleted by a prolonged and rapid centrifugation. An additional purification of mitochondria can be obtained with continuous or discontinuous gradients of sucrose or with Percoll density gradient (or other products). The pellet of mitochondria obtained before the gradients is called "crude mitochondrial fraction" (which includes membrane fragments, lysosomes, and peroxisomes) whereas it is reputed a "pure mitochondrial fraction" after sucrose gradients.

Contamination by nuclear DNA is attributable to the adhesion of the chromatin to the outer organelle membrane. A treatment of the suspension of mitochondria with DNase can help to destroy the chromatin. This treatment is rather dangerous for mtDNA if mitochondrial membranes are not intact. Another way is to treat mitochondria with digitonin. The effect of this drug, a detergent specific of cholesterol, is to destroy at least partially the external membranes (the only one which contains cholesterol) where the nuclear chromatin is fixed (the resulting peeled organelle is called a mitoplast).

Remember that each additional step represents a progress towards a higher purity of the mitochondria extract and thus of the mtDNA but lowers the final amount of mtDNA (frequently less than 1 ug of mtDNA per g tissue or organism). Degree of purity and amount required for RFLP depend on the method used for visualization and on the size of the restriction fragments to be observed.

2) MtDNA EXTRACTION

The pellet of mitochondria (crude or pure) is lysed with SDS and when the solution is clear, proteins can be removed by phenol extraction. Sometimes, DNA obtained in this way is good enough for analysis. It is however possible to perform an additional purification on CsCl gradient, either alone (based on the differential density of mitochondrial and nuclear DNA) or using a dye such as ethidium bromide (EB), 4',6-diamidino-2-phenylindole (DAPI), bisbenzimide, or propidium iodide (PDI). The dye allows to separate the DNAs according to their physical properties. In presence of EB or PDI supercoiled mtDNA can be banded separately from the linear one, of nuclear origin. Supercoiled DNA admit less intercaling dye than linear DNA between the two strands, and consequently is less lightened by EB. The efficiency of this method depends on the proportion of mtDNA still in a covalently closed circular form at this step. DAPI and bisbenzimide may help to separate DNAs upon their average base composition (it binds preferentially to DNAs rich in A+T).

3) PURIFICATION OF MtDNA FROM TOTAL DNA

Other methods tend to extract total DNA (or DNA only enriched in mtDNA) and to separate mtDNA and nuclear DNA on the basis of their differential properties. One way for the preparation of the mtDNAs is an adaptation of the method of separation of plasmid DNA from the bacterial chromosome by denaturation / renaturation. An alkali is used to denature the DNA (Birnboim and Doly 1979). Then, the mixture is neutralized: the mtDNA reanneals rapidly whereas nuclear (linear) DNA partially renatures in a complex network and forms, with proteins and SDS, an aggregate which can be removed by centrifugation. Another way involves an ultracentrifugation, as indicated above, in CsCl with intercaling dye. As the concentration of mtDNA relative to nuclear DNA is usually low in somatic cells (1% of the total DNA), it is useful either to enrich the initial preparation in mtDNA and to repeat the step(s) of purification.

4) WORKING WITH TOTAL DNA

Instead of purifying mitochondria or mtDNA, total DNA can be extracted, digested, transferred, and the position of mitochondrial DNA bands is revealed with a probe (cold or radioactive). The DNA used as probe can be a highly purified mtDNA (better from the same species or from a related species) or a total genome cloned or, for special purposes, parts of the genome (either restriction fragments extracted from gels or cloned restriction fragments).

The choice of the strategy (purify mitochondria, purify mtDNA, work with total DNA) and the choice of the technics to be used to prepare the extracts depend on numerous parameters: equipment of the laboratory, method of visualization, size of the smallest fragments (function of the mtDNA and on the enzymes used), number of strains analysed...

I. PURIFICATION OF MITOCHONDRIA BY CELLULAR FRACTIONATION. CsCl GRADIENTS OPTIONAL

This preparation comprises three steps: pellet mitochondria; purify mitochondria; purify mtDNA on CsCl gradient. See at the end of the protocol the simplified possibilities. This protocol is described for *Drosophila* (adult flies) but is also convenient for other stages, other tissues or other materials. Recommanded only if few strains have to be studied or to get few micrograms of pure mtDNA to prepare probes or to establish physical maps or if you plan to use four cutter enzymes. Only few ug of mtDNA are obtained from several grams of flies or tissue.

I A. The complete protocol

1) Several grams of flies are homogenized in 15% sucrose in TE. Debris are eliminated by a low speed centrifugation in a SS34 Sorvall rotor or J 20.2 Beckman rotor at 1000 g (3,500 rpm) for 5 min. Work with large volumes of solution (50 ml of solution per gram

of flies). Spin the supernatant for 10 min at 2500 g (4,500 rpm). Preserve supernatant. Pellet mitochondria at 10,500 g (9,500 rpm) for 20 min (crude pellet).

2) During the last 20 min centrifugation, prepare a discontinuous sucrose gradient in clear 50 ml polycarbonate tubes: 15 ml sucrose 42.5%, 10 ml sucrose 30% and 10 ml sucrose 20%. This gradient is prepared begining either with the highest concentration solution (layer the other ones on a 5 mm thick slice cut in a cork of the appropriate diameter with a clearance of 2mm) or the lowest one (use a long and thick needle to gently deposit the solutions at the bottom of the tube). The pellet of mitochondria is resuspended with a Pasteur pipette in 1 ml of 15% sucrose. Layer this suspension on the discontinuous sucrose gradient. Run for 55 min in a swinging rotor Sorvall HB4 or Beckman 13.1 at 25,000 g (12,500 rpm). The band of mitochondria is located between the 30 and 42.5% beds. Take out the mitochondria with a 5 or 10 ml pipette and a propipette.

A continuous sucrose gradient 1M - 2M can also been used, prepared either with a gradient maker or from diffusion of a discontinuous gradient (sucrose concentrations 1M to 2M with 2 or three intermediate concentrations, allowed to diffuse for several hr).

3) To the 10-15 ml of this suspension of mitochondria, add an equal vol of a 7.5% sucrose solution and mix (in order to obtain a final concentration of about 15%). Centrifuge for 20 min in fixed angle rotor at 9 500 rpm. The pellet is now very enriched in mitochondria (pure pellet).

4) Lyse the pellet with 1 or 2 ml of 1% SDS in TE. Complete the volume to 3.3 ml and add 3.96 g of solid cesium chloride. The final vol (4.25 ml) is suitable for the tubes of the Beckman SW50 and Kontron SP55.5 rotors (use tubes with small volumes; avoid vertical rotors). A pre-run (10 min 10,000 rpm) allows to eliminate proteins-SDS complex floating on top. Then, adjust the refractive index to 1.399-1.400 (density 1.7 g/cm^3).

5) Spin for 40 hours at 35 000 rpm, 20°C. Pierce the bottom of the tube and develop the gradient in 1 fraction of 1 ml and 12 - 13 fractions of about 250 ul (count the drops). The separation of mitochondrial and nuclear DNAs is based on density differences and on differential configuration of the molecules.

6) Load 5 ul of each fraction on a gel. Nuclear DNA forms a band just above the largest restriction fragments of lambda DNA restricted with *Hin*dIII; mtDNA forms two bands on each side of nuclear DNA: the slower is the circular, relaxed DNA; the faster is the linearized mtDNA. High concentration of CsCl slows down the migration of DNA.

7) Keep the right fraction(s). In *Drosophila*, nuclear DNA is found in fractions 5 to 7 and mitochondrial DNA in fractions 7 to 9. Because *Drosophila* mtDNA is very A+T rich it is lightest than nuclear DNA and found above it in gradients. Fractions from several gradients in which mtDNA is contaminated by nuclear DNA can be pooled and centrifuged again for further purification.

8) Remove CsCl by dialysis or ethanol precipitation.

Microdialyse 10 to 50 ul of solution in an eppendorf tube: cut the cap and close the aperture with a dialysis membrane maintained with a indiarubber.

Precipitation of DNA: to 1 vol of CsCl solution add 3 vol water. Polysaccharides (which co-band with DNA in CsCl gradients) can be first precipitated by adding 1/10 vol of ethanol. Keep 1 hr at -20°C. Spin for 15 min at 6,000 rpm. Add 2 vol ethanol to the supernatant. Keep at least 3 hr (one night) at -20°C. Spin DNA.

TE: Tris 10 mM pH 8.0, EDTA 1 mM.

Sucrose: percentage indicates w/w (i.e. 15% sucrose is 15 g sucrose for a final weight of 100 g). Use fresh solutions (less than one week) or frozen solutions. Before use, shake vigorously the solutions that have been kept frozen. The concentration of sucrose solutions can be controled with a refractometer.

The previous protocol can be interrupted at an intermediate stage: after the first mitochondrial pellet (step 1: crude mitochondria pellet) or after the second one (step 3: pure mitochondria pellet). In both cases, mtDNA is extracted by SDS lysis of mitochiondria followed by phenol extraction (see below for a simplified preparation of phenol; see also in section III A the minipreparation of mtDNA with a similar principle). If mtDNA is contaminated, the optional step described above can be added or an alternative solution can be chosen: DNase or digitonin treatment can be tried. These various protocols are described below. A protocol similar to the sucrose gradient but involving Percoll density gradient, often of better efficiency, is described first in the next section.

Prepare phenol as followed: 250 g ultrapur phenol (Merk): add directly in the bottle 0.3 g hydroxyquinolin, 75 ml water. Stir for the night under an extractor hood. Neutralize with 30 ml Tris 0.5 M, NaCl 0.15 M, EDTA 1 mM, pH 7.6.

I B. Purification of mitochondria on a Percoll^{R} density gradient

1) Prepare Percoll (Pharmacia) 50% for a final volume of 80 ml (for two centrifuge tubes of 38 ml each). To 40 ml Percoll add 5 ml of a 2M sucrose solution: add this solution drop after drop with magnetic stir at the lowest speed.

2) Add to the Percoll mixture 35 ml of the solution for isolation of mitochondria: sucrose 0.15 M, mannitol 0.1 M, Hepes 20 mM, pH 7.5, EGTA 1 mM. Mix gently.

3) Preform the gradient: in a **fixed** angle rotor (not swinging rotor), spin two polyallomere tubes, each with 38 ml of the previous mixture for 60 min at 60,000 g.

4) Layer the suspension of mitochondria obtained with the protocol described in the section I A 1 (maximum 10 mg of mitochondria resuspended in 1 or 2 ml) onto the Percoll gradient with a Pasteur pipette.

5) Centrifuge in a fixed angle rotor at 60,000 g for 5 min when the correct speed is reached.

6) Mitochondria form several thin bands at the superior third of the tube. Pierce the tube with a large needle (on a syringe) under the mitochondrial band and puncture. Stop the pricking at equal distance between mitochondria and contaminant (a floculous and diffuse region located just above).

7) Wash mitochondria: add 5 vol of solution of isolation, mix, centrifuge 10 min at 10,000 g.

8) The mitochondrial pellet is not very hard. Consequently, pipette gently the supernatant, add 2 ml of solution of isolation, mix, centrifuge for 10 min at 10,000 g. If the pellet is still not hard, wash again.

Mitochondria prepared in this way can be used as in section IA.

I C. Phenol extraction of the mitochondrial pellet

As stated above, the protocol can be interrupted either after the obtainement of the first (step 1) or the second (step 3) pellet of mitochondria. If the crude fraction is used, it is recommanded to resuspend the mitochondria and centrifuge them again prior the lysis. The pellet is gently lysed with 0.5 ml of SDS 1% or 2% at room temperature for 10 min. The clear lysate is transferred into a 1.5 ml microcentrifuge tube. Proteins are extracted with 0.5 ml of neutralized phenol. Spin for 5 min at 12,000 g. Adjust the aqueous supernatant to a final concentration of 0.3 M NaAc. Add 2.5 vol of absolute alcohol, mix (by inversion of the tube), keep at -20°C overnight. Centrifuge mtDNA, wash the pellet with 70° ethanol, dry under vacuum, dissolve in appropriate vol of TE.

For some organisms (we have experience with young honeybees) this protocol, applied to the pure mitochondrial fraction (step 3) allows to produce a mtDNA purified enough to visualize small restriction fragments with the silver staining method (section VI B).

I D. DNase treatment

The mitochondrial pellet is resuspended in 5-10 ml sucrose 15%. Add a solution 0.1 mg/ml pancreatic deoxyribonuclease solution up to a concentration of 20 - 50 ug/ml and adjust to 20 mM $MgCl_2$. Keep for 30 min at 37°C. Add EDTA up to 20 mM and centrifuge mitochondria.

Alternatively, use 150 units/ml of the micrococcal endonuclease with 1 mM $CaCl_2$. Incubate for 15 min at 20°C and stop the reaction by addition of EDTA up to 5 mM.

I E. Digitonin treatment

The pellet of mitochondria is suspended in 15% sucrose (minimal volume) and incubated with digitonin (0.15mg of digitonin/mg of proteins) for 15 min at 0°C. The reaction is stopped by addition of 2 vol of 15% sucrose and the mitoplasts are pelleted as mitochondria are.

I F. Special materials. Some biological groups are more difficult than other ones. Molluscs, and more precisely terrestrial Gastropods are among them. A protocol is given in this section for landsnails, according to Stine (1989).

1) Mince the foot, genitalia, and hepatopancreas of a single individual.

2) Homogenize the tissue in 5 ml of ice-cold 0.25 M sucrose in TEK (50 mM Tris-HCl, pH 7.5, 10 mM EDTA, 1.5% KCl) containing 140 mg/ml of ethidium bromide (EB) in a Dounce homogenizer. The EB must be added to inhibit nuclease activity.

3) Remove the mucopolysaccharides by centrifugating the sample through a layer of 1.1 M sucrose in TEK at 13,000 g for 50 min at 4°C.

4) Resuspend the pellet containing the mitochondria in 0.25 M sucrose in TEK and repeat the centrifugation through the dense sucrose (the repetition of this step is necessary).

5) Disrupt the mitochondria by resuspending the pellet in 1.4 ml of ice cold 2% NP40 (Sigma) in TEK (NP40 lyses all membranes except the nuclear membrane). Place the sample on ice for 10 min and then centrifuge at 13,000 g for 10 min at 4°C.

6) Extract the supernatant first with phenol and then with chloroform. Recover the DNA by ethanol precipitation.

II. PURIFICATION OF mtDNA ON A CsCl GRADIENT WITH DYE

Separation of mtDNA from nuclear DNA is achieved through gradients of CsCl mixed with a dye, generally ethidium bromide (EB), rarely 4',6-diamidino-2-phenylindole (DAPI) or propidium iodine (PDI).

II A. Ethidium bromide-CsCl protocol. According to Lansman et al., 1981

1) Minced <10 g fresh tissues with scissors or disrupt in Waring blendor. Homogenize in 2 or 3 ml of MBS-Ca^{++} per gram using 8-10 strokes with a tight-fitting, motor-driven glass teflon homogenizer. Add Na_2EDTA to final concentration of 10 mM.

2) Remove cell debris and nuclei by centrifugation at 700 g for 5 min in a swinging rotor. Repeat with supernatant.

3) Pellet the mitochondria: centrifuge for 20 min at 20,000 g. Wash the pellet: resuspend in 10 ml MBS-EDTA and centrifuge at 20,000 g for 20 min. (An additional purification can be made with a sucrose gradient, see section IA).

4) MtDNA lysis. Mitochondria are resuspended in 3 ml of STE and lysed by addition of 0.15 ml 25% SDS (few min at room temperature). Add 1.1 mg solid CsCl per ml solution. Add 0.2 ml of a 10 mg/ml solution of ethidium bromide in STE. Adjust the refractive index to 1.391-1.392 by adding either solid CsCl or STE.

5) Centrifuge for 30-40 hr at 20°C in the Beckamn SW50.1 rotor (160,000 g) using either cellulose nitrate or polyallomer tubes.

6) Visualize the gradients under UV light: covently closed circular (supercoiled) DNA should be visible as a sharp band approximately 0.5 cm below a strong nuclear DNA band. The lower band can be collected either by puncturing the bottom of the tube or by inserting a hypodermic through the side of the tube under UV illumunation. The mtDNA band is often not visible: puncture under the nuclear band.

7) MtDNA is generally contaminated by RNA and nuclear DNA. However, it is purified enough to be used for RFLP. If mtDNA must be used as probe, it should be rebanded in a new CsCl-EB gradient.

8) Adjust the vol of the sample to 0.5 ml. Samples are extracted 2 or 3 times with butanol saturated with 5 M NaCl to remove EB. Add 1.1 ml of water, 3.2 ml of ethanol, mix. Incubate at -70°C for 60 min. Spin at 30,000 g for 30 min. Dry the mtDNA pellet and resuspend in an appropriate vol. of TE.

MBS: 0.21 M mannitol, 0.07 M sucrose, 0.05 M Tris-HCl, pH 7.5.
MBS-Ca^{++}: MBS 3 mM $CaCl_2$.
MBS-EDTA: MBS 10 mM EDTA.
TE: 10 mM Tris, 1 mM EDTA, pH 8.0.
STE: 100 mM NaCl, 50 mM Tris-HCl, 10 mM EDTA, pH 8.0.

Instead of EB, bisbenzimide or DAPI can be used: they binds preferentially to DNAs rich in A+T, lowering their density; they form a band, visualized under UV light, above nuclear DNA.

DAPI (adapt the volumes for centrifuge tubes): 4.9 g solid CsCl for 6 ml **final** vol; add 40 ul of DAPI solution 10 mg/ml.

Bisbenzimide: per ml of the DNA aqueous solution add 1.15 g of CsCl and 120 mg of bisbenzimide. Centrifuge at 100,000 g for 24 hr. Remove bisbenzimide by four extraction with pentan-2-ol saturated with CsCl (20 min for phase separation).

II B. Extraction with propidium iodide. According to Wright et al. 1983; Carr and Griffith 1987 have adapted this method for a miniultracentrifuge which strongly reduces (to 10 hr) the time required for extraction.

1) Mince approximately 1-2 g of soft tissue (fresh or frozen) on a chilled glass plate on ice.

2) Tranfer the tissue to a chilled Dounce homogenizer and make up the volume to 15 ml with ice-cold buffer containing 0.25 M sucrose, 10 mM Tris (pH 7.0), 5 mM Na_2EDTA. Adjust the chilled buffer to pH 7.6 with NaOH just prior the use. Homogenize tissue by hand.

3) Transfer the homogenate to a chilled 16 ml centrifuge tube. Centrifuge (Sorvall SM-24, Beckman JA-20.1) at 1000 g for 5 min at 5°C to pellet nuclei and cellular debris.

4) Recover the supernatant and centrifuge it at 23,500 g for 20 min at 5°C to pellet the mitochondria.

6) Pour off the supernatant and resuspend the pellet in about 1.5 ml of 10 mM Tris (pH 7.4), 1 mM Na_2EDTA (TE buffer). Measure the volume of the resuspension and adjust the total volume to 2.4 ml with the same buffer.

7) Add 1/4 vol. (600 ul) 10% (w/v in H20) SDS, mix gently to lyse the mitochondria. Wait 5 min.

8) Add 1/6 vol (500 ul) 7 M CsCl in TE buffer. Mix gently. Store on ice or at 5°C for at least 1 hr to precipitate high weight (nuclear) DNA and proteins.

9) Remix. Centrifuge at 17,500 g for 10 min at 5°C.

10) Recover 2.9 ml of the supernatant. Add 600 ul 2mg/ml propidium iodide (PDI) in TE buffer. Mix. Adjust the sample density to 1.58 g/cm3 by adding 2.52 g CsCl. Mix thoroughly to dissolve salt completely and homogeneously. (4.2 ml sample final).

11) Centrifuge in a Beckman SW60Ti rotor at 36,000 rpm, 21°C for 24 - 36 hr.

12) Collect the lower mitochondrial band under UV light (305 nm) by bottom puncture. Add to the fraction (0.3-0.5 ml) an equal vol of TE.

13) Layer the mixture onto CsCl-PDI step gradient consisting of 2.5 ml of d=1.40 g/ml (density) solution underlayered with 0.7 ml of d=1.70 g/ml solution (PDI at 100 ug/ml in each) in a Beckman SW60Ti clear centrifuge tube.

14) Centrifuge in a Beckman SW60Ti rotor at 45,000 rpm, 21°C, for 3 - 5 hr. Collect the first (i.e. bottom) 1.4 ml of the gradient by bottom puncture. Add 0.8 ml of CsCl d=1.57 g/ml with 100 ugPDI/ml in TE.

15) Centrifuge to buoyant equilibrium for 18 - 24 hr at 36,000 rpm, 21°C in a Beckman SW60Ti rotor.

16) The faint mtDNA band (usually the only band, the lower one if two bands are present) is collected as above.

17) Remove PDI and CsCl by dialysis at 4°C: first again 1 M NaCl in TE for 4 hr; next against 10% (v/v) Dowex AG50-X8 in 1 M NaCl, TE for 12 - 18 hr; and finally against 10 mM Tris, 0.1 mM EDTA pH 8 with two buffer changes.

Adaptation for a more rapid procedure with a mini-ultracentrifuge. Steps 1 to 10 are identical.

11) Divide the volume between two 2.0 ml quick seal tubes for the TLA-100.2 rotor. Centrifuge at 100,000 rpm (436,000 g) for 5-10 hr at 20°C in the Beckman TL-100 ultracentrifuge.

12) Two bands will be visible under 302-nm ultraviolet light: a strongly fluorescent upper band (nuclear DNA) and a weaker lower band (mtDNA). Collect the lower band in minimum volume, ca. 200 ul. Pool the bands from the duplicate tubes.

13) Make up the volume to just over 2.0 ml with a CsCl solution at density 1.55 g/cm3 (80 g CsCl, 5,5 ml 2 mg/ml PDI, 83.5 ml TE buffer). Load the sample in a single tube.

14) Repeat centrifugation step 11. The mtDNA band is now more intense. Remove PDI and CsCl as above.

II C. Rough purification of mitochondria, DNase and EB-CsCl. According to Watanabe et al. 1985.

1) Homogenize 30g wet weight of liver in 8 vol 0.25 M sucrose.

2) Centrifuge the homogenate for 15 min at 300g and 4°C.

3) Centrifuge the supernatant for 30 min at 5,000 g to sediment mitochondria.

4) The mitochondrial fraction is treated with 0.1 mg/ml DNase at 37°C for 30 min in 30 ml of 0.25 M sucrose containing 50 mM $MgCl_2$.

4) After the addition of EDTA up to 50 mM, the suspension is centrifugated as in step 3.

5) Mitochondria are resuspended in an equal vol of 0.15M Tris-HCl at pH 8.0 and 2 ml of a solution of SDS.

6) Nucleic acids are extracted by adding an equal vol of phenol and are precipitated with 2 vol of ethanol at -20°C overnight.

7) MtDNAs are purified by CsCl-ethidium bromide ultracentrifugation at 85,000 g for 36 hr at 20°C. Fractions containing closed circular mtDNA molecules are collected, extracted four times with H_2O-saturated butanol to remove the dye, diluted with an equal vol of H_2O and precipitated with 2 vol of ethanol at -20°C.

III. MINIPREPARATIONS OF MITOCHONDRIA WITH PHENOL EXTRACTION

The following methods use the simplified steps described section I, adapted to minipreparations. They are suitable for screening numerous strains. The methods described in B and C are given in the litterature for vertebrates.

III A. DNase and phenol. According to Powell and Zuniga 1983, described for *Drosophila.*

1) Using a Dounce homogenizer with a lose-fitting pestle, thoroughly homogenize tissue in Homo buffer; continue until the homogenate can be pipetted with a Pasteur pipette. About 1ml/100 flies.

2) Spin homogenate for 10 min at 4°C at 1,000 g to pellet nuclei. Save supernatant. Repeat 3 times, saving supernatant each time. The final supernatant is pipetted into a 1.5 ml disposable microfuge tube.

3) Spin final supernatant at 12,000 g for 30 min at 4°C.

4) Discard supernatant and wash organelle pellet with 1.2 ml Homo buffer. Repeat step 3.

5) Remove supernatant and resuspend pellet in 350 ul Lysis buffer.

6) Add DNase-free RNase (5ul of a 20 mg/ml stock). Incubate at 37°C for 10 min.

7) Add about an equal volume of redistilled buffer saturated phenol. Mix thoroughly and incubate at 65°C for 10 min, mixing occasionally.

8) Let cool tube on ice. Spin in a microcentrifuge at 4°C for 20 min. Save the upper aqueous layer.

9) Extract with phenol:chloroform:isoamyl alcohol, 50:48:2.

10) Add about twice the volume of ether to the final aqueous phase, shake. Spin for 2 min. Eliminate ether (upper phase). Repeat and evaporate remaining ether 15 min.

11) Add 3 vol. absolute ethanol, mix gently, freeze on dry ice or in freezer at -70°C for at least 15 min.

12) Spin frozen microfuge tubes for 15-30 min. Remove the supernatant, wash the pellet with 70°C ethanol and centrifuge. Drive off residual alcohol under a vacuum about 5 min. Do not overdry. Take pellet in an appropriate volume of sterile water.

Homo(geneizing) buffer: 0.25 M sucrose, 1 mM EDTA, 2.5 mM CaC12, 30 mM Tris-Hcl, pH 7.5, 0,015% (v/v) Triton X-100.

Lysis buffer: 0.5 M NaAc, 10 mM EDTA, 0,5% sarkosyl, 10 mM Tris-HCl, pH 8.0.

III B. Modification of the Powell-Zuniga protocol. According to Yamagata et al. 1988, for mammals.

1) Liver or kidney are homogenized in a 10 vol solution pH 7.5 of 0.25 M sucrose, 1mM EDTA, 2.5 mM $CaCl_2$, and 30 mM Tris HCl.

2) The homogenate is centrifuged for 30 min at 600 g at 4°C.

3) The supernatant is centrifuged for 30 min at 13,000 g at 4°C to sediment mitochondria.

4) The mitochondrion pellet is resuspended in a 10 vol solution (pH 8.0) of 0.5 Na Ac, 10 mM EDTA and 10 mM Tris-HCl. Sarcosyl is added at a final concentration of 0.5%.

5) MtDNAs are extracted from the mixture by adding an equal vol of water saturated phenol and are precipitated with 2.5 vol of ethanol at -20°C overnight.

6) MtDNAs are resuspended in an appropriate vol of water and are stored at 4°C.

III C. Proteinase and phenol extraction. According to Jones et al. 1988: improvement of the Powell-Zuniga method: the preparation is reputed to contain less nuclear DNA (larger volumes, additional centrifugations, another homogenizer).

1) Wash fresly chilled tissue (0.3 g) in chilled distilled water to remove fat and blood. Transfer in homogenizing tubes with at least 10 ml prechilled buffer (30 mM tris-HCl, 1 mM EDTA, 2.5 mM $CaCl_2$, 0.25 M sucrose).

2) Homogenization: use, instead of Dounce homogenizer, a motor-driven glass teflon homogenizer. The best clearance between tube and pestle is 0.2 mm. Use the minimum strokes necessary (five to eight) at low speed (200 rpm) in a large volume of buffer.

3) Spin the homogenate at 1,000 g at 4°C for 15 min in a Beckman centrifuge JA 20.1 rotor. Carefully pipette the supernatant.

4) Resuspend the nuclear pellet in 5 ml of fresh buffer and respin at 1300 g for 10 min. Add the supernatant to the original one and spin the pooled supernatants at 1,300 g for 10 min. Repeat the spin until there is no remaining pellet to be seen.

5) Spin the final supernatant at 15,000 g at 4°C for 30 min to pellet the mitochondria.

6) Discard the supernatant, resupend the mt pellet in 2 ml STE buffer (0.05 M Tris-HCl, 0.1M NaCl, 0.01M EDTA, pH8.0) warmed to 37°C prior to use.

7) Add a small amount (20-25 ul for kidney or heart, 100 ul for liver) of 25% SDS solution until the supernatant clears, indicating the complete lysis of the mitochondria.

8) Add 30 ul RNase A (20 ug/ml solution DNase-free, preboiled for 10 min and cooled on ice) and incubate for 30 min at 37°C.

9) Add 20 ul of proteinase K (20 mg/ml solution) and incubate for 30 min at 37°C.

10) Shake the lysate with an equal volume of Tris-equilibrated phenol and centrifuge for 30 min at 20,000 g (the longer and harder the lysate is spun, the more effective this step is).

11) Pipette off the aqueous supernatant and add an equal volume of phenol chloroform mixture, mix, spin at 12,000 g for 10 min. Repeat this step until there is no proteins at the interphase.

12) Extract the supernatant twice with an equal volume of chloroform and twice with an equal volume of ether. Shake the tube until the interphase between the ether and the supernatant is sharp (i.e. no bubbles or opaque layer).

13) Discard the ether (top layer) and allow any residual ether to evaporate (at least 15 min).

14) If necessary, add distilled water to the supernatant to bring the volume up to 2 ml. Add 6 ml of absolute ethanol, covers the tube with parafilm, mix by inversion. store the sample at -70°C for at least 2 hr or overnight.

15) Remove sample from -70°C, allow to stand until it reaches room temperature, mix again (any salts precipitated will be redissolved). Spin for 30 min at 20,000g at 4°C to pellet the DNA.

16) Discard the supernatant and vacuum dry the mtDNA pellet. Dissolve in 40 ul TE buffer (10 mM Tris-HCl, 0.5 mM EDTA) and transfer to a sterile eppendorf tube for storage at -20°C.

III D. Minipreparations of mtDNA from virgin eggs. According to Solignac et al. 1984. The protocol of extraction is the simplest of all methods but harvesting virgin eggs can be laborious and limited to few suitable materials. Convenient for numerous strains, this protocol furnishes an excellent mtDNA.

1) Virgin eggs are obtained from unimpregnated females (separated from males at the pupal stage or as young adults). About 50 females are introduced in a plastic box 4x3x7 cm and a disc of corn medium covered with fresh yeast where they can lay their eggs is changed every day. Yeast (containing eggs) is removed with a paint-brush and water. Virgin eggs can be kept in water at +4°C during several weeks.

2) Eggs are washed in a large vol of water and decanted 4 to 6 times. Remnant contamination by medium or filamentous fungi can be eliminated by a short centrifugation (5 min at 5,000 rpm) of the eggs suspended in a 40% sucrose solution.

3) Eggs are homogenized in the solution: Tris 10 mM, EDTA 1 mM, pH 8, glucose 1%. Work with half a pea of eggs (about 5,000) for 0.5 ml solution.

4) Add 50 ul of a 10% SDS solution. Mix. Keep 5 min at room temperature.

5) Add 0.5 ml of neutralized phenol and keep on ice for at least 15 min. Shake periodically.

6) Centrifuge the microfuge tube for 5 min at 5,000 rpm. Pipette the supernatant (Pasteur pipette or Pipetman P200). Adjust to a final concentration of 0.3 M NaAc, add 3 vol of ethanol and mix. Precipitate nucleic acids for one night at -20°C.

7) Centrifuge for 15 min at 10,000 rpm. Wash the pellet (i.e. add 1 ml 70% ethanol and centrifuge 5 min at 10,000 rpm; eliminate the ethanol, add 1 ml of 90% ethanol and centrifuge 5 min at 10,000 rpm). Eliminate the ethanol and dry the pellet.

8) Dissolve the pellet in 20 to 100 ul TE or sterile water. Keep at -20°C. The extract contains mtDNA and nuclear DNA (one half of each in *Drosophila*). RNA has to be removed only in protocols using fluorescence to visualize fragments less than 900 bp or if silver staining is used: in this case add 1 to 5 ul per sample of preboiled RNase 1 mg/ml after the digestion by restriction endonucleases.

NB. Extraction of mtDNA is very easy and rapid. The longest step is to collect unimpregnated females. In some favourable cases the following procedure can be followed: if the female of the species under study can be easily crossed with males of a related species and if among the hybrids, females are fertile and males are sterile, sexes have not to be separated in the F1.

III E. Simplified protocol for adult flies. Modified from Coen et al. 1982 (a method of D. Ish-Horowicz). This method is very rapid and convenient for a lot of strains. Although not described originally for the study of mitochondrial DNA, this protocol gives sometimes clear mitochondrial bands after digestion (i.e. the smear due to nuclear DNA is not tto strong). Convenient for one observation in flurescence. The same protocol is used with an additional step in section IV B.

1) Homogenize 20 flies (better females with well developed ovaries) in 1 ml homogenization buffer (10 mM Tris, 60 mM NaCl, 5% sucrose, 10 mM EDTA).

2) Add 1 ml lysis buffer (Tris 0.3 M,pH 9.0, 100 mM EDTA, 5% sucrose, 1.25% SDS, 8 ul/ml diethylypyrocarbonate = DEPC) and incubate at 65°C for 30 min.

3) Add 300 ul 8 M KAc and incubate on ice for 45 min.

4) Centrifuge at 10,000 rpm (Sorvall, SS-34 rotor) for 10 min and take supernatant.

5) Precipitate DNA with 1 vol isopropanol: 5 min at room temperature, spin at 12,000 g for 10 min.

6) Dissolve the pellet in 500 ul H20.

7) Two versions here: 30 min at room temperature, then step 8. or phenol extraction then step 8.

8) Precipitate DNA with 1/10 vol 3M NaAc and 2 vol ethanol.

IV. PLASMID-LIKE mtDNA MINIPREPARATIONS

IV A. Minipreparation of mitochondria. According to Tamura and Aotsuka 1988. This alkaline lysis procedure is analogous to that used for the preparation of plasmid, covalently closed, DNA (Birnboim and Doly 1979). A similar protocol has been described for human tissues by Welter et al. 1989.

1) 50 mg of adult Drosophila (50 flies) in 1 ml of a chilled homogenizing buffer: 0.25M sucrose, 10 mM EDTA, 30 mM Tris-HCl, pH 7.5 using a microhomogenizer.

2) Transfer the homogenate to a chilled 1.5 ml disposable centrifuge tube and centrifuge at 1 000 g for one min at 4°C in order to pellet the nuclei and cellular debris.

3) Recover the supernatant and recentrifuge at 12,000 g for 10 min at 4°C, thereby pelleting the mitochondria.

4) Discard the supernatant and resuspend the mitochondrial pellet in 10 mM Tris EDTA buffer pH 8 containing 0.15M NaCl and 10 mM EDTA. The total volume should be 50 ul.

5) Add 100 ul of freshly prepared 0.18 N NaOH containing 1% SDS. Vortex gently. Store on ice for 5 min.

6) Add 75 ul of an ice-cold solution of KAc (3M potassium, 5M acetate, Maniatis et al., 1982). Vortex gently and store on ice for 5 min.

7) Centrifuge at 12,000 g for 5 min. at 4°C. Recover the supernatant and add an equal volume of phenol:chloroform:isoamyl alcohol, 50:48:2. Mix thoroughly by vortexing.

8) Centrifuge at 12,000 g at room temperature. Tranfer the aqueous phase to a fresh tube and add 2 vol. of ethanol. Mix by vortexing and let stand at room temperature for 15 min.

9) Centrifuge at 12,000 g at room temperature. Wash the resulting pellet with 1 ml of 70% ethanol and dry briefly under reduced pressure.

10) Add an appropriate vol. of 10 mM Tris-HCl buffer pH 8.0 containing 1 mM EDTA and 20 ug/ml of DNase-free pancreatic RNase.

11) Store the solution of mtDNA at -20°C.

The same procedure can be followed for vertebrates with 25 mg of fresh tissue (liver or kidney).

IV B. Miniprep on crude DNA. Also for midi and maxiprep. According to Afonso et al. 1988. This protocol is essentially that of Coen et al. (section III E) followed by an alkaline denaturation (Birnboim and Doly 1979).

1) Grind flies very gently in 2 ml precooled solution I (+4°C) using a Potter-Elvehjem tube for 1 min at 500 rpm.

2) Add 2 ml of solution II, mix and incubate at 65°C for 30 min.

3) Add 960 ul of KAc 5M, mix well, and incubate at -20°C for 8 min. or on ice for 45 min.

4) Centrifuge at 12,000 rpm for 5 min. at +4°C. Extract DNA-containing supernatant with a pipette and discard pellet.

5) To supernatant add 2 vol ethanol or 1 vol isopropanol (+4°C) and mix; let stand at room temperature for at least 5 min.

6) Centrifuge as in 4. Discard supernatant.

7) Dehydrate pellet for at least 5 min at room temperature and resuspend in 1 ml TE buffer.

8) Add 2 ml of alcaline solution. Mix by gentle 6 to 10 tube inversions.

9) Incubate on ice for 4 to 5 min.

10) Neutralize pH by adding 1.5 ml of NaAc, pH 5.2. Mix by gentle tube inversions.

11) Incubate for 15 min at -20°C or 60 min on ice.

12) Centrifuge at 12,000 g for 5 min. Extract DNA-containing supernatant.

13) Add two vol of phenol chloroform to the supernatant. Mix by shaking. Incubate on ice for 5 min.

14) Centrifuge at 3000 g for 5 min. Collect aqueous phase (another extraction step can be inserted here for sensitive restriction enzymes). Add 2 vol absolute ethanol or 1 vol isopropanol (+4°C).

15) Incubate at room temperature for 5 min.

16 Centrifuge at 12,000 g for 5 min. Discard supernatant.

17) Wash the pellet twice with 70% ethanol (-20°C).

18) Dry the pellet under vacuum. Resuspend in the appropriate vol of TE and store at +4°C.

Solution I: 10 mM Tris, 60 mMNaCl, 5% sucrose (W/v), 10 mM EDTA (pH 8.0). Bring all to pH 7.5 with HCl and autoclave.

Solution II is 300 mM Tris, 5% sucrose (w/v), 10 mM EDTA. Autoclave. Add 1.25% SDS and 0.8% diethylpyrocarbonate - DEP (freshly mixed).

TE is 25 mM Tris, 10 mM EDTA, pH 8.0.

Alcaline solution is 0.1% SDS and 0.2N NaOH, prepare fresh from 20% SDS, pH 9.0 and 2N NaOH stocks.

Mini-, midi-, maxi-preparations: 15 ml centrifuge tubes for 80 flies (described here); 1.5 ml microcentrifuge tubes for 5-10 flies; 30 ml tubes for 300 flies.

V. TOTAL DNA EXTRACTION

The following protocols are developed to prepare crude extracts of DNA. This DNA is convenient for digestion, transfer and hybridization. The expected proportion of mtDNA is about 1%; however, the actual proportion is frequently higher. The extract can be performed from a single fly (for example for the determination of an individual genotype with a diagnostic restriction enzyme) or from a lot of flies when several enzymes have to be used for the study of variability. In principle, these methods are convenient for large restriction fragments the size of which is above 300-500 bp. Obviously, all the previous methods of extraction can also be used to prepare pure and more or less contaminated mtDNA for hybridization technics.

V A. One of the numerous recipes. DNA preparation for individual grasshoppers according to Marchant 1988.

1) Grind the insect to a fine powder in a mortar with LN_2. The LN_2 is allowed to evaporate.

2) The powder was added to 5ml of a solution of 8 parts 0.05 M Tris, 0.1 M NaCl, 0.1 M Na_2EDTA, pH 7.0 with HCl; 1 part 5% SDS; 1 part 2mg/ml protéinase K (freshly prepared). Incubate at 37°C for 1-3 hr.

3) Extract with 3 ml phenol.

4) Precipitate DNA by adding 2 vol of a solution of 2.9 M sodium perchlorate in 80% ethanol. Mix. Cool to -20°C for 1 hr. Centrifuge for 15 min at 11,000 rpm in Sorval HB4 rotor. Wash the pellet, centrifuge, dry.

V B. Protocol for individual flies. An adaptation for individual flies of the protocol given section III E which can be adaptated for as many flies as wished.

Individual flies are homogenized in 50 ul of cold 0.1 M TE (0.1 Tris, pH 9, 0.1 M EDTA) in a microcentrifuge tube with a plastic pestle. Add 50 ul of a preheated 2% SDS solution. Incubate at 65°C for 1 hr. After cooling at room temperature for a few min, add 13 ul 8M KAc. Mix. After 30 min on ice, samples are spun for 10 min in an eppendorf centrifuge. Add isopropanol (60 ul) to the supernatant. After 10 min at room temperature, the samples are spun for 10 min. The pellet is washed with 70% ethanol, centrifuged, dried under vacuum and resuspended in 10 ul 10 mM Tris - 1 mM EDTA pH 8.

V C. The "PCR" recipe. This recipe is proposed by Kocher et al. 1989 to prepare mtDNA (in fact total DNA) for PCR amplification. However, used with honeybees, this method allow to visualize by ethidium bromide staining the largest mtDNA restriction fragments (> 2 kb).

1) Grind one bee (head+thorax) in 5 ml of the solution: Tris-HCl 100 mM pH 8.0, EDTA 10 mM, NaCl 100 mM, SDS 0.1%, dithiothreitol (DTT) 50 mM, proteinase K 0.1 to 0.5 mg /ml.

2) Incubate at 37°C for 2 to 4 hr.

3) Centrifure for 5 min at 6,000 rpm to eliminate debris.

4) Extract with 2 phenols, 1 phenol chloroform, 1 chloroform.

5) Precipitate with alcohol or concentrate with Centricon.

V D. Isolation of "cytoplasmic nucleic acids". According to Lansman et al. 1981.

1) For 0.5 to 2 g tissue: mince and homogenize in 5 ml of 2X TMS (60 mM Tris-HCl pH 7.4, 300 mM NaCl, 3 mM $MgCl_2$).

2) Add 0.1 ml of a 10% P40 (BRL) solution and incubate on ice for 5 min (P40 lysis all membranes except the nuclear one).

3) Pellet intact nuclei by centrifugation in 15 ml conical tubes fro 5 min at 1,500 g.

4) Transfer the supernatant in a 30 ml Corex tube, add 5 ml of phenol and incubate at 65°C for 5 min with periodic thorough mixing.

5) Spin for 10 min at 6,000 g in a swinging bucket rotor.

6) Reextract the upper aqueous phase with 5 ml phenol. Spin.

7) Nucleic acids are precipitated by addition of 2 vol ethanol and incubation at -70°C for 30 min or -20°C for 2 hr.

8) The precipitate is pelleted by centrifugation at 6,000 g for 15 min. The pellet is washed with 75% ethanol and the centrifugation is repeated. Dry. Dissolve.

VI. VISUALIZATION OF mtDNA FRAGMENTS

Detection limits.

Silver staining: around 10-30 pg of DNA (i. e. enabling application of 10-40 ng of mtDNA digested with tetranucleotide restriction endonucleases).

Ethidium bromide: 1 to 5 ng DNA i.e. 100 to 300 ng have to be applied on the gel to detect a 250 bp band. DAPI is a litlle more sensitive than EB but expensive.

Hybridization of Southern blot: 10 pg per band.

End labelling: 5 to 10 ng allow to load 3 gels.

VI A. Fluorescence

Stain with ethidium bromide added to the gel buffer (1mg/ml)and migration buffer (1mg/ml) or by incubation of the gel after electrophoresis. In this last case incubate for 10 min in a 2ug/ml EB solution and wash with distilled water; stain and wash in a refrigerator (dark preserve EB and cold reinforce fluorescence). For small amounts of DNA, stain overnight at +4°C. EB stock solutions must be kept in dark. Manipulate with gloves. Caution with waste.

VI B. Silver Staining with polyacrylamide gels. According to Tegelström 1986. "Fast, cheep, sensitive". **High purity water is essential (freshly redistilled water); solutions freshly made; works only with acrylamide gels.** Separate the mtDNA fragments on 5% **polyacrylamide** gels (0.7 mm tick). Run the gels on a vertical electrophoresis apparatus for 3 hr (200 V; 35 mA) using 1 x TBE buffer, pH 8.3 (10 x stock solution: 89 mM Tris-base, 89 mM boric acid, 2 mM EDTA).

1) 20 min of presoaking in 0.1% cetyltrimethylammonium bromide (CTAB, Sigma).

2) 20 min of rinsing in distilled water.

3) 15 min in 0.3% ammonia (stock sol is 28%).

4) 15 min in ammoniacal silver prepared as follow: 0,4g of silver nitrate disolved in 2 ml of distilled water, then 1 ml of freshly prepared 1M sodium hydroxyde, then 1 ml of 25% ammonia, then 248 ml water.

5) The gel is developped for 10-20 min in 2% sodium carbonate with 0,02% formaldehyde. Photograph the gel. Dry the gel.

VI C. End labeling of restriction fragments. According to Brown 1980 and Wright et al 1983. The same protocol is used whatever the DNA ends generated by the restriction enzymes (5' single -stranded extension, blunt end, 3' single-stranded extension).

1) Use 5 or better 10 ng mtDNA per sample. Digest with restriction enzyme in 20 ul in a 1.5 ml eppendorf tube.

2) Add 0.5 uCi (3.7×10^4 Bq) of each of the four radioactive nucleotides* and 0.25 unit of Kleenow enzyme. Kleenow fragment works with H and M restriction enzyme buffers. For L buffer, add NaCl to 50 mM. Mix. Keep for 75 to 90 min on ice.

3) Precipitation: add 340 ul of NaAc 0.3 M containing 20 ug of *E. coli* tRNA and 900 ul ethanol.

4) Incubate for 2 hr at -70°C or overnight at -20°C.

5) Spin for 15 min at 12,000 g. Supernatant are radioactive and must be carefully collected. Edge of the tubes can be dried on Kleenex. Wash with 1 ml 75% ethanol and centrifuge as above.

6) Dissolve directly the pellet in 30 ul of loading buffer containing Bromophenol blue. Load 10 ul on a 1% agarose gel and 10 ul on a 5% acrylamide gel; the remaining 10 ul are available for a new experiment if necessary.

7) Dry and autoradiograph the gels. Appropriate DNA size standards are prepared in the same way as mtDNAs.

*A mixture of the four nucleotides is recommanded for site mapping because various single and double digests are used. For screening variability with a given enzyme, a single radioactive nucleotide can be used. Replace any lacking radioactive nucleotide by 1 ul of a 1 mM solution of the corresponding (cold) nucleotide. Use 0.5 uCi of each radioactive nucleotide when three or four are used and 1 uCi when one or two are used. If a single nucleotide is used, it must be complementary of a particular nucleotide (the nucleotide replaced below by *) depending on the structure of the restricted DNA:

5’ o o o o	5’ o o o o o	5’ o o o o o o o o o
3’ o o o o * o o o	3’ o o o o *	3’ o o o o *

Theoretically, all the fragments, whatever their size, are equally labelled.

VI D. Hybridization of transferred restriction fragments. The technics for probe labeling, DNA transfer, and hybridization are not peculiar to mtDNA. Only, the protocol of hybridization is simplified.

1) Label 200 ng of purified or cloned mtDNA (nick-translation or random priming)

2) Prehybridization: 2 hr 30 to 3 hr at 55°C in Solution A.

3) Hybridization: 16 hr at 55°C in Solution A + the probe previously denatured (5 min in boiling water; 5 min on ice).

4) Wash 6 times 30 min at 55°C in Solution A.

Solution A is Denhardt 5x, SSC 6x and SDS 0.1%. To prepare this solution, mix:

200 ml of a 10g/l of bovine serum albumine (BSA)

200 ml of a 10g/l Ficoll solution,

200 ml of a 10g/l Polyvinyl Pyrolidone solution

600 ml 20x SSC

10 ml SDS 20%

H_2O to 2 l.

For the visualization of small restriction fragments (300 to 1,500 bp), use 3% Nusieve gels with TBE buffer and transfer DNA on Hybond N membranes (Amersham).

VI E. Hybridization of dried gels. Instead of transferring DNA, the gel itself can be used, after drying, as a membrane filter.

1) Run, stain and photograph gel as usual.

2) Denature gel for 20 min in 500 mM NaOH, 150 mM NaCl.

3) Neutralize for 20 min in 500 mM Tris pH 8 150 mM NaCl.

4) Place gel on Whatman 3MM paper or on a piece of dialysis membrane.

5) Dry gel completely (gel drier).

6) When ready for hybridization, separate dry gel from support. This is easily done by briefly soaking the gel which will then peel off from the 3MM paper or the dialysis membrane.

7) Treat dry gel exactly as a nitrocellulose filter.

VI F. Quantitation

This section concerns the determination not of the absolute quantity of mtDNA but that of the relative amount of two or several molecular types present in a mixture. This ratio is calculated through the comparison of the intensity of two or several restriction fragments. This quantitative study is useful in case of mixture of different mtDNA types (for example when the relative proportion of two or several mtDNA types present in a population have to be determined through a mass extraction, or in cases of site or length heteroplasmy).

Negatives of the photos (Polaroid 665 for example) of the gels stained with EB or X-ray autoradiographies can be scanned with a microdensitometer. In both cases, avoid an overexposure.

Caution: thanks to the action of DNases, vortex, phenol and alcohol, a variable proportion of mtDNA molecules is linearized in the course of extraction. If the breaks occur at random, long restriction fragments are under-representated (whatever the enzyme used). If the breaks are non random, the restriction fragment including the region in which they occur preferentially will be under-representated.

Fluorescence of a band under UV light, after EB staining, is proportional to the quantity (weight) of DNA. As the number of molecules is the same for each band, providing that the molecules are homogeneous in length and sequence, as it is generally the case for individual or isofemale mtDNA extracts, the intensity of fluorescence is proportional to the size of the fragment in the restricted DNA. In case of mixture, quantitation measured through fluorescence must be corrected for the length. The comparison of fragments whose size is as close as possible is always preferable.

Theoretically, end labeled restriction fragments give signals with the same intensity, whatever their size. In fact, this is rarely the case, even for fragments sharing the same ends (digestion with a single enzyme which has no ambiguity in the recognition site).

If a hybridization technics is used, and if a total mtDNA is used as a probe, the signal is expected proportional to the size of the fragments. However, this is not always the case. If a quantitation has to be made, avoid overexposure and use homologous probe: gene order is not strictly conserved in animal mtDNA (distantly related species have different gene order) and the different genes evolve at different rates.

Aknowledgements. It is a great pleasure for me to thank Monique Monnerot, Nicole Dennebouy, Jean-Claude Callen who initiate me to prepare mitochondrial DNA and to experiment with it in the lab of Jean-Claude Mounolou .

REFERENCES

Afonso JM, Pestano J, Hernandez M (1988) Rapid isolation of mitochondrial DNA from *Drosophila* adults. Biochem Genet 26:381-386

Birnboim HC, Doly J (1979) A rapid alkaline extraction procedure for screening recombinant plasmid DNA. Nucleic Acids Res 7:1513-1523

Brown WM (1980) Polymorphism in mitochondrial DNA of humans as revealed by restriction endonuclease analysis. Proc Natl Acad Sci USA 77:3605-3609

Carr SM, Griffith OM (1987) Rapid isolation of animal mitochondrial DNA in a small fixed angle rotor at ultrahigh speed. Biochem Genet 25:385-390

Coen ES, Thoday JM, Dover G (1982) Rate of turnover of structural variants in the rDNA gene family of *Drosophila melanogaster*. Nature 295:564-568

Giles RE, Blanc E, Cann HM, Wallace DC (1980) Maternal inheritance of human mitochondrial DNA. Proc Natl Acad Sci USA 77:6715-6719

Jones CS, Tegelström H, Latchman DS, Berry.RJ (1988) An improved rapid method for mitochondrial DNA isolation suitable for use in the study of closely related populations. Biochem Genet 26:83-88

Kocher TD, Thomas WK, Meyer A, Edwards SV, Pääbo S, Villablanca FX, Wilson AC (1989) Dynamics of mitochondrial DNA evolution in animals: amplification and sequencing with conserved primers. Proc Natl Acad Sci USA 86:6196-6200

Lansman RA, Shade RO, Shapira JF, Avise JC (1981) The use of restriction endonucleases to measure mitochondrial DNA sequence relatedness in natural populations. III. Techniques and potential applications. J Mol Evol 17:214-226

Maniatis T, Fritsch E F, Sambrook J (1982) Molecular Cloning, a Laboratory Manual, Cold Spring Harbor Laboratory, Cold Spring Harbor New York

Marchant AD (1988) Apparent introgression of mitochondrial DNA across a narrow hybrid zone in the *Caledia captiva* species-complex. Heredity 60:39-46

Powell J P, Zuniga M C (1983) A simplified procedure for studing mtDNA polymorphisms. Biochem Genet 21:1051-1055

Solignac M, Génermont J, Monnerot M, Mounolou J-C (1984) Genetics of mitochondria in *Drosophila*: mtDNA inheritance in heteroplasmic strains of *D. mauritiana*. Mol Gen Genet 197:183-188

Stine OC (1989) *Cepaea nemoralis* from Lexington, Virginia: the isolation and charaterization of their mitochondrial DNA, the implications for their origin and climatic selection. Malacologia 30:305-315

Tamura K, Aotsuka T (1988) Rapid isolation method of animal mitochondrial DNA by the alkaline lysis procedure. Biochem Genet 26:815-819

Tegelström H (1986) Mitochondrial DNA in natural populations. An improved routine for the screening of genetic variation based on sensitive silver staining. Electrophoresis 7:226-230

Watanabe T, Hayashi Y, Ogasawara N, Tomoita T (1985) Polymorphism of mitochondrial DNA in pigs based on restriction endonuclease cleavage patterns. Biochem Genet 23:105-113

Welter C, Dooley S, Blin N (1989) A rapid protocol for the purification of mitochondrial DNA suitable for studying restriction fragment length polymorphism. Gene 83:169-172

Wright JW, Spolsky C, Brown WM (1983) The origin of the parthenogenetic lizard *Cnemidophorus laredoensis* inferred from mitochondrial DNA analysis. Herpetologica 39:410-416

Yamagata Y, Ishikawa A, Tsubota Y, Namikawa T, Harai A (1987) Genetic differentiation between laboratory lines of the musk shrew (*Suncus murinus*, Insectivora) based on restriction endonuclease cleavage patterns of mitochondrial DNA. Biochem Genet 25:429-446

Hybridization of DNA probes to filterbound DNA using the "DIG - DNA Labeling and Detection System Nonradioactive" (Boehringer Mannheim)

Bachmann, L. and Sperlich, D.
Universität Tübingen, Lehrbereich Populationsgenetik, FRG

Hybridization of labeled DNA to filterbound DNA is stillan important method in molecular biology but is furthermore introduced and applied as a simple, rapid and powerful instrument in classical fields of biology like taxonomy, too.

Traditionally probe DNA was labeled radioactively, most commonly by means of ^{32}P labeled triphosphates. The quite short half-life of the ^{32}P isotope and the requirement of security devices favours the use of nonradioactive labeling and detection techniques, especially for those, who do not apply such methods very often.

During the last years various nonradioactive DNA labeling and detection system have been elaborated. One of them, the "DIG - DNA Labeling and Detection System" of Boehringer Mannheim is emploid in our laboratory routinely in order to probe Southern blots and plaque lifts as well as to screen bacterial colonies for recombinant plasmids.

The principle of this procedure will be described briefly:

- The probe DNA is labeled by incorporation of the thymidine analog digoxigenin-11-dUTP by the random primed labeling technique of Feinberg & Vogelstein (1983).

- Hybridization of the labeled DNA to filterbound DNA

- Detection of the hybridization with an anti-digoxigenin-antibody

- Visualization of the antibody-antigen conjugate by a colour reaction via an alkaline-phosphatase linked to the anti-digoxigenin-antibody

Protocol

The protocol for labeling of DNA, hybridization to filterbound DNA and detection of the signal described here follows mainly the standard protocol of the Boehringer Mannheim applications manual. Reagents: Hexanucleotide mixture, dNTP labeling mixture, Klenow enzym, blocking reagent, anti-digoxigenin-antibody-conjugate, NBT solution and X-phosphate solution are taken from the "DIG DNA Labeling and Detection Kit Nonradioactive" (Boehringer Mannheim).

Preparation of filters for hybridization:

- Transfer of digested, electrophoretically separated DNA to both Nitrocellulose- or Nylonmembranes according to Southern (1975)

- Plaque lifts are performed as described by Benton & Davis (1977)

- Bacterial colonies are transfered to Nitrocellulosemembranes following the protocol of Grunstein & Hogness (1975)

NATO ASI Series, Vol. H 57
Molecular Techniques in Taxonomy
Edited by G. M. Hewitt et al.

Digoxigenin-labeling of probe DNA:

Usually 100 - 500 ng of probe DNA are labeled. The DNA is denaturated in 15µl of destilled water by boiling for 10 min. After a short chilling on ice 2µl of a Hexanucleotide mixture, 2µl of a dNTP labeling mixture and finally 1µl Klenow enzyme (2 units) are added. The reaction is incubated for one to several hours at 37°C. In order to precipitate the DNA 10µl 3M NaAc, 70µl water and 300µl ethanol are added and kept 30 min at -20°C. After centrifugation for 15 min at 13.000 rpm the pellet is washed briefly with 70% ethanol, dried and redissolved in 100µl TE - buffer (10 mM Tris-HCl; 1mM EDTA; pH 7.8).

Hybridization of DIG - labeled probe DNA to filterbound DNA:

Hybridization is performed at temperatures of 60 to 68°C. After 1h of prehybridization at hybridization temperature chosen according to stringency conditions in 5 x SSPE; 0.1% N-lauroylsarcosine, Na-salt; 0.02% SDS; 0.1% blocking reagent the denaturated probe DNA (10 min boiling) is added and hybridized overnight. The filters are washed 2 x 10 min at room temperature with 2 x SSPE; 0.1% SDS and then 2 x 30 min at hybridization temperature with 0.1 to 2 x SSPE according to stringency conditions; 0.1% SDS. The hybridization solution can be stored after hybridization at -20°C and can be re-used several times.

Detection and visualization of hybridized probe DNA:

The filters are washed briefly in 100 mM Tris-HCl; 150 mM NaCl; pH 7.5 (buffer 1) followed by 30 min in buffer 1; 0.5% blocking reagent and again briefly in buffer 1 only. The anti-digoxigenin-antibody-conjugate (diluted 1:5000 to a concentration of 150 mU/ml in buffer 1) is added for approximately 60 min. Again the filters are washed 2 x 15 min in buffer 1 and equilibrated in 100 mM Tris-HCl; 100 mM NaCl; 50 mM $MgCl_2$; pH 9.5 (buffer 2). The filters are now ready for the enzyme reaction. During incubation in colour solution (45µl NBT solution and 35µl X-phosphate solution in 10ml buffer 2) a colour precipitate is formed within minutes up to one day.

Discussion

The method described above is an interesting alternative to other techniques using radioactively labeled probe DNA. The procedure is simple and can be easily performed routinely in any laboratory equiped for molecular work. Sensitivity and specifity of the DIG system are almost identical with those worked out for ^{32}P labeled probe DNA, especially the detection of single-copy DNA or of very short restriction fragments is no problem. Two advantages have to be emphasized, first, the hybridization solutions can be stored for long time until they will be used again, and second, the routine work is not depending on the availability of ^{32}P-labeled triphosphates of high specific activity, what may be critical sometimes. We will present examples from our applications of probing Southern blots, plaque lifts and for the screening of bacterial colonies for recombinant plasmids.

References

Benton, W. D. & Davis, R. W., 1977. Science, 196: 180-182

Boehringer Mannheim GmbH, 1989. DIG DNA Labeling and Detection Nonradioactive. Applications manual

Feinberg, A. P. & Vogelstein, B., 1983. Anal. Biochem., 132: 6-13

Grunstein, G. & Hogness, D. S., 1975. Proc. Natl. Acad. Sci. USA, 72: 3961-3965

Southern, E. M., 1975. J. Mol. Biol., 98: 503-517

DNA Fingerprinting

Royston E. Carter
Department of Genetics
Queens Medical Centre
Nottingham NG7 2UH

Laboratory protocols used at Nottingham for producing DNA fingerprints from birds, mammals and invertebrates.

DNA EXTRACTION

High quality, high molecular weight genomic DNA is essential for successful DNA fingerprint analysis, and is conveniently prepared from avian erythrocytes. For mammalian studies DNA is usually prepared from isolated leucocytes, alternatively homogenized tissues may be used (e.g. liver).

Fifteen microlitres of avian whole blood is suspended in 600μl of isotonic buffer (1 x SET; 0.15M NaCl, 1mM EDTA, 50mM Tris, pH 8.0). The cells are lysed by the addition of 7.5μl of 25% w/v sodium dodecyl sulphate solution. Lysis causes the release of nucleases, which are inactivated by incubating with 15μl of proteinase K solution (10mg ml^{-1}) overnight at 55°C.

Proteins contaminating the DNA are denatured and removed during a series of extractions with immiscible organic solvents.

To the DNA solution is added 500μl of buffered phenol (pH 8.0), the phases are mixed for 15-30 minutes then separated by centrifugation. Proteins partition to the lower organic phase or precipitate at the solvent interface. The aqueous phase is recovered and additionally extracted either with repeated phenol or phenol/chloroform/isoamyl alcohol (24:23:1 w/v) until no further precipitation occurs at the interface. The final traces of phenol are removed by a brief chloroform/isoamyl alcohol (23:1 w/v)) extraction. The DNA is then recovered by the addition of 2 x volumes of cold (-20°C) absolute ethanol, for 30 minutes at -20°C which causes the precipiation of the DNA. This is collected by centrifugation. Traces of ethanol are removed by

NATO ASI Series, Vol. H 57
Molecular Techniques in Taxonomy
Edited by G. M. Hewitt et al.

drying *in vacuo* and the DNA is dissolved in TE buffer (10mM Tris 1mM EDTA pH 8.0) at 55°C overnight.

DNA RESTRICTION

The genomic DNA isolated is a suitable substrate for digestion with a variety of restriction enzymes. Ideally a restriction enzyme is chosen which cuts frequently in most genomic DNA but not within tandemly repeated minisatellite sequences, a variety of such "four base pair" restriction enzymes exist, e.g. Hae III, Alu I and Hinf I, although their suitability varies with different species.

An aliquot (10μl) containing an excess (> 10μg) of genomic DNA is digested to completion with 10 U of the chosen "four base pair" restriction enzyme according to the manufacturers instructions, usually overnight and in the presence of 4mM Spermidine HCl.

The extent of the digestion is monitored by a "minigel assay", a small aliquot is electrophoresed through an 0.8% agarose minigel, stained with ethidium bromide and the resulting smear is examined. All samples are quantitatively assayed fluorometrically, and are adjusted to 0.15μg μl^{-1} with 2 x B.P.B.

(10 x B.P.B.; 20% w/v Ficoll$^{(R)}$ 400, 0.2M EDTA, 0.25% w/v Bromophenol blue, 0.25% Xylene cyanol FF).

ELECTROPHORESIS

DNA fragments are separated by electrophoresis according to size by molecular sieving through an agarose gel under the influence of an applied electrical field.

A 0.7 - 1.0% w/v agarose gel is prepared by dissolving by microwave the appropriate mass of LE agarose into 375ml 1 x TAE buffer (1 x TAE, 0.04M Tris Acetate, 1mM EDTA, pH 8.0). The agarose solution is cooled to 55°C and poured into a gel-mould to set. The gel is placed into an electrophoresis tank containing 2.5l of 1 x TAE electrophoresis buffer. The samples and appropriate molecular weight markers (e.g. bacteriophage lambda DNA digested with the restriction enzyme Hind III) are heated to 65°C for ten minutes then rapidly quenched on ice to dissociate fragments which have joined by their restriction generated cohesive termini. They are micropipetted into the preformed sample wells of the gel. The samples are allowed to equilibrate with the electrophoresis buffer for ten minutes prior to commencing electrophoresis. Electrophoresis is necessarily long and slow (40 - 72 hours at 40 V depending on the species and chosen enzyme), in order to minimize "band smiling".

BLOTTING

For ease of handling the DNA fragments are transferred from the gel to a solid support matrix by capillary blotting, thus maintaining their relative positions.

Two routine methods of preparing blots are employed depending upon the membrane chosen: Southern transfer[1] must be used for nitrocellulose and can be used for all nylon varieties. Alternatively, some nylon membranes are amenable to "alkali transfer"[1].

Southern transfer

Large DNA fragments (> 10 Kb) retained within the gel are further fragmented *in situ* by brief acid hydrolosis by soaking the gel in 0.2M HCl for 10 minutes. The double stranded DNA is then separated by alkali in 1.5M NaCl, 0.5M NaOH for 35 minutes, followed by a gel neutralization in 3M NaCl, 0.5M Tris pH 8.0 for 45 minutes. The DNA fragments are then transferred to the membrane by blotting. The gel is placed on a wick in contact with a reservoir of the high ionic strength buffer 20 x SSC (20 x SSC; 3M NaCl, 0.3M sodium citrate), a membrane is then placed onto the gel surface. DNA fragments are eluted from the gel and deposited onto the membrane surface, as the 20 x SSC is absorbed into paper towels above the filter membrane.

Alkali transfer

Acid hydrolosis and denaturation are performed as above, but the neutralization step is omitted and transfer is in either 0.4M NaOH or 0.25M NaOH, 1.5M NaCl.

DNA is then fixed to the membrane by drying *in vacuo* for 2 hours at 80°C.

PREHYBRIDIZATION, HYBRIDIZATION AND WASHING

A probe capable of recognizing and binding to minisatellite DNA is prepared and is used to wash the filter, in order that it may bind to homologous sequences immobilized on the membrane surface. The required match between a probe and target can be regulated by controlling temperature and/or the ionic strength (stringency) during the hybridizing step or may occur during the post hybridization washing.

Prehybridisation

Non specific hybridization of the labelled probe to positively charged sites on the membrane surface is prevented by a prehybridization step. The membrane is washed with proteinaceous "blocking" agents at the desired stringency, e.g. 1% BLOTTO, 1 x SSC, 1.0% SDS at 65°C for several hours (typically 8 - 10). This is done either in bottles or cake boxes depending upon the number of filters to be processed.

Hybridization

In our procedures stringency is regulated during hybridization, usually a moderate stringency (1 x SSC at 65°C) is employed - it is easier to prevent hybrids forming than to dissociate them during the washing stage.

Post Hybridization Washing

Non bound probe is removed from the membranes by washing them at 65°C in several changes of wash solution (1 x SSC, 0.1% SDS).

PREPARATION OF A HYBRIDIZATION PROBE

A hybridization probe is a piece of nucleic acid which can be hybridized to specific target sequences, and which is "labelled" to allow their detection. A probe may be DNA or RNA, and can be labelled either radioactively or chromogenically. For genetic fingerprint analyses a high activity label is necessary and for this reason we routinely employ ^{32}P RNA probes.

The minisatellite region from the multilocus fingerprint probes 33.6 and 33.15[2] were sub- cloned into a transcription vector[3].

Radiolabelled ($\alpha^{32}P$) CTP is incorporated into multiple RNA copies using a commercial kit transcribing from a T_7 RNA polymerase promoter[4].

One microgram of plasmid containing the minisatellite region that has previously been linearzied distal to the polymerase promoter is labelled in a reaction containing unlabelled UTP, GTP and ATP, transcription buffer, DTT, T_7 RNA polymerase and $\alpha^{32}P$ CTP. The probe is separated from unincorporated nucleotides by spun column chromatography[1], and is added (120,000 counts ml^{-1}) after prehybridization is complete.

AUTORADIOGRAPHY

Labelled probe hybridized to target sequences on the membrane may be detected by autoradiography. β radiation emitted from ^{32}P labelled nucleotides incorporated in the probe will expose X-ray film. Intensifying screens are used for initial exposures to amplify the image, exposures without amplification offer improved resolution but require much longer exposures even when high specific activity probes are used.

After washing is complete the filters (while still damp) are wrapped in Saran wrap and exposed to pre-flashed X-ray film for 4 hours at -80°C in a cassette with two tungsten intensifying screens. This autoradiograph is developed photographically and used to gauge additional exposures with screens, if necessary, or exposures without screens (3-10 days at room temperature).

The Jeffreys' probes 33.6 and 33.15 and pSPT derivatives are the subject of patent No. GBA 2166445 and worldwide patents (pending) for commercial diagnostic use. All enquiries regarding the probes should be directed to ICI Diagnostics, Gadbrook Park, Northwich, Cheshire, UK.

REFERENCES

1. In: *Molecular Cloning - A Laboratory Manual*, 2nd Ed. Cold Spring Harbor Laboratory Press. 1989. Fritsch-Maniatis.
2. Jeffreys, A.J., Wilson, V. and Thein, S.L. 1985. Hypervariable 'minisatellite' regions in human DNA. *Nature*, 314: 67-73.
3. Carter, R.E., Wetton, J.H. and Parkin, D.T. 1989. Improved genetic fingerprinting using RNA probes. *N.A.R.* 17: 5867.
4. Little, P.F.R. and Jackson, I.J. Application of plasmids containing promoters specific for phage-encoded RNA polymerases. In: *DNA Cloning*, Vol. III. Ed. Glover, D.M. IRL Press, 1987.

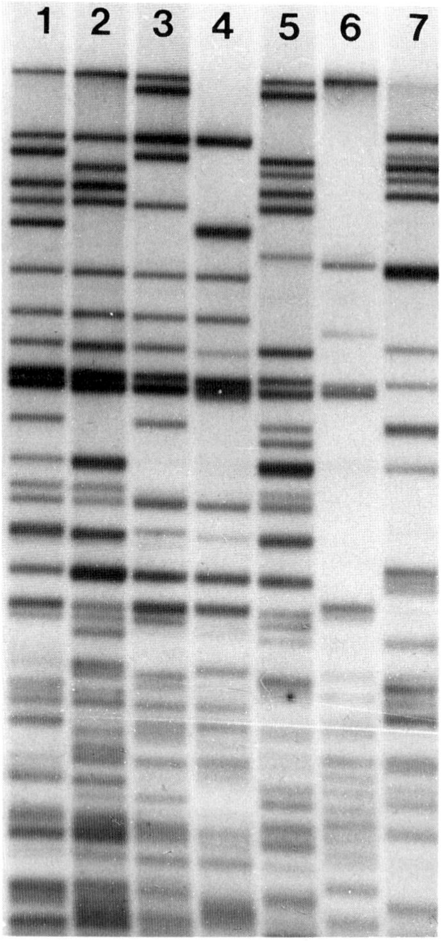

Genetic fingerprints of seven House Sparrows (*Passer domesticus*) revealed by RNA probe pSPT 19.6. Three micrograms of Hae III restricted DNA from three sibs (1-3), their parents (4 and 5) and two unrelated adults (6 and 7).

THE POLYMERASE CHAIN REACTION: DNA EXTRACTION AND AMPLIFICATION

Chris Simon[1,2], Adrian Franke[2], and Andrew Martin[2]

Ecology and Evolutionary Biology
University of Connecticut
Storrs, CT 06269

Zoology Program
University of Hawaii
Honolulu, HI 96822

The polymerase chain reaction (PCR) is a method of cloning DNA without the use of microorganisms (Saiki et al. 1985, Mullis et al. 1986). In five short years since the development of PCR, this technology has been modified for many uses (Innes et al. 1990) and has essentially revolutionized molecular biology (Guyer and Koshland 1989). PCR allows the rapid selection, isolation and amplification of DNA regions of interest from small amounts of tissue and can be used to help prepare DNA for sequencing. Because it works well for small amounts of tissue and for small pieces of DNA, PCR allows the examination of nucleotide sequences from ancient preserved specimens which have been dried, frozen, hidden in anaerobic sediments, or soaked in alcohol or formalin (Paabo 1990). The greatly increased speed of extraction, amplification, and sequencing has made nucleotide sequence data available on the large scale necessary for population biological and systematic studies. If sequencing studies turn up consistent differences in nucleotide sequence among populations or taxa, PCR can be combined with RFLP analysis to rapidly screen for these known differences. Dot blot analysis with allele specific probes or allele specific PCR primers can be used for a similar purpose (Innes et al. 1990).

PCR makes use of 18-25 bp pieces of DNA called primers which flank the region to be copied and create starting points for strand synthesis. These primers must be homologous or nearly homologous to segments of the sample mtDNA bordering the region to be cloned. If the nucleotide sequence of the sample mtDNA is not known, potential primers must be chosen based on sequences of close relatives. For taxa for which no close relatives have been sequenced, highly conserved regions of DNA can serve as "universal" primers for DNA amplification (Kocher et al. 1989, Simon et al. 1990). For

NATO ASI Series, Vol. H 57
Molecular Techniques in Taxonomy
Edited by G. M. Hewitt et al.

amplification of protein coding regions for unstudied species, degenerate primers can be used which are designed to include a mixture of possible primers all matching a known conserved amino acid sequence. The substitution of nucleotides by the non-specific base analog inosine in degenerate positions increases the universality of a primer (Linz et al. 1990). Tips for constructing PCR primers are provided in Appendix 1. References useful for designing primers are provided in Appendix 2. Useful PCR primers with varying degrees of conservation are provided in Appendix 3. A discussion of the choice of particular DNA regions appropriate for various taxonomic levels of investigation is given in part 1 of this book (Simon, this volume).

The PCR reaction takes place in three basic steps: template denaturation, primer annealing, and new strand extension (Figure 1). The reaction mixture contains template DNA, two primers each complementary to opposite strands, buffer, the four nucleotides in equal proportions, and a thermo-stable polymerase enzyme (_Taq_ polymerase). In the first step, the temperature is raised rapidly to 92-96° C and held there for 15 seconds to 1 minute. This causes the double stranded template DNA to dissociate. An initial denaturation of 3-4 minutes prior to the first cycle improves the efficiency of the reaction (Adding the _Taq_ polymerase after this initial denaturation step will prolong the life of the enzyme). For the second step, the temperature is dropped rapidly to approximately 50° C. This allows annealing of the oligonucleotide primers. The excess of primers assures that they will compete successfully with the double strands of the original template molecules for annealing. We use a range of annealing temperatures from 45 to 72° C; 50° C is a good starting point. The lower the temperature, the less specific the annealing, therefore we recommend using as high a temperature as possible to avoid mispriming. In the third, extension, step of the PCR cycle, the temperature is raised to 72-74° C for maximum efficiency of the _Taq_ polymerase. At this temperature, _Taq_ has been found to add between 35 and 100 nucleotides per second such that an extension time of one minute is sufficient to amplify up to 2 Kb (Innis and Gelfand 1990), although empirically we have found 2 minutes to work well. The cycle ends when the temperature is again raised to 92° C and the strands dissociate. The number of cycles necessary varies between 25 and 40 depending on the number of target molecules in the initial reaction.

Figure 1. Diagrammatic representation of the PCR Cycle. A) In the first cycle of PCR, dissociation at 94 degrees C is followed by annealing of the primers to the complementary strands, and finally by extension at 74 degrees C. Long products are produced from the dissociated circular mtDNA strands because extension is terminated only by a new cycle of denaturation or the Taq enzyme falling off. B) In the second cycle of a PCR reaction a second set of long products is generated for each molecule of starting template DNA. This is not shown because it is identical to the diagram in A. In addition, one set of short products is produced from the long products. Extension of the short product stops when the end of the long product is reached. C) In all subsequent cycles of PCR, one set of long products is produced from each molecule of starting template DNA (not shown) and short products are produced from long products and from existing short products. Thus, short products increase exponentially while long products accumulate linearly.

A.

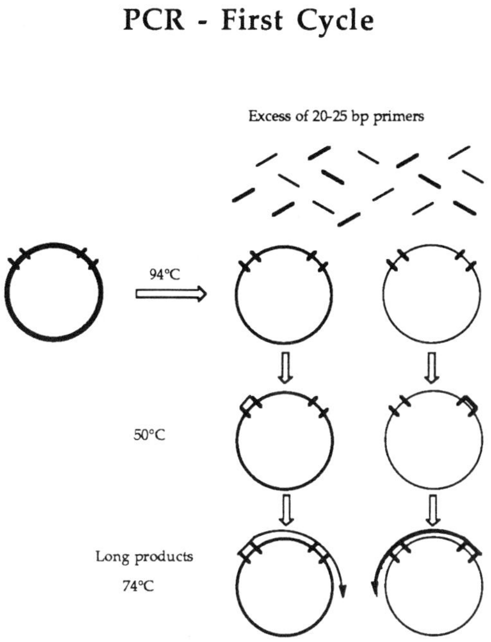

B.

PCR - Second Cycle

Short Products Produced From Long Products

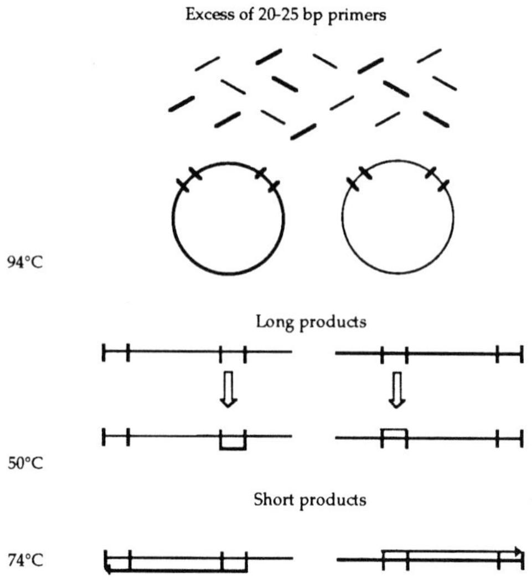

C.

Short Products of Third Cycle

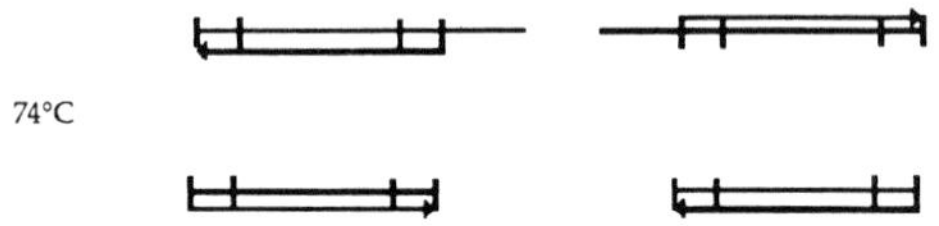

Short products: multiply exponentially

Long products: one set produced each cycle per molecule of starting template

Short Products of Fourth Cycle

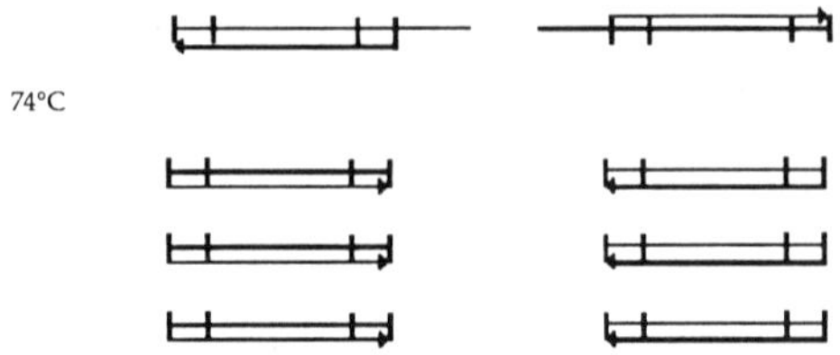

In addition to the standard PCR reaction which produces a double stranded product, PCR can be modified to produce single stranded DNA (Gyllensten and Erlich 1988). In this "asymmetric amplification", one primer, the "limiting primer" is added in 50 to 100 times less concentration. Amplification proceeds exponentially until the limiting primer is exhausted. At this point, the reaction becomes linear and only one strand is produced. Two amplification mixtures are prepared each containing a limiting amount of one of the two primers. The limiting primer can be used for sequencing or a third, internal, primer can be used.

The protocols presented below were developed for extracting and amplifying mitochondrial DNA (mtDNA) from insect tissue but they vary little from the standard protocols used in PCR. The book, "PCR Protocols: A Guide to Methods and Applications" (Innes et al. 1990) provides detailed descriptions of basic PCR methodology as well as a wide variety of modifications. The protocols provided here emphasize simplicity, efficiency, and applications to mitochondrial DNA, especially that of invertebrate taxa.

EXTRACTION PROTOCOLS

Most extraction protocols share the same basic steps: homogenization, destruction of extraneous or harmful proteins (such as nucleases), lysis of membranes containing DNA, and the separation of DNA from cell debris, digested proteins and other extraneous materials. Because PCR primers are specific, simple genomic DNA preps (containing mitochondrial and nuclear DNA) work well for mtDNA amplification. Thus, time consuming ultracentrifugation of DNA in a cesium gradient can be avoided. Below we present two protocols: A standard genomic protocol and a simplified genomic protocol for use with small amounts of tissue. In the standard protocol, we have noted where amounts of certain components can vary. Obviously, these amounts are not critical. Success of each recipe may vary depending on the sample organisms. In the second protocol, the ability of PCR to work from small amounts of tissue allows the omission of centrifugation, unpleasant phenol/chloroform extractions, and ETOH precipitation. We have just begun using this technique and have obtained excellent amplifications for periodical cicadas for the 12A & B primer set (Appendix 3).

Extraction of Genomic DNA (Periodical cicada used as an example) for Mitochondrial DNA Amplification.

1. Homogenize entire insect (trim wings, legs, excess chitin): Flight muscles and egg masses are excellent sources of mitochondrial DNA. Amount of homogenization buffer will depend on the amount of tissue used.

HOMOGENIZATION BUFFER:	10 mM NaCl	.584 g	
	50 mM Tris HCl	6.05 g	(pH 7.5 in 1M soln.)
	100 mM EDTA	37.22 g	
	250 mM Sucrose	85.57 g	
	Sterile Distilled Water	to 1 L	

2. Centrifuge 10 minutes at 3.5K rpm. Transfer supernatant to new tube discard pellet (mostly nuclei and cell debris). This and the following step increase the ratio of mitochondria to nuclei and are optional.
3. Centrifuge 20 minutes at 13K rpm. At this speed the mitochondria will pellet. Save pellet and dissolve in 700 microliters sterile distilled water.
4. Transfer to 1.5 ml eppendorf and add: 70 µl 10.0% SDS (to make 1%), and Proteinase K (to make 100 µg/ml; extraction protocols are highly variable in amount of Proteinase K from 1 µg/ml to 500 µg/ml).
5. Incubate at 60° C overnight. (Other protocols: 37° C for 10 - 20 hours; or 55 - 65° C for 2-3 hours).
6. Extract once with 500 µl Phenol (equilibrated with Tris HCl pH 8 and containing 0.1 % 8-Hydroxy quinolin), then extract 3 times with 500 µl Chloroform/Isoamyl Alcohol (24:1).
7. Pipette aqueous phase into new 1.5 ml eppendorf and ppt. DNA with 2 volumes 95% ETOH. Spin at 10K rpm for 2 minutes.
8. Wash pellet with 70% ETOH and dry
9. Resuspend in 300 µl sterile distilled water.
10. Use 5 µl in PCR reaction. Note: mtDNA content varies among species and among individuals (see "Sample Concentration," below).

Simplified Extraction Protocol for Small Amounts of Tissue Modified from Kawasaki (1990).

1. Remove three eggs from one periodical cicada (or small amount of other tissue) and place in 35 µl lysis buffer:

LYSIS BUFFER:	10 mM	Tris HCl pH 7.5-8.0
	1 mM	EDTA
	1 %	NonIdet (preferentially lyses mitochondria)
	100 µg/ml	Proteinase K

2. Add an equal volume of sterile distilled water.
3. Heat to 95° C for 3 min.
4. Use 1 µl in PCR reaction.

AMPLIFICATION PROTOCOL

10 X PCR Buffer:	670 µl	1M Tris HCl pH 8.8
	20 µl	1M $MgCl_2$
	83 µl	2M $(NH_4)_2SO_4$
	227 µl	sterile distilled water
	1000 µl	total

In a .5 µl eppendorf tube, make a 100 µl rxn using:

Volume			Final Concentration
1-5 µl	extracted DNA		unknown (see below)
10 µl	10X PCR Buffer:		67.0 mM Tris HCl 2.0 mM $MgCl_2$ 16.6 mM $(NH_4)_2SO_4$
5 µl	primer 1	10µM stock	0.5 µM
5 µl	primer 2	10µM stock	0.5 µM
10 µl	dNTP's	10mM stock	1.0 mM total or .25 mM each
.5 µl	*Taq*		1 unit

Optional:

	Bovine Serum Albumin	2 - 100	µg/ml (nuclease free)
or	Autoclaved Gelatin	100	µg/ml
	β-Mercaptoethanol	10	mM

Overlay with 2 drops mineral oil and amplify 30-35 cycles (Denature: 94° C; Anneal: 50° C; Extend: 72° C) as illustrated.

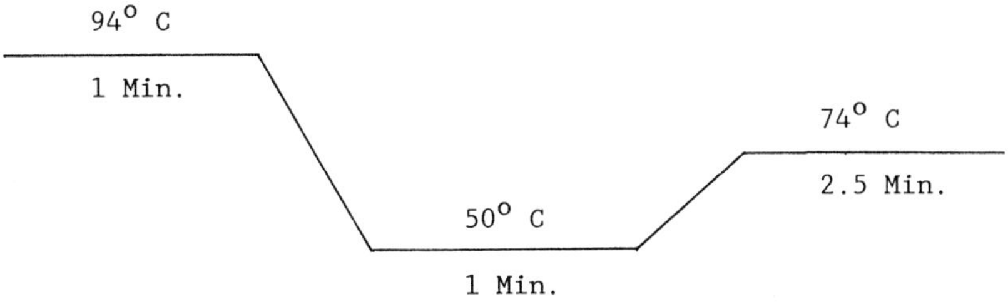

Following the reaction, run 10 µl of product with 4 µl dense dye on a minigel, with a known size standard (Phi-X HAE III). Stain with EtBr. A water control (all rxn components except template DNA) should always be run because contamination is a serious problem in PCR (see Kwok 1990 for procedures to minimize contamination). Minigels can be run with TBE or TA buffer.

		Amount	10X Concentration
10X TBE Buffer:	Tris base	215.60 g	890 mM
	Boric Acid	110.05 g	890 mM
	EDTA	14.80 g	20 mM
	Sterile dH_2O to 2.0 L		

Template Concentration. When extracting total DNA it is difficult to know the concentration of mitochondrial versus genomic DNA. Probing of each sample with a mitochondrial probe on a southern blot would provide an estimate of relative mtDNA concentration but is too time consuming for routine procedures. We have found considerable variation in the amount of total DNA from individual to individual. Total DNA concentrations can be equalized by eye by running 15 µl of each extraction on a minigel, staining with EtBr, comparing staining intensities among individual samples and diluting more concentrated samples. Although PCR is supposed to work from one or two molecules of starting DNA, often when comparing samples of the same species extracted in parallel, some samples simply will not amplify. A sample may be either too concentrated or too dilute to work. We recommend titration of sample concentration in both directions.

PCR Buffer. The PCR buffer may contain the salt $(NH_4)_2SO_4$, or KCl or no salt other than $MgCl_2$ (e.g. Pääbo 1990). Magnesium concentration effects primer annealing and enzyme activity. It may have to be re-adjusted for each primer pair for optimum amplification. Cetus recommends $MgCl_2$ titration over the range 1.5 - 4.0 mM and that the final magnesium concentration be at lest 0.5 to 1.0 mM higher than the total dNTP concentration (Oste 1989). We have found that it is useful to titrate each primer separately for the single stranded reactions.

Primer Concentration. If primer concentrations are too high, non-specific priming can occur (resulting in multiple products). High primer concentration can also increase the incidence of primer-dimers (see primer construction tips).

dNTP Concentration. The four dNTPs should be in equal concentrations. Innis and Gelfand (1990) recommend comparing nucleotide concentrations between 20 and 200 µM each dNTP. Low concentrations minimize mispriming.

Taq Concentration. *Taq* concentrations can be varied between .5 units per reaction to 2.5 units per reaction with good success. Concentrations necessary will vary with template and primers. Because *Taq* is expensive, it pays to try a titration series. High concentrations of *Taq* may result in non-specific products.

Asymmetrical amplifications. As described above, asymmetrical (single stranded) amplifications can be performed using the same protocol as for symmetrical (double stranded) reactions and varying the concentration of one of the primers to 1/100 or 1/50 that of the other. The number of cycles is increased from 35 to 40 to increase the amount of single stranded product produced during the linear phase of the reaction. Template DNA for the asymmetrical amplification can come directly from the original genomic extraction, from a double stranded PCR reaction, or from a band of double stranded product cut or melted (diluted 100X in water) from a 2-4% NuSieve agarose gel slice. If the template DNA for the asymmetrical amplification is contaminated with primers from the double stranded reaction or if the starting amount of amplified double stranded DNA is high enough, it may not be necessary to add limiting primer in order to obtain sufficient single stranded DNA for sequencing. We have tried all of these techniques and found that the relative success of each varies across species and primer pairs. It is not uncommon that in single stranded reactions, one primer will amplify better than its opposing partner. If this is the case, there are several options: sequence double stranded rather than single stranded template; kinase one primer and then digest one strand of a double stranded PCR reaction to produce single stranded DNA (see Kreitman, this volume); or add a third primer, external to the problem primer, for single stranded amplification. Amplify double stranded with the three primers. The problem strand will thus increase more than exponentially. Use 1 µl of this solution for single stranded reactions with the each of the two original primers, respectively (use no limiting primer). Sequence the single strands with the appropriate original primers. We have found that the problem primer as well as its opposing partner will sequence well with this technique.

To Prepare DNA for Sequencing: To purify DNA for sequencing, the reaction mixture can be concentrated (to approx. 30-40 µl) using a Centricon 30 microconcentrator[1] (Amicon, Danvers, MA) or an Ultrafree Microcentrator 30,000 (Millipore, Bedford, MA) which remove salts, excess nucleotides and primers. Both methods retain more DNA than ethanol precipitation and the latter is less expensive, faster to use, and works well. In our laboratory sequencing is performed using approximately 5-7 µl of the purified, concentrated single stranded product in a dideoxy sequencing reaction using the Sequenase enzyme system (U.S. Biochemical Co, Cleveland, OH).

ACKNOWLEDGEMENTS

Ideas presented in this paper benefited from discussions with and/or contributions from the following people: R. Cann, H. Croom, R. DeSalle, R. Harrison, D. Irwin, B. Kessing, T. Kocher, C. McIntosh, E. Metz, O. McMillan, C. Orego, S. Paabo, S. Palumbi, A. Phillips, A. Sidow, M. Stoneking, K. Thomas, and A.C. Wilson. This work was supported by NSF BSR 88-22710 to CS.

REFERENCES

Guyer RL, Koshland Jr. DE (1990) The molecule of the year. Science 246:1543-1544

Gyllensten UB, Erlich HA (1988) Generation of single-stranded DNA by the polymerase chain reaction ad its application to direct sequencing of the HLA-DQA locus. Proc. Natl. Acad. Sci. USA 85:7652-7656

Innis MA, Gelfand DH, Sninsky JJ, White TJ (1990) PCR Protocols. A guide to methods and applications. Academic Press, NY

Innis MA, Gelfand DH (1990) Optimization of PCRs. In: Innis MA, Gelfand DH, Sninsky JJ, White TJ (eds) PCR Protocols. Academic Press, NY, p 3

Kawasaki ES (1990) Sample preparation from blood, cells and other fluids. In: Innis MA, Gelfand DH, Sninsky JJ, White TJ (eds) PCR Protocols. Academic Press, NY, p 146

Kocher TD, Thomas WK, Meyer A, Edwards SV, Pääbo S, Villablanca FX, Wilson AC (1989) Dynamics of mitochondrial DNA sequence evolution in animals: Amplification and sequencing with conserved primers. Proc. Natnl. Acad. Sci. USA 86:6196-6200

---

1 Amicon now recommends their new 100,00 MW filter which cleans up PCR products faster and more efficiently and retains pieces up to 125 bp with 90% efficiency. Millipore still recommends their 30,000 MW filter for PCR clean up stating that their filter is "not as tight."

Kwok S (1990) Procedures to minimize PCR-product carry-over. In: Innis MA, Gelfand DH, Sninsky JJ, White TJ (eds) PCR Protocols. Academic Press, NY, p 142

Linz U, Rubsamen-Waigmann H, Roesch R, and Seliger H. (1990) Inosine-substituted primers prevent false-negative PCR results due to 3' mismatches. Amplifications 4:14-15

Mullis K, Faloona F, Scharf S, Saiki RK, Horn G, Erlich H (1986) Specific enzymatic amplification of DNA in vitro: the polymerase chain reaction. Cold Spring Harbor Symposium on Quantitative Biology. 51: 263-273

Oste C (1989) Optimization of magnesium concentration in the PCR reaction. Amplifications 1:10-11

Pääbo S (1990) Amplifying ancient DNA. In: Innis MA, Gelfand DH, Sninsky JJ, White TJ (eds) PCR Protocols. Academic Press, NY, p 159

Saiki RK, Scharf S, Faloona F, Mullis KB, Horn GT, Erlich HA, Arnheim N (1985) Enzymatic amplification of B-globin genomic sequences and restriction site analysis for diagnosis of sickle cell anemia. Science 230: 1350-1354

Simon C, Pääbo S, Kocher T, Wilson AC (1990) Evolution of mitochondrial ribosomal RNA in insects as shown by the polymersase chain reaction. In: Clegg M, O'Brien S (eds) Molecular Evolution. UCLA Symposium. Alan R. Liss, Inc., NY, p 235

Appendix 1. Tips on primer construction. The first version of these primer tips was compiled by CS during a research visit to the laboratory of A.C. Wilson in July of 1988 and includes input from D. Irwin, T. Kocher, C. Orego, S. Pääbo, M. Stoneking, and K. Thomas. This updated version incorporates information from our subsequent experiences plus insights compiled in Innis et al. (1990).

1. Oligonucleotide primers should be approximately 18-25 bp long. Above 25 bp there is not much increase in primer annealing specificity.

2. To choose primers for unsequenced organisms, try regions which are known to be highly conserved. Remember to take into account possible gene rearrangements and differences in direction of transcription when choosing opposing primers. Use references in Appendix 2 for these purposes.

3. A convenient size DNA segment for PCR amplifications and sequencing is 400-500 bp. Amplifications of up to 10 Kb have been performed successfully and produce enough product to be visible on a EtBr stained gel. Pieces of 20 kb have been amplified and visualized by probing (Dan Clutter, pers. comm.). Once a reference sequence is obtained for the species under investigation, A useful strategy is to choose universal primers 1000 bp apart for the initial amplification and then to sequence with internal primers which are more specific to your particular organism.

4. The 3' end of the primer is most critical. The first three bases should be complementary. It has been shown that a match at the 3' end of a primer is essential for the success of PCR because Taq polymerase lacks a 3'-5' exonuclease activity (Linz et al. 1990). For this reason, it is best not to end a primer on the third position of a codon. A "T" at the 3' end of a primer has been shown to pair with any base (Kwok et al. 1990).

5. The 5' end of a primer is less critical. Some people add a restriction site to the 5' end (for later cloning or expression vector analysis). This site, of course, does not anneal in the first round of polymerization. Thereafter it is copied. In fact, any mismatches in the primer will be copied such that in later runs primers will match the newly created attachment sites perfectly. As a result, annealing temperature can be increased after about 5 cycles of PCR to increase priming specificity.

6. The base composition of a primer should be roughly equivalent for all four bases. G's and C's form stronger bonds that A's and T's. The annealing temperature should be as high as possible to prevent non-specific priming. For standard conditions annealing temperature should be 10 degrees C less than the melting temperature.

$$T_{melting} \text{ (for 20bp or less)} = (4 \times (G+C)) + (2 \times (A+T)),$$

where G+C = the total number of G's and C's in the primer, etc. Computer programs are available for calculating bonding energy of oligonucleotide primers (see Kreitman, this volume).

Appendix 1 (continued)

7. Runs of nucleotides should be avoided because they can pair with runs of nucleotides outside of the target sequence or complementary runs within the same primer. Avoid palindromic base sequences which could cause the primer to fold and pair with itself. Secondary structure in the template can also interfere with priming. Substitution of 7-deaza-2'-deoxyGTP for dGTP can help solve this problem (Innis and Gelfand 1990).

8. Primer pairs must not have complementary 3' ends. Primers with complementary ends can pair with each other and amplify to form what is known as primer artifacts or primer dimers (in multiples of the size of the original primers). During amplification, primer dimers compete for nucleotides, primers and enzymes and if they form early, can take over a reaction. Watson (1989) reports that, "primer artifacts may form even in the absence of obvious complementarity." He suggests that the Taq enzyme may hold the primers together long enough for extension to take place. Higher annealing temperatures can reduce primer dimers.

9. Primers can be made less specific to match variable positions in the target DNA by substituting inosine (which will pair with any base) or by making degenerate primers--a mixture of primers derived from an amino acid sequence containing all possible combinations of bases at degenerate codon positions (Compton 1990). Degenerate primers should not be degenerate on the 3' end (use the single-codon amimo acids methionine and tryptophan for this purpose).

10. If at first you don't succeed, repeat the same amplification. If the reaction fails, lower the annealing temperature. If still no product appears, dilute the template DNA. If this fails, try a different primer combination. Two oligonucleotides that look equally good sometimes are not.

Additional References:

Compton T (1990) Degenerate primers for DNA amplification. In: Innis MA, Gelfand DH, Sninsky JJ, White TJ (eds) PCR Protocols: A Guide to Methods and Applications. Academic Press, Inc. NY

Kwok S, Kellog DE, McKinney N, Spasic D, Goda L, Levenson C, Spinsky JJ (1990) Effects of primer template mismatches on the polymerase chain reaction: Human Immunodeficiency Virus Type I model studies. Nucleic Acids Research 18:999-1005

Watson R (1989) The formation of primer artifacts in polymerase chain reactions. Amplifications 2:5-6

Appendix 2. References useful for constructing PCR primers for mitochondrial DNA. Comparative gene order, direction of transcription, and level of conservation of nucleotides across species must be taken into account when designing primers.

Complete Mitochondrial DNA Sequences

HUMAN — Anderson S, Bankier AT, Barrell BG, de Bruijn MHL, Coulson AR, Drouin J, Eperon IC, Nierlich DP, Roe BP, Sanger F, Schreier PH, Smith AJH, Staden R and Young IG (1981) Sequence and organization of the human mitochondrial genome. Nature 290:457-465

COW — Anderson S, De Bruijn MHL, Coulson AR, Eperon IC, Sanger F and Young IG (1982) Complete sequence of bovine mitochondrial DNA. Conserved features of the mammalian mitochondrial genome. J Mol Biol 156:683-717

RAT — Gadaleta G, Pepe G, DeCandia G, Quagliariello C, Sbisa E, and Saccone C (1989) The complete nucleotide sequence of the *Rattus norvegicus* mitochondrial genome: cryptic signals revealed by comparative analysis between vertebrates. J Mol Evol 28:497-516

MOUSE — Bibb MJ, Van Etten RA, Wright CT, Walberg MW, and Clayton DA (1981) Sequence and gene organization of mouse mitochondrial DNA. Cell 26:167-180

CHICKEN — Desjardins P, and Morais R (1990) Sequence and gene organization of the chicken mitochondrial genome. J Mol Biol 212:599-634

FROG — Roe BA, Ma DP, Wilson RK, and Wong JFH (1985) The complete nucleotide sequence of the *Xenopus laevis* mitochondrial genome J Biol Chem 260(17):9759-9774

URCHIN — Jacobs HT, Elliott DJ, Math VB, and Farquharson A (1988) Nucleotide sequence and gene organization of sea urchin mitochondrial DNA. J Mol Biol 202:185-217

URCHIN — Cantatore P, Roberti M, Rainaldi G, Gadaleta MN, and Saccone C (1989) The complete nucleotide sequence, gene organization, and genetic code of the mitochondrial genome of *Paracentrotus lividus*. J Biol Chem 264:10965-10975

DROSOPHILA — Clary, DO and Wolstenholme DR (1985) The mitochondrial DNA molecule of *Drosophila yakuba*: Nucleotide sequence, gene organization, and genetic code. J Mol Evol 22:252-271

NEMATODE — Wolstenholme, et al. in prep. Gene order for *Ascaris suum* given in: Wolstenholme DR, MacFarlane JL, Okimoto R, Clary DO, Wahleithner JA (1987) Bizarre tRNAs inferred from DNA sequences of mitochondrial genomes of nematode worms. Proc. Natl. Acad. Sci. USA 84:1324-1328

Appendix 2 (cont.)

Compilations

srDNA Neefs J-M, Van de Peer Y, Hendriks L, and DeWachter R (1990) Compilation of small ribosomal subunit RNA sequences. Nucl Acids Res 18 supplement:2237-2318

lrDNA Gutell RR, Schnare MN, and Gray MW (1990) A compilation of large subunit (23S-like) ribosomal RNA sequences presented in a secondary structure format. Nucl Acids Res 18 supplement:2319-2330

tRNA Sprinzl M, Moll J, Meissner F, and Hartman T. (1987) Compilation of tRNA sequences and sequences of tRNA genes. Nucl Acids Res 15 supplement:r53-r188

Partial mtDNA Sequences for Invertebrate Taxa

CICADA: Simon C, Pääbo S, Kocher T, and Wilson AC (1990) Evolution of the mitochondrial ribosomal RNA in insects as shown by the polymerase chain reaction. pp 235-244 In M Clegg & S O'Brien (eds.), Molecular Evolution. UCLA Symposia on Molecular and Cellular Biology, New Series, Vol. 122. Alan R. Liss, Inc., NY

LOCUST: Haucke H-R and Gellissen G (1988) Different mitochondrial gene orders among insects: exchanged tRNA gene positions in the COII/COIII region between an orthopteran and a dipteran species. Curr Genet 14:471-476

McCracken A, Uhlenbusch I, Gellissen G (1987) Structure of the cloned Locusta migratoria mitochondrial genome: restriction mapping and sequence of its ND-1 gene. Curr Genet 11:625-630

Uhlenbusch I, McCracken A, and Gellissen G (1987) The gene for the large (16S) ribosomal RNA from the Locusta migratoria mitochondrial genome. Curr Genet 11:631-638

CRICKET: Rand DM and Harrison RG (1989) Molecular population genetics of mitochondrial DNA size variation in crickets. Genetics 121:551-569

BEE: Crozier RH, Crozier YC, and Mackinlay AG (1989) The CO-I and CO-II region of honeybee mitochondrial DNA: Evidence for variation in insect mitochondrial evolutionary rates. Mol Biol Evol 6:399-411

Vlasak I, Burgschwaiger S, Kreil G (1987) Nucleotide sequence of the large ribosomal RNA of honeybee mitochondria. Nucl Acids Res 15:2388

Appendix 2 (cont.)

MOSQUITO: HsuChen C-C, Dubin DT (1984) A cluster of four transfer RNA genes in mosquito mitochondrial DNA. Biochem Int 8:385-391

HsuChen C-C, Kotin RM, Dubin DT (1984) Sequences of the coding and flanking regions of the large ribosomal subunit RNA gene of mosquito mitochondria. Nucl Acids Res 12:7771-7785

Dublin DT, HsuChen C-C, Tillotson (1986) Mosquito mitochondrial transfer RNAs for valine, glycine, and glutamate: RNA and gene sequences and vicinal genome organization. Curr Genet 10:701-707

DROSOPHILA: Clary DO, Wolstenholme DR (1987) Drosophila mitochondrial DNA: conserved sequence in the A+T rich region and supporting evidence for a secondary structure model of the small ribosomal RNA. J Mol Evol 25:116-125

DeSalle R, Freedman T, Prager EM, Wilson AC (1987) Tempo and mode of sequence evolution in mitochondrial DNA of Hawaiian Drosophila. Proc Natl Acad Sci USA 83:6902-6906

Garesse R (1988) Drosophila melanogaster mitochondrial DNA: Gene organization and evolutionary considerations. Genetics 118:649-663 (extends known sequence to 90% of coding regions)

Satta Y, Ishiwa I, Chigusa SI (1987) Analysis of nucleotide substitutions of mitochondrial DNAs in Drosophila melanogaster and its sibling species. Mol Biol Evol 4:638-650

URCHINS: Thomas WK, Maa J, and Wilson AC (1989) Shifting constraints on tRNA genes during mitochondrial DNA evolution in animals New Biologist 1:93-100

STARFISH: Jacobs HT, Asakawa S, Araki T, Miura K-I, Smith MJ, Watanabe K (1989) Conserved tRNA gene cluster in starfish mitochondrial DNA. Current Genetics 15:193-206

Smith MJ, Banfield DK, Doteval K, Gorski S, Kowbel DJ (1989) Gene arrangement in sea star mitochondrial DNA demonstrates a major inversion event during echinoderm evolution. Gene 76:181-185

FLATWORM: Garey JR and Wolstenholme DR (1989) Platyhelminth mitochondrial DNA: Evidence for early evolutionary origin of a $tRNA^{ser}$ AGN that contains a dihydrouridine arm replacement loop, and of serine-specifying AGA and AGG codons. J Mol Evol 28:374-387

Appendix 3. Mitochondrial PCR Primers. The following primers were designed for use with mtDNA. Many were designed to be universal. This compilation was updated from Kessing B, Croom H, Martin A, McIntosh C, McMillan WO, Palumbi S. 1989. The Simple Fools Guide to PCR. Version 1.0. Available from the authors.

Primers are named for the gene in which they are located followed by a letter designation indicating the order in which they were created. All primers are written in the 5' to 3' direction. Codon spacing is used when appropriate. Arrows following primer designation indicate direction of priming (opposing primers are on opposites strands). Direction of priming will be consistent across genes only within the taxa for which the primers were designed (e.g. Insects, Echinoderms, or Vertebrates). If opposing primers are located in different genes and are used for a taxon other that that for which they were designed, direction of priming should be checked by locating the primers in the most closely related sequenced species. Primers are compared to the human, mouse, cow, frog (<u>Xenopus laevis</u>), urchin (<u>Strongylocentrotus purpuratus</u>), fly (<u>Drosophila yakuba</u>) and miscellaneous other mtDNA sequences with their 5'-end positions on the light strand (with reference to humans) numbered as in the references contained in Appendix 2 (unless unnumbered in publication). Approximate sizes of pieces to be amplified can be obtained by subtracting location numbers of opposing primers. The figure at the end of this appendix illustrates gene order and direction of transcription in various taxa.

Primers ending in "r" are the reversed complements of existing primers which someone has tried and found to work (sometimes modified slightly to avoid any lack of conservation which may have been present at the 5' end of its reverse analog). All primers could theoretically be reversed.

<u>12S RNA Gene Primers</u> (Summary figure with locations follows 12S primers)

12Sai (25mer) ➤	5'- AAACTAGGATTAGATACCCTATTAT -3'	Position
Fly	....	14588
Cicada	....	Simon & Franke, unpubl.
Cricket (<u>Gryllus</u>)	....	Rand & Harrison, unpubl.
Human	.....G.............C.C...	1067
Mouse, Rat	.....G.............C.C...	485, 486
Cow	.....G.............C.C...	843
Frog	.....G.............C.C...	2485
Urchin	....C...............G....	491

12Sair (20mer) ➤	5'- AGGGTATCTAATCCTAGTTT -3'	Position
Cicada, Drosoph.	....	as above

12Sbi (20mer) ➤	5'- AAGAGCGACGGGCGATGTGT -3'	Position
Fly	....	14214
Human	G..G.T......G.......	1478
Mouse	G..G.T......G.......	901
Cow	G..G.T......G.......	1262
Frog	G..G.T......G.......	2897
Urchin	G....T..............	855

Comments: 12Sai, 12Sair and 12Sbi were made for cicadas (insects) based on comparisons of Drosophila yakuba and Gryllus. The human versions of these primers are one set of the "universal" primers of Kocher et al. 1989. 12Sa has an 11 base pair core (in two pieces) which is conserved across eukaryotic, bacterial, plastid, and mitochondrial small ribosomal RNA genes in 103 out of 106 species examined. The 12Sb primer has a 14bp 3' end which is conserved across 96 of the same 106 species, with the exception of the A in the sixth position of D. yakuba (Simon et al. 1990) and S. purpuratus (Jacobs et al. 1988) The insect primer set 12Sai & 12Sbi will amplify vertebrate DNA and the vertebrate primer set 12Sa & b will amplify insect DNA, despite the mismatches near the 3' end in the 12Sa primers. Because of the 14 bp match of the 3' end of the 12Sb primer between nuclear and mitochondrial DNA, we had some concern that nuclear DNA might compete for this oligonucleotide (although we have obtained excellent sequences from this primer set, it does not work 100% of the time for cicadas). To avoid this potential problem, we designed a new primer, 12Sfi which was moved 8 bases upstream (3' end matches cicada):

			Position
12Sfi (20mer)◄	5'- CGGGCGATGTGTACATAATT -3'	Fly	14221
12Sbi (20mer)◄	5'- AAGAGCGACGGGCGATGTGT -3'	Fly	14214

Comments: 12Sfi works well for cicada, cricket, & planthopper.

12Sc (20mer)►	5'- AAGGTGGATTTGGTAGTAAA -3'	Position
Fly	....	14275
Cicada	..AC........AA......	Simon et al. 1990
Human	...........A.C......	1416
Mouse	....A......A........	839
Cow	....A......A.C......	1200
Frog	....C......A.C......	2834
Urchin	Does not exist!	----

Comments: Wilson lab primer modified to match Drosophila.

12Se (19mer)►	5'- ATTCAAAGAATTTGGCGGT -3'	Position
Cicada	....	Simon et al. 1990
Fly	.C.T...A...........	14521
Human	.C......G.CC.......	1154
Mouse	.C......G.C........	581
Cow	.C......G.C........	938
Frog	.CC.....G.C........	2573
Urchin	........G..........	547

Comments: The 12se primer is located at the beginning of Domain III of the 12S gene and works well for cicada.

12Sgi (18mer)►	5'- AAGTTTTATTTTGGCTTA -3'	Position
Fly	....	14939
Human	.G....GG.CC.A..C.T	651
Cow	.G....GG.CC.A..C.T	434
Mouse	.G....GG.CC....C.T	72
Rat	.G....GG.CC....C.T	71
Frog	.G....GC.CC.A..C.T	2205
Urchin	GGTCC.AGTCCCAATC.T	82

Comments: 12Sgi primer was made for cicadas based on a plant rRNA sequencing primer of Elizabeth Zimmer's. It is located only a few bases away from the beginning of Domain I of the small ribosomal subunit RNA gene. It is not universal but a vertebrate primer could be made at this site which would work for a wide variety of vertebrates.

12Sg could be used in combination with 12Sb to amplify most of the 12s gene (about 750 bp), then the other 12S primers could be used as internal sequencing primers operating from this one amplification product. The diagram on the following page shows the location of all 12S primers on the 13-year cicada sequence (Simon & Franke, unpubl.)

12Sh (21mer)►	5'- GACAAAATTCGTGCCAGCAGT -3'	Position
Cicada	....	Simon & Franke, unpubl.
Drosophila yakuba	...C.....G...........	14756
Drosophila virilis	...C.....G.........A.	
Human	.GTC..T...........CAC	877
Chimpanzee	.GTC..T.......T...CAC	Hixson & Brown, 1976
Gorilla	.GTC..T...........CAC	" " " "
Mouse, Rat	.GT...T...........CAC	296, 295
Cow	.GT...TC..........CAC	656
Frog	.GTC..TC......-...CGC	2294

Comments: 12Sh primer was made for cicadas. Removing three bases from the 3' end would make it "universal."

12Sj (21mer)◄	5'- TACAAAACAGATTCCTCTG -3'	Position
Fly	..........G........	14490
Cicada -Okanagana	..T.G..............	Simon, Paabo, McIntosh, unpubl.
-Dicerop.	..T.G..............	" " " "
-Magicicada	....	Simon et al. 1990
Human	....G.....GC......A	1195
Chimpanzee	....G.....GC......A	Hixson & Brown, 1976
Gorilla	....G.....GC......A	" " " "
Mouse, Rat	..T.G.....GC......A	614, 614
Cow	..T.G.....GC......A	971
Frog	....G.....GC......A	2606

Comments: The 12Sh primer was made for cicadas. This primer could be made more versatile by omitting the 3' base.

12S REGION IN MAGICICADA TREDECIM SHOWING PRIMERS

```
        12SG►
AAGTTTTATTTTGGCTTAAA~13 BASES AAAATTTTAATTATATTACATGCATAACTT 51


TAAATTGGTTGAAAATTATCTGAATGATAATTATCCAGAGCGCAGTATTAAGTATTATTA 111

                                                    12SH►
ATTATTAATTTTAGAGGTAATAAATTTATTTATAGAGGACAAAATTCGTGCAGCAGTTGC 171


GTTATACGATTTTCTAAATTTAACTATGTTAGTTTCAGTTAAAAAGTGTGTTGATATTCA 231


ATTTTTTTAAATTTTGGTGGAATAAAATATAAATATGTGTTTAATTTTATGTCTGAGAAA 291

                 12S A►
CTTTTATATAAAACTAGGATTAGATACCCTATTATTGAGAGTGTAAATAAAATAACTAGA 351
          TTTGATCCTAATCTATGGGA
                ◄12S Ar
                                  12S E►
TTATTAATAGTTATGATCTTTAAATTCAAAGAATTTGGCGGTAATTTATCTAATCAGAGG 411
                                                      GTCTCC

AATCTGTTTTGTAATTGATAATCCACGATAGATTCTATTTTAAATAAATTTGTATACCTC 471
TTAGACAAAACAT
◄12S J

TGTCAAGAATGTTTTATCAGAATAATTTTCATTTGTTTTATTAATAAAATGTCAGGTCAA 531


GGTGCAGTTAATTTTAAAGAAAAAATGGATTACATTATTGTAAAAAATGAATTGTTTTCT 591

                   12S C►
ATAATGAAAATCATGAAACTGGATTTGAAAGTAAATTTCATTAAATATGTGTTTTTGAAT 651


TTAGGTTCTAAATT                                                665
        ◄TTAATACATGTGTAGCGGGC  12S F
               ◄TGTGTAGCGGGCAGCGAGAA  12S B
```

<u>16s RNA Primers</u>

16sa (20mer) ◄	5'- ATGTTTTTGTTAAACAGGCG -3'	Position
Fly	....	13398
Locust	....	690
Human	.........G..........	2491
Mouse	.........G..........	1927
Cow	.........G..........	2306
Frog	.........G..........	3976
Urchin	.........G..........	5093

16sar (20mer) ►	5'- CGCCTGTTTAACAAAAACAT -3'	Position
Fly	....	13398
Locust	....	690
Human	..........C.........	2491
Mouse	..........C.........	1948
Cow	..........C.........	2306
Frog	..........C.........	3976
Urchin	..........C.........	5093

16sb (22mer) ►	5'- ACGTGATCTGAGTTCAGACCGG -3'	Position
Human	....	3058
Mouse	....	2501
Cow	....	2852
Frog	....	4573
Urchin	....	5662
Fly	..A.............A.....	12888
Locust	..A...................	163

16sbr (22mer) ◄	5'- CCGGTCTGAACTCAGATCACGT -3'	Position
Human	....	3058
Mouse	....	2501
Cow	....	2852
Frog	....	4573
Urchin	....	5662
Fly	.....T.............T..	12888
Locust	...................A..	163

<u>Comments</u>: The 16Sa primer was developed for cicadas from comparisons of crickets and <u>Drosophila</u>. The 16Sb primer is modified from the Wilson lab 16S1 primer. The 16Sar and 16Sbr primers face one another and can be used to amplify a 500-650 base fragment. 16Sa can be used with 12Sa in insects to amplify amplify approximately 1200 bp. These 16S primers work for urchins, vertebrates, insects, corals, gastropods and probably many other taxa as suggested by the extreme conservation illustrated above. In fact, there are many conserved sites in this gene which would make good universal primers and many labs have taken advantage of this (e.g. within 100 base pairs downstream of our 16Sa primer, the Wilson lab has their 16S2 primer and the Templeton lab has their 16SMid primer). Uhlenbush et al. (1987) illustrate these conserved regions in relation 16S rRNA structure.

Cytochrome oxidase I Primers

COla (21mer) ◄	5'- AGT ATA AGC GTC TGG GTA GTC-3'	Position
Human	G.. G.. T.. A.. G..	7227
Mouse	G.. G.. ... A..	6651
Cow	T.. G.. T.. A..	7010
Frog	T..	8720
Urchin	T.. ... G.. ... ... A.. ...	7108
Fly	... G.. ... A.. A.. ... A..	2791

COlc (22mer) ►	5'- TC GTC TGA TCC GTC TTT GTC AC-3'	Position
Human	.. .. C.A A.. ..	6454
Mouse	.T ... ...A C.. A.T ..	5878
Cow	.. ..A ... .A A.A A.T ..	6237
Frog	.. ..T ... ..A ..A ..A A.. ..	7947
Urchin	.. ..	6335
Fly	.T ..A ... ..A ..A G.. A.T ..	2018

COld (27mer) ◄	5'- GAA CAT GAT GAA GAA GTG CAC CTT CCC -3'	Position
Human	... T.. ... AGT ... A.. G.T T.. GG.	7258
Mouse	... ATA ... .GA ... ... GGA T.. TG.	6582
Cow	A.. T.. A.. TGC ... ... G.T T.. GG.	6940
Frog	A.. T.. T.C TCC A.. A.. G.T T.. T.G	8650
Urchin	... ... T.. ... ... A.. ... T..	7039
Fly	T.. ACA T.A T.. T.. A.T G.. T.. TTA	2723

Comments: The COld primer was designed for Echinometrid sea urchins, and works well only within this group.

COle (23mer) ◄	5'- CCA GAG ATT AGA GGG AAT CAG TG -3'	Position
Human	..T ... .A. ..G ..A	7110
Mouse	G.T ..A .A. .AT	6533
Cow	..T ... .A. ..T ... .A ..	6892
Frog	... .TA .A. .AC	8602
Urchin	... ... .AG ..G ..A ..C	6992
Fly	... .TA .A. .AT ... T..	2672

COlf (20mer) ►	5'- CCT GCA GGA GGA GGA GAY CC -3'	Position
Human	..C ..C ... ...C ..	6569
Mouse	..C ..T ... .G ..C ..	5990
Cow	..G ... ...C ..	6431
Frog	... ..C ... .T ..C ..	8061
Urchin	... ...G ... ..T ..	6451
Fly	..A .T ... ... .T. ..	2131

where Y = C or T

<u>Comments</u>: C01e and C01f make superb dsDNA amplification. ssDNA amplifications often smear for both primers. Sequencing is difficult with the C01f primer, perhaps because of its small size and degeneracy. Successful amplifications have been obtained for sharks, lamprey, fish, sea urchins (C01f only) and corals. Note that **C01f is a degenerate primer.**

<u>Cytochrome oxidase II Primers</u>

C02a (23mer) ➤	5'- GGG GCT AAC CAT AGA TTC ATG CC -3'	Position
Human	..A ..A ... ..C ..T	8189
Mouse	..A T.. ... .C ..T	7713
Cow	... T.A ... ..C ..T ... ..A ..	7974
Frog	..A ..A ... ..C ..C ..T ..A ..	9709
Urchin	... ...	8312
Fly	... .T ... ..T ..T	3682

<u>Comments</u>: This primer was based on the urchin sequence. It is in a region of high amino acid conservation, but is not very useful for anything but sea urchins.

<u>Cytochrome oxidase III Primers</u>

Co3a (20mer) ➤	5'- TTATTTATTGCATCAGAAGT -3'	Position
Locust	....	2038
Fly	.........TT.........	4995
Human	C.......A.C........	9459

Co3b (20mer) ➤	5'- TCAACAAAGTGTCAGTATCA -3'	Position
Locust	....	2503
Fly	........A...........	5460
Human	..T.....A..C........	9924

<u>Comments</u>: Co3a & b primers were made for cicadas using sequence information from Haucke and Gellissen (1988). These primers face each other. They should work well for most insects. Many more COIII primer sites can be found in this same reference as well as primers for COII and ATPase 6 & 8.

<u>ATPase 6 Primers</u>

ATP6 (22mer) ➤	5'- G TGC GCT TGG TGT TCC CTG TGG -3'	Position
Human	. G.G TG. A.. ... G.. T.. ...	8936
Mouse	. ..A AA. ... A.. ... T.. ...	8333
Cow	. ..G AG .G. ... ... .T.. ...	8698
Frog	A ..G TG. ... ... T.C A..	9709
Urchin	. ...	9039
Fly	A AAT TGC A.. ... A.. T.. A..	4478

<u>Comments</u>: Pretty poorly conserved at the 5' end, this primer works well only in Strongylocentrotid sea urchins, where it does a great job.

<u>NADH dehydrogenase Primers</u>

```
ND4c (18mer) ➤     5'- TAC TCC CTA TAC ATA TTT -3'        Position
   Human               ... ... ..C ... ... ...            11975
   Mouse               ... ..A A.. ... ... A..            11187
   Cow                 ..T ..T ... ... ..G C.A            11748
   Frog                ... ..T ..C ... ... ..C            13488
   Urchin              ... ... ... ATG ... ...            11888
```

```
ND5a (21mer) ◄     5'- GAA TTC TAT GAT CGA TCA TGT -3'    Position
   Human               ... ... ... ... G.. C.. ...        12650
   Mouse               ..G ..G ... A.. T.. ... ...        12055
   Cow                 ... ... ... A.. A.. ... ...        12420
   Frog                A.. ... A.. ... A.. ... ...        14170
   Urchin              ... C.. C.. ... ... ... ...        12486
```

<u>Comments</u>: NADH primers are not really good primers. Amplifications are inconsistent and should be performed at annealing temperatures of 45°C or lower. However, sequence has been obtained for approximately 250 bp from the ND4 primer. Successful amplifications have been obtained for fish.

<u>Cytochrome b Primers</u>

```
Cyb1 (26mer) ➤     5'- CCA TCC AAC ATC TCA GCA TGA TGA AA -3'   Position
   Human               ... ... ... ... ..C ... ... ... ..       14817
   Mouse               ... ... ... ..T ... T.. ... ... ..       14208
   Cow                 ... ..A ... ..T ... T.. ... ... ..       15753
   Frog                ... ..A ... ..T ... T.. .T. ... ..       16321
   Urchin              ..C ... ... C.T ..C ATT ..G ... ..       14581
```

```
Cyb2 (24mer) ◄     5'- CCC TCA GAA TGA TAT TTG TCC TCA  -3'     Position
   Human               ... ... ... ... ... ... ... ...          15175
   Mouse               A.. ... ... ... ... ... ... ...          14565
   Cow                 T.. ... ... ... ... ... ... ...          15753
   Frog                ... ... A.. A.. ... ... ... ...          16677
   Urchin              AG. ... A.. G.. C.. ... C.. C..          14937
```

<u>Comments</u>: These primers seem to work fairly well with most vertebrates. They are based on Kocher et al. (1989).

<u>tRNA Primers</u>: As illustrated in the figure at the end of this appendix, tRNA genes are rearranged considerably among major groups of animals (Jacobs et al. 1988, Wolstenholme et al. 1987). In addition, they may exhibit minor variation in location within animal classes (e.g. Insecta: Crozier et al. 1989, Haucke and Gellissen 1988; Mammalia: Paabo, pers. comm.). Thus even when tRNA sequences are highly conserved between two organisms, the same primer pairs may not function together due to their relative locations. Note also that there are instances where some tRNA's have lost their

function and evolve at higher than normal rates (e.g. Thomas et al. 1989, Cantore et al. 1987), that mitochondrial tRNA are degenerate in nematodes (Wolstenholme et al. 1987) and that *Chlamydomonas* and *Tetrahymena* lack some tRNAs and trypanosomes lack all tRNAs (Attardi and Schatz, 1988).

t-Arg (24mer) ◄	5'- CGAAATCAGAGGTTCTCCTTAAAC -3'	Position
Cow	........TTTA..T.AT..T..A	10203
Urchin	....	7380
Fly	.....CT.ACT.CAA.TAA.CGCT	6102

Comments: Made to match urchins. Clearly not a good primer for anything else.

t-Phe (20mer) ◄	5'- TCTTCTAGGCATTTTCAGTG -3'	Position
Human	C.G....AA...........	625
Mouse	C.A....A............	49
Cow	..A.................	411
Frog	..A...CA............	2182
Urchin	C...TG.A............	52

t-Pro (20mer) ►	5'- CTACCTCCAACTCCCAAAGC -3'	Position
Human	.C...ATT.G.A........	15980
Mouse	.AC.AC.AGG.A........	15701
Cow	TC...AT....C........	15753
Frog	.C..TATTG..C........	17510
Urchin	TACAT.G.............	

Comments: t-Phe and t-Pro go through the entire vertebrate D-loop (control region) based on Kocher et al. (1989). Successful amplifications have been obtained for Fish.

t-Iso (19mer) ►	5'- ATTT-ACCCTATCAAGGTAA -3'	Position
Fly	....-...............	42
Human	....-..T.......A....	4303
Mouse	....-..T.......A....	3728
Cow	....-..T.......A....	4079
Frog	....C..T.......A..G.	

Comments: The above primer can be used in combination with one of the 12S primers to amplify the A+T-rich region in insects It has worked for *Drosophila virilis*. It did not work for *Magicicada*. This may be because in this cicada, the t-RNA is transposed or because the A+T-rich region can be up to 4Kb (Martin and Simon 1990). Considerable length variation has been found in this region in invertebrates (Rand and Harrison 1989; Snyder et al. 1987); in weevils the A+T-rich region is known to vary between 9 and 13Kb (Boyce et al. 1989). Note that tRNA iso is transposed in vertebrates and *Xenopus* has an insertion and a substitution one base from the 3' end.

Appendix Figure 1. Mitochondrial gene order in Strongylocentrotus (urchin), vertebrates [except in chicken where tRNA (Glu) and ND6 lie between the D-loop and tRNA (Pro) (Desjardins and Morals 1990)], Drosophila (fly), and Ascaris (nematode). tRNA genes (dark bars) are indicated by the one letter code of the amino acid they correspond to. Arrows indicate direction of transcription. O indicates origin of replication (for heavy [H], light [L], or both [R] strands). The origin of replication of Strongylocentrotus remains unmapped. The hypervariable, non-coding regions are indicated by stippling in all but Ascaris where it is labeled "A T" for A+T-rich region. Redrawn from Wolstenholme et al. 1987 and Jacobs et al. 1988.

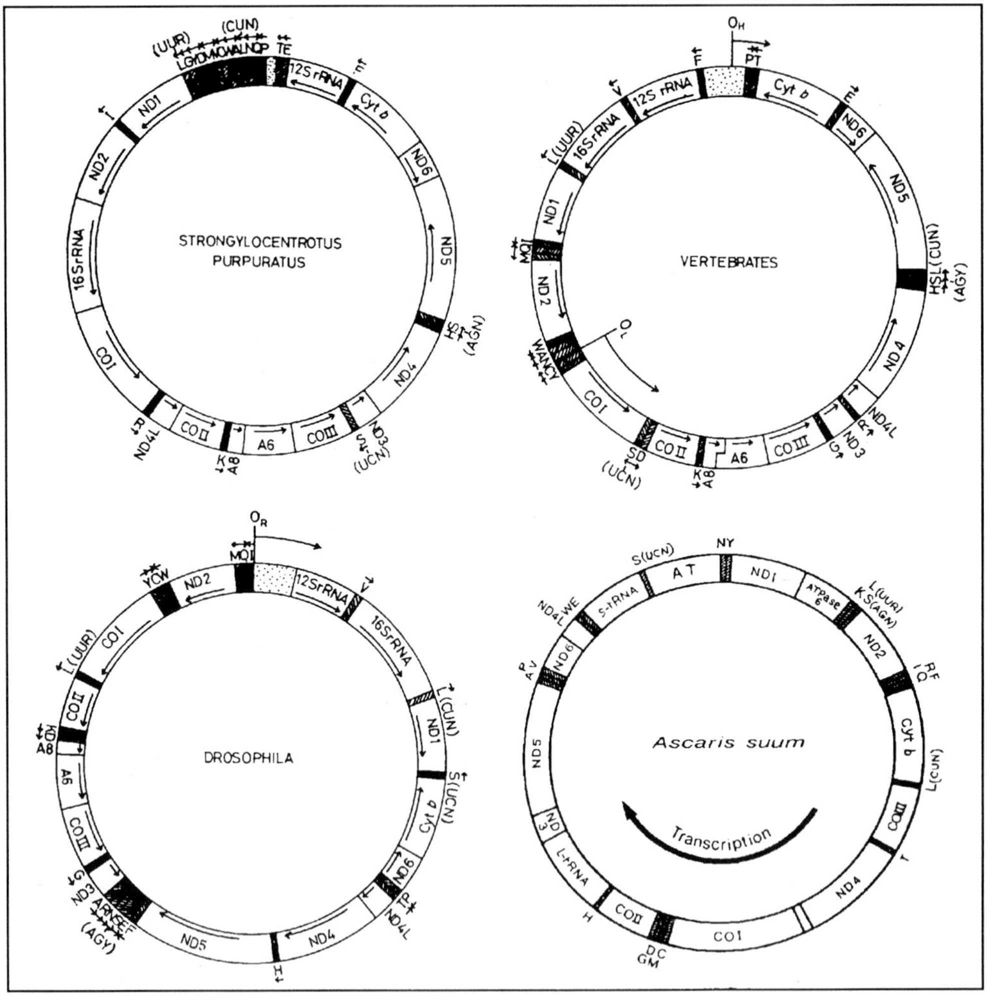

Additional References for Appendix 3:

Attardi G, Schatz R, (1988) Biogenesis of mitochondria. Ann. Rev. Cell Biol. 4:289-333

Boyce TM, Zwick ME, Aquadro CF. 1989. Mitochondrial DNA in the bark weevils: size, structure and heteroplasmy. Genetics 123: 825-836

Cantore P, Gadaleta MN, Roberti M, Saccone C, Wilson AC (1987) Duplication and remoulding of tRNA genes during the evolutionary rearrangement of mitochondrial genomes. Nature 329:853-855

Hixson JE, Brown WM (1986) A comparison of the small ribosomal RNA genes from the mitochondrial DNA of the great apes and humans: sequence, structure, evolution, and phylogenetic implications. Mol. Biol. Evol. 3:1-18

Snyder M, Fraser AR, LaRoche J, Gartner-Kepkay KE, Zouros E. (1987) Atypical mitochondrial DNA from the deep-sea scallop Placopecten magellanicus. Proc. Natl. Acad. Sci. USA 84:7595-7599

PROTOCOLS FOR 4-CUTTER BLOTS AND PCR SEQUENCING

Martin Kreitman
Department of Ecology and Evolution
Princeton University
Princeton, N.J. 08544

4-CUTTER DIGESTIONS

***See over for Micro-titre plate protocol**

-For 3 μg genomic DNA

Enzyme	Rxn. temp.	Tube color	Label Color	* NOTES
Alu I	37			
Ban I	37			
Dde I	37			
Hae III	37			
Hha I	37			
Hinf I	37			
Msp I	37			
Sau 3A	37			
Sau 96 I	37			
Taq I	65			

-5 μl gDNA (3μg) into Eppendorf tube.

-For each sample gently mix on ice:

sdH_2O	45 μl
10X Buffer	5.5 μl
Enzyme	5 Units
RNAase(10 mg/ml)	0.03 μl

-Dispense 50 μl of the above enzyme mixture to each tube.

NATO ASI Series, Vol. H 57
Molecular Techniques in Taxonomy
Edited by G. M. Hewitt et al.

-Incubate 6 hours to overnight at appropriate temperature. Incubate in oven rather than water bath to prevent condensation at top of tube.
-Quick spin samples in Eppendorf.
-For each sample mix on ice:

0.5 M EDTA	1µl
3 M NaOAc	10 µl
T.E	30 µl
cold 100% EtOH	260 µl

-Add 300 µl of the above EtOH mixture to each tube. Mix gently.
-Put into -20° C overnight or -80° C for 30 minutes.
-Spin in Eppendorf at r.t. for 10 minutes.
-Aspirate off supernatant with stretched pipette. Let samples sit a few minutes, then aspirate off as much supernatant as possible.
-Wash with 1 ml cold 70% EtOH. Aspirate. Vacuum dry.
-Resuspend in 3.0 µl formamide dye loading buffer. Spin, vortex, spin.
-Store at -70° C if not run immediately.
-Run 1.5 to 2.0 µl on a 5% DNA sequencing gel.

Four-cutter Digestions (Micro-titre Plates)
-50 µl digests as before (one micro-titre plate to one enzyme, one population).
-Seal plates with Dynatech plate sealers (*# 001-010-3501, Dynatech Laboratories Inc, 14340 Sullyfield Circle, Chantilly, Virginia, 22021. 1-800-336-4543),* mix gently on multi-tube vortexer and leave in incubator of appropriate temperature overnight.
-Add 6µl of 3M NaOAc + 44 µl of Isopropanol to each reaction. Seal plates again and mix gently on multi-tube vortexer. -20 ^{O}C for 30 min
-Spin in plate-holding rotor of IEC Centra-4B clinical centrifuge 4K rpm 10 min
-Aspirate off supernatants.
-Add 200 µl of cold 70% EtOH, seal again. Brief spin, aspirate dry. Air dry. Add 3 µl formamide loading dye. Seal with regular tape, mix thoroughly on multi-tube vortexer, spin down.

4-CUTTER ACRYLAMIDE BUFFER-GRADIENT GEL (5 % ACRYLAMIDE, 0.5x TBE)

Gel mix (1 l)

125 ml 40 % Acrylamide
460 g Urea, Ultra Pure
5 g Amberlite

-Add H_2O up to about 950 ml and stir for 20 min
-Filter through 500 ml sterile flask. Add 50 ml TBE (10x= 1M) and make up to 1 liter.

For each gel use:
(all at refrigeration temp; mix on ice)
70 ml gel mix 140 µl 25% AP 70 µl Temed
-Mix well
-Pour gel using 50cc syringe as quickly as possible
-Insert comb or top spacer.

Gel plates

30x40 cm plates. Clean and fully silanize one plate; clean and silanize *bottom half* of other plate.
-Clamp together with binder clips using 0.4mm spacers on two sides and on bottom.
-The bottom and side spacers are sealed with a tiny bit of plasticine dough.
PAY SPECIAL ATTENTION TO SEALING THE BOTTOM CORNERS - THEY LEAK!

Gel rig

-Allow gel to set up (15 min) and remove comb, clips, tape, and bottom center spacers.
-Set up on large gel rig using large clips. Seal sides of top reservoir between the front plate and the gel rig with water-proof neopreme gasket.
-Fill top reservoir with 0.5x TBE; check for leaks.

-Fill bottom reservoir with 2:1 1X TBE:3MNaOAc.

-Use hooked syringe to blow out air bubble in space left by the bottom center spacer.

-Use pipette to blow air bubbles out of pockets.

-Clamp aluminum heat-distributing plate to back of gel set-up (this prevents "smiling")

-Pre-run for about 25 min at 45W.

Loading samples

-Use Drummond Sequencing pipette.

-Pre-heat samples for about 3-5 min at 90°C (load 30 samples in heat block at a time if using eppendorf tubes; add the next 30 when the first lot has been taken out... or put microtitre plate directly on hot block).

-Clean out pockets thoroughly with Pasteur pipette. Do this repeatedly to prevent a build up of Urea in the pockets.

-Load 1.5-2μl of each sample.

-Lane markers: 123 bp ladder and HincII phi X-174 (1ng of each in formamide dye loading buffer)

-Start running gel as soon as loading is finished to prevent diffusion of samples.

Gel run

-Run at 45W (ie starting at about 1300V and finishing at a max of about 1700V) for $2^{1}/_{2}$ - 3 hr.'s (until the first dye front enters the bottom reservoir; the blue dye front should be about $^{2}/_{3}$ of the way down.)

ELECTROPHORETIC TRANSFER OF 4-CUTTER GELS

Preparing Transfer Apparatus

[Supplied by Polytech Products, 95 Properzi Way, Somerville, Mass 02143 (tel 617-666-5064). *This consists of a large transfer bath (vol about 25 L) with horizontal platinum wire electrodes at top and bottom. The blotting-paper/gel/filter sandwich is placed in the tank in a sandwich between two*

pieces of large "ScotchBrite" and two white plastic grids. We use a small, self-priming pump to route the buffer via an outlet at the bottom of the tank through a coil of copper tubing in a "cooler" containing ice, and back into the buffer bath. This is necessary in order to cool the buffer which otherwise heats up during the transfer; heating can severely detrimentally affect the transfer]

-Drain old buffer from tank, and drain ice-melt water from ice container.
-Disconnect cooling system (copper tube coil in ice container).
-Stop buffer tank outlet and add 21 L of 0.5x TBE. Bury copper tubing in ice container with ice, and re-connect tubing to buffer tank outlet. Unstop outlet.
-Pre-run pump for 30 mins to pre-cool buffer. Make sure that buffer input tube is away from buffer tank outlet to ensure that all buffer is cooled.

Transfer

-Separate gel plates carefully with Knife. Optionally use spatula or metal ruler.
-Place appropriately cut filter paper (usually *Whatman* 1Chr) on top of gel and notch corner with the first lane marker. Trim off excess gel material.
-Place filter paper gel-side up on top of maroon Scotch-Brite pads sitting on first white grill of transfer apparatus and wet liberally with .5x TBE.
-Wet appropriately cut and labeled transfer membrane (*GeneScreen* works fine but can get a little expensive if you're doing a large amount of this kind of work. We use bulk-supplied MSI nylon membrane in transfer tank and place label-side up on top of the wet gel. With gloved finger, squegee out excess buffer and bubbles. Notch corner as before. (We order MSI nylon membrane through Fisher Cat # NO4 HY 00100. It comes in a 100' roll (12" across) which costs $1124.)
-Place other maroon Scotch-Brite pads on top along with second white grill and turn upside down so that labeled transfer membrane faces down and filter paper is on top. Rubber band each side.
-Place whole set-up in tank. Then place piece with wire facing such that hook-up to the electrode is facing up.
-Weigh this down with some appropriate weight.
-Place cover on apparatus and hook up electrodes.

-Transfer for 30-40 minutes. Transfer no more than 2 at a time. *{We have done it with up to four at one time, but this doesn't always work - and a fourfold catastophe can be bad news!} We use a* Skare Powerpak Model 1[James Skare, 665 North Str., Tewksbury Ma 01876 (tel 617-638-7086)] *which gives a 108V dc output at about 500ma.*

-Wrap in Saran Wrap with folds on label side. Make sure filter is well wet (in 0.5x TBE) for storage.

-UV crosslink for 3 minutes with non-labeled (ie DNA side) side facing UV source. We use an unprotected UV light-box approx. 9" above the filter.

DIRECTIONS FOR DNA SEQUENCING

We do all our annealing and sequencing steps in microtiter plates (Falcon 3911) and I would urge you to do the same. We spin samples in a IEC Centra 4b with microtiter plate carriers. (By the way, this is a great general purpose centrifuge).

ANNEALING STEP

DNA*	3 μl
5x sequenase buffer	2 μl
sd water	4.4 μl
20'mer primer (10ng/μl)	0.6 μl
Total volume	10 μl

*: 6ng of a 20'mer primer is approximately 1p<u>M</u>. Use about the same amount of template DNA (roughly 1μg of a 3000bp plasmid or fragment).

-Seal microtiter plate with waterproof tape (stretchable polyvinyl tape is good). Heat to 60 $^{\circ}$ C for 30 minutes in oven. Cool on ice; quick spin. May be put into freezer until a later date if needed. If not, then continue with next step.....

Set up the following items:

Gel plates: Set up buffer gradient gel(see 4-cutter protocol). Prerun gel @ 40-45W for 30x40x0.04cm gel (should be approx. 1300V).

Clinical centrifuge: Set up microtiter centrifugation apparatus. Balance with empty microtiter dish if needed.

Microtiter dish: Put dideoxy's in bottom of wells 2.5 μl each.

Heat block: 85° C

Water bath: 37° C

-To each annealed DNA at room temp. add the following Labeling Reaction:

DTT	1.0
Diluted Labeling Mix	2.0
^{35}S dATP (Amersham)	0.5

(These three items are mixed together on ice, then added to DNA).

-Add to DNA+mix:

Diluted Sequenase*	2.0

**Dilute just before use* (1 seq'ase: 7 cold T.E).

-Quick spin and mix gently.

-Set timer to 5 minutes. Try to complete the following within the 5 min. time limit: to sides of microtiter dish containing dideoxy mixes add 3.5 μl of Labeling Reaction. Prop the microtiter dish up so that the Labeling Reaction does not run down into the dideoxy mix. As long as 2-3 or more min. have gone by spin in clinical centrifuge until it hits 2000 rpm; put into 37° C water bath. (May be helpful to cut V's in sides to make dish sink a little to insure good contact of the wells with the water). Leave for 5 minutes.

-Add 4 μl of Stop Buffer.

- Heat to 85° C for 3-5 minutes and load onto gel with shark's teeth comb. Run at 45W (1900V Maximum) until bromophenyl blue has run out.

**NOTE: Have to take cap off of the geiger counter to check for ^{35}S contamination.

GEL DRYING AND AUTORADIOGRAPHY

-Take apart gel (don't worry about bubbles,etc.) and soak for 15 minutes in about 1 liter of 10% CH_3OH: 10% Acetic Acid.

-Lift out very gently and evenly so gel doesn't slip. Stand gel plate up at an angle and allow to air dry for a few minutes.

-Cut a piece of 3mm paper to fit gel and place on top of gel. Cover with Saran Wrap and cut to fit gel exactly or it won't be smooth. Can even cut into the gel on all edges.

-Warm up gel dryer to 80° C and make sure the two dry ice traps are very full.

-When placing gel in gel dryer, make sure all pieces of dryer are perfectly centered . Gel itself is placed paper side down on top of white piece. Start dryer. Pull gray piece over to set a seal and dry for 15 minutes.

-Take off of dryer. Remove Saran Wrap. The Saran should come straight off or the gel isn't dry enough.

-Put into a cassette overnight straight against the film and clamp cassette shut and leave at room temperature overnight.

PREPARATION OF SINGLE-STRANDED DNA FROM PCR AMPLIFIED DNA USING LAMBDA-EXONUCLEASE

Kinasing an oligo

For 10 µg of oligo in 30 µl rxn:

10x kinase buffer (NEBL)	3 µl
10mM ATP	3 µl
20 U) kinase	2 µl

-30 min at 37 C

-10 min at 95 C to heat kill the enzyme. Phenol/Chloroform, chloroform extr.'s

-Clean up with NAP-5 column (Pharmacia) in 1/10 T.E. Conc. of oligo is now 10ng/µl.

PCR

Regular double-stranded amplification using 100 ng of each primer and 100ng of genomic DNA in a 100 µl rxn for 25 - 35 cycles.

lambda-Exonuclease Digest

-Add 10µl lambda-Exo supplement buffer (see below)

-Add 0.5µl lambda Exonuclease (approx. 2U)

-30 min at 37 C

-10 min at 65 C to heat kill the enzyme

-Phenol, Phenol/chloroform extractions (Doing this carefully is very important)

-Add 1/2 vol. 7.5M NH_4OAc

-2 vol EtOH ppt (10 min). Wash 70% EtOH, dry.

-Resuspend pellet in 10 µl of water; use half to sequence

-Use the phosphorylated oligo as the sequencing primer or an internal primer corresponding to the same strand. If only a little dsDNA is seen on the mini-gel used to visualise the PCR products, resuspend in 5 µl and sequence all of it.

Lambda-Exo supplement buffer:

775mM glycine

278mM KOH

5.8mM $MgCL_2$

STOCK SOLUTIONS

40% Acrylamide (19:1 Acrylamide:bis-acrylamide)

Acrylamide	190g
Bis-acrylamide	10g

diH_2O to <500mL

MB-1 Amberlite resin (Sigma)	10-15g

Stir until dissolved

Adjust volume in graduated cylinder to 500mL with diH_2O

Filter through 0.2 or 0.4 micron sterilizing filter (Nalgene) to remove particulates. (Be sure to first remove the cotton plug to prevent fibers from being pulled into the clean acryl. solution.)

Store at 4^oC.

6% 0.5X TBE Acrylamide Mix

40% Acryl. sol'n	75mL
Urea (ICN UltraPure)	230g

diH_2O to about 450 mL (not more than 475!)

MB-1 Amberlite resin	2g

Stir on low heat until dissolved. Do not let sol'n get hot.

Vacuum filter twice through Whatman #1H filter paper to remove MB-1 resin

10X TBE	25mL

Adjust volume in graduated cylinder to 500 mL with diH_2O

Filter through 0.2 or 0.4 micron sterilizing filter

Store at 4°C.

6% 1X "Marathon" Acrylamide Mix

Same as above but substitute 50mL 10X Marathon buffer for 10X TBE buffer.

10X TBE

Tris base	432g
Boric Acid	220g
Na_2EDTA	37.2g

make up to 4L with diH_2O.

10X "Marathon" TBE

Tris base	654g
Boric Acid	111.3g
Na_2EDTA	37.2g

make up to 4L with diH_2O.

10X PCR Buffer (Perkin Elmer Cetus) "2.0m*M* Mg^{++}"

1M TrisHCl, pH8.3 (100m*M*)	100µl
1M KCl (500m*M*)	500µl
1M $MgCl_2$ (20m*M*)	20µl
1% gelatin (Difco or Sigma) (0.1%)	100µl
sdH_2O	280µl

Make sure stock solutions contain no extraneous DNA!

10X PCR Buffer (Innis, et al.)

1M TrisCl, pH8.0 (100m*M*)	100µl
1M $MgCl_2$ (30m*M*)	30µl
Tween20 (0.5%)	5µl
Nonidet P-40 (0.5%)	5µl
sdH_2O	860µl

<u>40X TAE Buffer</u>

Tris base (2*M*)	242g
diH_2O 800mL	
HOAc (glacial) to pH8.0	
NaOAc (anhydrous) (0.8*M*)	65.6g
Na_2EDTA (0.04*M*)	16.5g
diH_2O to 1L	

<u>T.E</u>

1M TrisCl, pH8.0 (10m*M*)	1mL
500m*M* EDTA pH8.0 (0.1m*M*)	20µl
sdH_2O to 100 mL	
autoclave	

<u>10X Kinase buffer (New England Biolabs, 1987)</u>

1*M* TrisCl pH8.0 (700m*M*)	700µl
1*M* $MgCl_2$ (100m*M*)	100µl
1*M* DTT (50m*M*)	50µl
sdH_2O	150µl

RIBOSOMAL RNA SEQUENCING

Michel Solignac

Laboratoire de Biologie et Génétique évolutives
CNRS, 91198 Gif-sur-Yvette Cedex, France

Ribosomal genes can be cloned or PCR amplified and then sequenced as other DNA sequences. However, the products of the genes, ribosomal RNAs, can be used as template for direct sequencing with the reverse transcriptase as polymerase following a protocol derived from the Sanger's technics for DNA (dideoxynucleotide chain termination). Universal primers can be used to sequence the chosen region of the rRNA.

I. RNA EXTRACTION

The main problem with RNA extraction is the rapid degradation by RNases. These enzymes are very resistant and more care has to be taken for RNA than for DNA extraction. All solutions and small items are sterilized. Work with gloves. Homogenization must be rapid. Guanidine thiocyanate which destroys proteins (and RNases) very rapidly is used in several protocols. The protocols of extraction are described for *Drosophila.*

I A. Protocol 1 using an ultracentrifuge: RNA banded. According to Fyrberg et al. 1983. Use a swing or a vertical rotor.

1) Work with disposable gloves. 3-5g of flies are ground in liquid nitrogen (LN_2) with porcelane mortar and pestle until a fine powder is obtained.

2) Work under an extractor hood, at room temperature. The powder is transferred into a large potter. Add 50 ml of the solution: Guanidine thiocyanate 4M, ß mercaptoethanol 1M, Na acetate (pH 4.5) 50mM, EDTA 10 mM.

3) Filter the homogenate onto nylon (100 micrometer). Add 1g cesium chloride per ml of the homogenate.

NATO ASI Series, Vol. H 57
Molecular Techniques in Taxonomy
Edited by G. M. Hewitt et al.

4) In a 50 ml quick seal tube for a Beckman VTi50 rotor, put 17 ml of a CsCl solution 1.8 g cm^{-3} (45 g CsCl in 33 ml water, adjust the refractive index to 1.4085). The homogenate is carefully layered onto the cesium bed.

5) Centrifuge for 15 hours, 40,000 rpm at 20°C. The RNA forms one or two opalescent bands in the middle of the tube. The band is punctured with a syringue (large needle). Cesium chloride is diluted with 3 vol. water. Polysaccharides are ethanol precipitated by adding 1/10 vol. ethanol. Shake and keep 1 hr at -20°C. Centrifuge for 40 min at 8,000 rpm in Corex tubes. The supernatant is adjusted to 0,3M NaAc, 2 vol. alcohol are added and RNA is precipitated for the night at -20°C. Centrifuge 20 min at the maximum speed allowed for Corex tubes. Dry the pellet and dissolve it in TE (10mM TRis, 1 mM EDTA).

I B. Protocol 2 using an ultracentrifuge: RNA pelleted. According to Perbal 1988.

1) Work with disposable gloves. 3-5 g flies are ground in LN_2 with porcelane mortar and pestle until a fine powder is obtained.

2) Transfer the powder into a Potter and add 10 ml of guanidium thiocyanate*.

3) Put 3 ml cesium chloride 7.5 M** in a Beckman SW 41 tube.

4) Layer 9 ml of the thiocyanate preparation onto the CsCl.

5) Centrifuge for at least 15 hr at 30,000 rpm at 4°C.

6) The RNA is pelleted whereas DNA forms a viscuous layer between the pellet and the CsCl interface.

7) Draw off the supernatant CsCl and DNA with a Pasteur pipette. Leave some solution on the pellet.

8) Pour off the remaining solution and cut off the bottom of the tube with a razor blade.

9) Resuspend the pellet in 400 ul sterile distilled water.

10) Heat for 5 min at 68°C.

11) Add 2.6 ml guanidine-HCl solution***. Mix.

12) Add 75 ul 1N acetic acid and 1.5 ml ethanol.

13) Let stand at -20°C for at least 3 hr.

14) Centrifuge at 15,000 rpm in a SS34 rotor (Sorvall) for 30 min.

15) Resuspend the pellet in 1 ml sterile water.

16) Centrifuge as in step 14. Save supernatant.

17) Resuspend the pellet in 0.5 ml water. Repeat 16.

18) Repeat 17. Pool the three supernatant (total 2 ml).

SOLUTIONS:
*Guanidine thiocyanate solution. For 100 ml: dissolve 50 g guanidine thiocyanate and 0.5 g n-lauryl sarcosine in 30 ml of sterile water. Heat and stir to help dissolution. Add 2.5 ml 1M sodium citrate buffer (pH 7.0). Add 0.7 ml ß 2-mercaptoethanol. Bring the pH of the mixture to 7.0 with few drops of 0.1 N NaOH. Bring up to 100 ml with sterile distilled water. Filter the solution on a millipore filter. Do not keep more than 2 weeks.
**Cesium chloride solution: dissolve 63 g CsCl in 49 ml water and add 1.25 ml 1 m NaAc (pH 5.0).
***Guanidine-HCl solution. For 100 ml: dissolve 72 g guanidine-HCl in 36 ml sterile distilled water; heat and stir to help dissolution. The pH of the solution is approximately 4.5. Add 2.5 ml 1M sodium citrate buffer (pH 7.0). Add 5 ml 0.1 M dithiothreitol solution. Bring up the pH of the solution to 7.0 with 0.1 N NaOH. Add water to 100 ml. Filter the solution on a 0.45 um pore size membrane. Make fresh every month.

Recently, extractions using cesium trifluoroacetate (CsTFA) have been recommanded. This salt is expensive but is more effective than CsCl in deproteinization and inhibition of RNases. RNA can be pelleted or banded by adjustement of the density. CsTFA is far from essential to purify total RNA for sequencing.

I C. The lithium chloride protocol. According to Maccecchini et al. 1979 and Qu et al. 1983. Work with gloves. Sterilize all solutions. Wash the potters with alcohol.

1) Homogenize 50 to 100 flies in a Potter in 1 ml Tris 100 mM, EDTA 100 mM, pH 7.4. Add 1 ml of 10% SDS. Mix. Transfer the homogenate in 3 microfuge tubes (1.5 ml).

2) Spin for 5 min at 8,000 rpm at +2°C.

3) Pipette the supernatant (about 500 ul) and add 1 vol of phenol:chloroform:isoamyl alcohol, 50:48:2. Shake vigorously.

4) Spin for 25 min at +5°C at 9,000 rpm.

5) Repeat step 3 and 4 at least 3 times. The final volume of the aqueous phase is about 300 ul.

6) Add 0.1 vol NaCl 3M to the supernatant and 2.5 vol of absolute ethanol. Let stand the tubes for 10 min in LN_2.

7) Spin for 15 min at 12,000 rpm at +2°C

8) Dry the pellet in a dessicator.

9) Resuspend the pellet in 0.6 ml lithium chloride 4M and 0.2 ml Tris 10 mM pH 7.4. Do not vortex. Resuspend the pellet by inversions of the tube.

10) Keep for 15 hr (overnight) at +4°C

11) Spin for 15 min at 12,000 rpm at +2°C and discard the supernatant.

12) Add to the pellet 0.4 ml of LiCl 4M + 0.4 ml Tris 10 mM pH 7.4. Resuspend the pellet by inversion of the tube several times.

13) Centrifuge for 15 min at 12,000 rpm at +2°C.

14) Wash the pellet with 75% ethanol.

15) Spin for 15 min at 12,000 rpm at +2°C. Dry the pellet in a dessicator. Resuspend the pellet in 30 ul Tris 10 mM pH 7.4.

I D. Combination of Guanidine and LiCl protocols. According to Savouret 1987.

1) Grind 200 mg flies in 2 ml TEG*

2) Add 200 ul SDS 10%. Homogenize.

3) Spin for 10 min at 8,000 rpm.

4) Pipette the supernatant and add to it 1 vol TEG Guanidine**.

5) Incubate at room temperatutre for 2 hr with periodic vortexing.

6) Centrifuge 10 min at 5,000 g (7,200 rpm).

7) Pipette the supernatant (1-2ml) in 15 ml Corex tubes. Add 6 vol LiCl 4M.

8) Vortex and incubate for 24 hr at +4°C.

9) Spin for 90 min at 8,200 rpm at 4°C.

10) Add 6 ml buffer 3*** and shake until the pellet is entirely resuspended.

11) Extract 2 times with phenol:chloroform:isoamyl alcohol, 50:48:2. Spin for 30 min at 8,200 rpm.

12) Add 1/10 vol NaAc 3 M and 2.5 vol ethanol. Incubate 2 hr at -20°C or 1 hr at - 70°C.

13) Spin for 15 min at 8,000 rpm.

14) Wash the pellet with ethanol 75%. Repeat step 16.

15) Dry the pellet and resuspend in Tris 10 mM pH 7.4.

SOLUTIONS:
*TEG: for 100 ml: 0.6 g Tris-HCl and 0.38 g EDTA. Sterilize.
**TEG-Guanidine: to 100 ml TEG add 59 g thiocyanate guanidine, 1 g SDS, and 5 ml mercaptoethanol just before use.
***Buffer 3: Tris-HCl 10 mM, pH 7.5, EDTA 1 mM.

I E. Quiagen^{R} protocol. According to "The Quiagenologist. Application protocols", Diagen 1988. This is only an example of the use of columns to purify macromolecules.

1) Grind 100 mg flies in a Potter washed with ethanol. Add 1 ml TE 100 mM (Tris 100 mM, EDTA 100 mM), pH 7.4. Add 1 ml 10% SDS. Transfer the homogenate in 3 microcentrifuge tubes.

2) Spin for 5 min at 9,000 rpm at +2°C.

3) Add an equal vol of phenol:chloroform:isoamyl acohol, 50:48:2. Mix vigorously.

4) Spin for 25 min at +5°C at 5,000 rpm.

5) Repeat 3 times steps 3 and 4. Treat with chloroform:isoamyl alcohol and spin for 25 min at +5°C at 5,000 rpm. This preparation must not contain SDS which interferes with the binding of nucleic acids to the column.

6) Adjust the binding conditions by adding to the supernatant 1/30 vol MOPS 1M pH 7.0 and 1/20 vol NaCl 5M (control carefully the volumes); the final concentration is approximately 250 mM.

7) Adjust a P1000 Pipetman to 600 ul and equilibrate a Quiagen-tip 20 (capacity for RNA: 40 ug) by pipetting 300 ul of buffer A in and out.

8) Adsorb the sample (solution of RNA) onto the Quiagen by pipetting it in and out 4 times. Repeat the same operation for the two other tubes (see step 1).

9) Wash the Quiagen tip with buffer A: adjust the Pipetman to 1 ml and pipette in and out about 750 ul of washing buffer (buffer A). Repeat with fresh buffer 5 times.

10) Adjust the Pipetman to 600 ul, mix equal vol of buffer C and E, and elute the RNA by pipetting in and out 3 times 200 ul of this mixture. Repeat 3 times with 200 ul fresh C+E mixture. Pool the 4 eluates.

11) Precipitate the eluted RNA by adding 0.8 vol isopropanol. incubate on ice for 20 min. Wash the pellet with 70% ethanol

SOLUTIONS:
Buffer A: 400 mM NaCl, 50 mM MOPS, 15% ethanol, pH 7.0
Buffer C: 1 M NaCl, 50 mM MOPS, 15% ethanol, pH 7.0
Buffer E: 1.1 M NaCl, 50 mM MOPS, 15% ethanol, 4 M Urea, pH 7.0

II. RNA CONTROL

Minigels (80 x 40 x 6 mm) for RNA: prepare a 0.8 to 1% agarose gel in TBE buffer (Tris base 90 mM, Boric acid 90 mM, EDTA 2 mM). Load 2 to 3 ul RNA heated for 3 min at 100°C + 1 ul of the loading mixture (100 ml deionised formamid, 100 mg Xylene cyanol, 100 mg Bromophenol blue, 4 ml EDTA 500 mM pH 8). Run 1 hr at 100 V. Ribosomal RNAs form two bands in the middle of the gel (a single band for *Drosophila* where the 28S rRNA is fragmentated)

Optical density. The measurement of RNA concentration is performed at 260 nm: 1 unit of DO corresponds to 20 mg/ml of RNA. The purity of a RNA solution is judged through the ratio of the values 260 nm/280 nm which is 2 for pure RNA and 1.8 for pure DNA. This ratio is changed if proteins or phenol are present.

If necessary, RNA can be concentrated in a dessicator, in a speedvac, or by ethanol precipitation. The ideal concentration is 4 ug/ul.

III. RNA SEQUENCING

The direct sequencing of rRNA has been developed by Qu et al. (1983) for the 28S rRNA and adapted to the study of 16S-like rRNAs by Lane et al. (1985). Radioactive labelling of the neosynthetized cDNA is achieved through a kination of the primer or through incorporation of the radionucleotide during the elongation reaction.

III A. Kination of the primer

1) Mix in a microfuge tube:
3 ul Tris 1 M pH8.0
1.5 ul 0.2 M $MgCl_2$
1.5 ul 0.1 M DTT (or beta- mercaptoethanol)
2 ul polynucleotide kinase (20 units)

25 pM (picomole) of the primer
20 uCi of **gamma** ^{32}P dATP
water to a final vol of 30 ul.

2) Incubate at 37°C for 45 min, then at 65°C for 5 min to stop the reaction.

3) Separate the primer from unincorporated nucleotides on a column of G 50 Sephadex suspended in 50 mM triethylamine. The stock solution of triethylamine is 0.5 M and neutralized (to pH 7.5) by CO_2 bubbling. Wash the column with triethylamine. The primer is in the first fractions. Alternatively, separate the primer and unincorporated nucleotides with a column Select-D G25 (5prime > 3prime, Inc).

4) Aliquot in fractions of 500 counts (G.M. minimonitor). Evaporate the solvant. Keep the pellet dry. This preparation can work for as much as 40 RNA sequences.

III B. Sequencing reactions

1) Hybridization: Add 3 ul water and 8 ul RNA (4 ug/ul) to the dried primer. Vortex. Heat for 5 min at 55°C (use aluminium blocs). Add 2.5 ul of the reverse transcriptase buffer (BRL 5x). Heat 5 min at 55°C. Allow to cold slowly down to 35°C (about 15 min in the aluminium bloc). Transfer on ice.

2) Prepare the sequence reaction tubes: 15 ul of the sequence solutions* in the appropriate tubes

3) Add to the hybridization solution: 2.5 ul reverse transcriptase buffer (BRL 5x), 3.0 ul reverse transcriptase (600 units of the cloned MLV reverse transcriptase, BRL, 40,000 units per tube), 7 ul water (total vol: 26 ul). Distribute 6ul of the mixture to each of the sequence tubes. Incubate 35 min at 37°C. Transfer on ice.

4) Precipitate nucleic acids: add 3 ul NaCl 2.5 M and 75 ul precooled ethanol per tube. Incubate 10 min in LN_2 (or 2 hr at -20°C). Centrifuge 20 min at 12,000 g and dry the pellet.

5) Alkaline hydrolysis. Resuspend the pellet in 60 ul 0.3 N NaOH. Incubate 15 hr at 37°C (overnight) or 1 hr at 65°C. Neutralize with a mixture of 8 ul acetic acid, 25 ul NaAc 2 M, pH 7.5 and 50 ul water (prepare the mixture for several tubes). Precipitate immediatly with 450 ul ethanol, 10 min in LN_2 (2 hr at -20°C). Centrifuge 20 min, wash the pellet with 200 ul 70% ethanol, centrifuge, dry the pellet. Keep dry.

6) Resuspend in 2.5 ul water. Mix carefuly. Add 2.5 ul mix blue (mix with the tip of a P20 Pipetman). Heat (denature) for 3 min at 90°C (tubes closed) and transfer on ice. Load 3 ul on sequence gels for long run. Add 2 ml mix blue. Load on the sequence gel for short run.

*Sequence solutions

	C	A	T	G
ddC (1mM)	80ul			
ddA (1mM)		80ul		
ddT (1mM)			80ul	
ddG (1mM)				80ul
dNTP**	160ul	160ul	160ul	160ul
BRL buffer	60ul	60ul	60ul	60ul

**dNTP = 1vol dC, 1 vol dA, 1 vol dT, 1 vol dG, all 1 mM and 1 vol water (final concentration: 0.2 mM for each nucleotide.

III C. Sequencing reaction with unlabeled primer.

1) Hybridization: in the reaction tube, to 11 ul of the 4 ug/ul RNA solution, add 0.2 ul of the unlabeled primer (#13 ng). Heat for 5 min at 55°C. Add 2.5 ul of the reverse transcriptase buffer and keep for 5 min at 55°C. Allow to cold slowly down to 35°C (about 15 min) in an aluminium bloc. Transfer on ice.

2) Prepare sequence tubes with 15 ul of each of the four sequence solutions (see above IIIB).

3) Return to the hybridization tubes now at 35°C. Add 7 ul H20, 2.5 ul of reverse transcriptase buffer, 3 ul of reverse transcriptase (600 units of cloned MLV BRL reverse transcriptase, 40,000 units per tube) and 1ul (10 uCi) of alfa- ^{35}S dATP. Centrifuge. Vortex.

4) Elongation. Distribute 6 ul of this mixture to each of the four sequence tubes. Mix. Incubate for 30-45 min at 37°C.

5) Precipitate: add 3 ul of 2.5 m NaCl and 75 ul ethanol. Incubate 120 min in LN_2. Spin 20 min at 12,000 g. Dry the pellet.

6) Alkaline hydrolysis. Resuspend the pellet in 60 ul NaOH 0.3 N. Incubate 1 hr at 65°C or 15 hr at 37°C (overnight).

7) Neutralize: add 8 ul acetic acid, 25 ul of 2M NaAc, pH 7.5, 50 ul H_2O per tube (prepare the mixture for several tubes).

8) Precipitate immediatly with 450 ul ethanol, incubate for 10 min in LN_2. Centrifuge for 20 min, wash the pellet with 400 ul ethanol 70%. Centrifuge. Dry the pellet. Keep it dry at -20°C.

9) For electrophoresis, resuspend (with the tip of a P20 Pipetman) the pellet in 2.5 ul water. Mix. Add 2.5 ul mix blue. Mix. Centrifuge. Denature for 3 min at 90°C (tube closed). Transfer on ice. Load 3 ul on a sequence gel for the long run. Add 2 ul blue, mix, load 3 ul on a sequence gel for the short run.

IV. PRIMERS

Each primer is convenient for a 300 nucleotide sequence. If several sequences are performed along the molecule, they do not need to be overlapping. It is sufficient to sequence the same regions for the species studied.

Primers used by Lane et al. (1985) for the 16S-like rRNA:
GWATTACCGCGGCKGCTG (positions 519-536 in *E. coli*)
CCGTCAATTCMTTTRAGTTT (positions 907-926 in *E. coli*)
ACGGGCGGTGTGTRC (positions 1392-1406 in *E. coli*)
K = G or T, M = A or C, R = A or G, W = A or T.

Primers used for *Drosophila*, developed in Bachellerie's laboratory for sequencing the divergent domains of the 28S rRNA:
D1: GCTGCATTCCCAAGCAACCCGACTC
D2: CCTTGGTCCGAGTTTCAAGACGGG
D8: ATTCCCCTGGTCCGCACCAGTT

REFERENCES

Fyrberg EA, Mahaffey JW, Bond BJ, Davidson N (1983) Transcripts of the six *Drosophila* actin genes accumulate in a stage and tissue specific manner. Cell 33:115-123

Lane DJ, Pace B, Olsen GJ, Stahl DA, Sogin ML, Pace NR (1985) Rapid determination of 16S ribosomal RNA sequences for phylogenetic analyses. Proc Natl Acad Sci USA 82:6955-6959

Perbal B (1988) A practical guide to molecular cloning. 2nd ed. Wiley, New York.

Qu LH, Michot B, Bachellerie J-P (1983) Improved methods for structure probing in large RNAs: a rapid "heterologous" sequencing approach is coupled to the direct mapping of nuclease accessible sites. Application to the 5' terminal domain of eukaryotic 28S rRNA. Nucleic Acids Res 11:5903-5919

Savouret JP (1987) Isolement d'ARN traductibles. Le Brin Complémentaire (Boehringer, Mannheim France) 2:1-2

QUANTITATIVE DNA:DNA HYBRIDIZATION AND HYDROXYAPATITE ELUTION

Brion D W Jarvis, George Ionas and John C Clarke

Department of Microbiology & Genetics,
Massey University
Palmerston North, New Zealand

PROTOCOL

1) **Labelling.** For each hybridization mixture digest 1μg of purified reference DNA with *Hae*III restriction enzyme and label this with dCTP[^{32}P] by nick translation according to the manufacturer's instructions.

2) **Removal of unincorporated dCTP[α^{32}P].** Load the labelling reaction mixture on a "mini-spin column" containing 0.9-1.0ml of compacted, pre-swollen Sephadex G40-50. Follow this with 200μl of 0.01M Tris buffer containing 0.1mM EDTA. Place the column and a 1.5ml microfuge tube in a 50ml plastic centrifuge tube and collect the DNA solution in the microfuge tube by centrifuging for 5 min at 3000g.

3) **Sonication of unlabelled DNA.** Sonicate 5ml of purified DNA (200μg/ml) in 0.28M PB (phosphate buffer, pH 6.8) with a Soniprep 150 sonicator fitted with a 90mm probe at 20k/c for 80 sec. or equivalent.

4) **Hybridization.** To a 10x100mm screw-capped Kimax tube add 0.75ml; of sonicated unlabelled DNA solution (150μg) and 0.1ml labelled reference DNA (1μg). Make up to 1.0ml with 0.28M PB. Denature the DNA in a boiling waterbath (10min at 100°C) and immediately plunge the tube into melting ice. Hybridize in a waterbath at T_m-25°C for 16hr. Terminate hybridization by adding an equal volume (1.0ml) of distilled water and place the tube on ice. Set up a positive control containing labelled reference DNA with homologous unlabelled DNA and a negative control containing labelled DNA alone.

5) **Batch separation of ssDNA from dsDNA.** For each hybridization mixture place 0.7g of DNA-grade hydroxyapatite (BioRad) (HA) in a 13x125mm thick walled test-tube and add 8ml of 0.10M PB containing 0.4% SDS. Mix with a variable speed overhead stirrer and raise the temperature of the mixture to T_m-25°C in a circulating waterbath. Immediately place the tube in a bench centrifuge (Sorvall Type A is suitable) held at T_m-25°C and centrifuge at 1000g for 60 sec. Discard the supernatant and re-wash the HA in 0.10M PB containing 0.4% SDS.
Add 0.5ml of diluted reassociated DNA solution and 8ml of 0.10M PB containing

NATO ASI Series, Vol. H 57
Molecular Techniques in Taxonomy
Edited by G. M. Hewitt et al.

0.4% SDS. Mix and repeat the above washing procedure exactly. Collect the supernatant which contains ssDNA in a scintillation vial. Repeat the washing procedure three more times and collect each wash in a separate scintillation vial. To elute dsDNA from HA add 8ml of 0.4M PB (without SDS) mix, and hold at room temperature for 5 min. Centrifuge as before and collect the supernatant in a scintillation vial. Repeat this process three more times.

6) **Assay of radioactivity.** Assay all 8 eluants by Cerenkov counting in a Beckman LS7000 liquid scintillation counter (or equivalent) at ^{32}P settings.

7) **Estimation of relatedness.**

$$\%\ dsDNA = \frac{\text{cpm eluted from HA with 0.4M PB}}{\text{total cpm eluted from HA}} \times 100$$

$$\%\ relatedness = \%\ dsDNA - \%\ \text{self-hybridized reference DNA (negative control)}$$

$$\text{Relative}\ \%\ relatedness = \frac{\%\ \text{relatedness of heterologous DNA}}{\%\ \text{relatedness of homologous DNA}} \times 100$$

The last of these figures permits the comparison of results obtained with different reference DNAs.

PRINCIPLE OF THE METHOD

Hydroxyapatite (HA, calcium phosphate hydroxide) can be used to separate double-stranded (ds) from single-stranded (ss) DNA, and hence to measure the extent of DNA:DNA hybridization. This is an important criterion in the taxonomy of bacteria, in particular, as it is often critical for the definition of species and genera. The DNA from a reference strain is radioactively labelled and hybridized in phosphate buffer solution with an excess of unlabelled DNA from each of the strains to be compared with it. Hybridization goes essentially to completion. No membrane-bound DNA is involved. HA in phosphate buffer of suitable molarity is used to separate the hybrid dsDNA from ssDNA, and both column and batch procedures have been used for this purpose. The percentage relatedness between the reference strain and each of the other strains is obtained by expressing the radioactivity associated with dsDNA as a percentage of the total radioactivity in solution.

QUESTIONS AND ANSWERS CONCERNING THE BASIC METHOD

How much DNA is required?

When labelled DNA is mixed with homologous unlabelled DNA, denatured and allowed to re-hybridize, the conditions must be such that substantially all the labelled DNA becomes hybridized. This requires a high concentration of unlabelled DNA.

At the same time self-hybridization of labelled DNA strands with each other must be minimised. This requires a low concentration of labelled DNA.

C_0t is the product of DNA concentration and incubation time. In the hybridization mixture a C_0t of approximately 100 is required to ensure complete hybridization of unlabelled DNA and a C_0t of 0.1 to ensure negligible self-hybridization of labelled DNA.

In a 1.0ml hybridization mixture this is achieved by using 150μg of unlabelled DNA for an adequate number of hybridization mixtures 1-2mg of DNA at about 200μg/ml will be required.

How pure must this DNA be?

We have found it essential to use highly purified DNAs for quantitative hybridization. In particular the labelled reference DNA must be free from contaminating cellular material. Conventional phenol-chloroform, RNAse treatments yield DNA which is free from contamination with proteins and phenol and has good spectrophotometric ratios (258:230 and 258:280nm) but it may contain as much polysaccharide as it does DNA!

We have used cesium chloride density gradient centrifugation to clean up phenol-chloroform extracted, ethanol precipitated DNA and measured the "hyperchromic shift" when dsDNA denatures to ssDNA as an indication of DNA purity. A "hyperchromic shift" of more than 40% indicates that the DNA is of sufficient purity for quantitative hybridization.

What molarity of phosphate buffer should be used for hybridization?

A 1.4 molar sodium phosphate buffer, pH 6.8, (1.4M PB) is diluted to 0.28M PB. Do not adjust the pH after dilution. The hybridization mixture will be diluted to 0.14M PB before separation on HA but hybridization is more rapid (and hence more complete) at the higher molarity ($\times$3.5).

Why do you sonicate the DNA?

Hybridization of unsonicated DNA tends to have a "zippering" effect and may include as dsDNA sequences which are heterologous. Sonication breaks the DNA into short lengths (4-500bp) in which the sequences are more likely to be mainly homologous or mainly heterologous.

How is the reference DNA labelled?

Reference DNA is usually labelled with ^{32}P. A specific activity of 400-800,000 cpm/μg is satisfactory. This can be achieved by *in vivo* labelling and the extraction of labelled DNA but it is more convenient to use one of the current *in vitro* labelling methods such as nick-translation.

At what temperature should hybridization be carried out?

The optimum temperature for hybridization is 25°C below the mid-point of the DNA melting curve (T_m). T_m values can be calculated from published % GC values since:

$$\% \, GC = 2.44 \, (T_m - 69.4)$$

This will give T_m in 1 × standard saline citrate (1 × SSC), whereas these hybridizations are carried out in 0.28M PB, but the difference in T_m is not significant.

How long should hybridization mixtures be incubated to ensure that hybridization goes to completion?

A C_ot of at least 100 is required (as described above). Hybridizations between heterologous DNAs are reported to go more slowly. We have found that a hybridization time of 16hr is sufficient and convenient. Hybridization mixtures can be kept for several days at 4°C before analysis.

What controls should be set up when each batch of DNAs is compared with a reference DNA?

1) A positive control consisting of homologous labelled and unlabelled DNA from the reference strain. This is used to standardize the results and can also indicate radiation damage to the reference DNA.

2) A negative control consisting of labelled DNA only. This indicates the extent to which labelled DNA hybridizes with itself or alternatively the extent of non-specific DNA binding by HA.

What conditions are required to bind dsDNA to HA and selectively elute ssDNA?

This depends on the physicochemical state of the HA and thus on the maker, the batch and the age of the HA suspension. HA binds DNA selectively at temperatures above 60°C. The phosphate buffer molarity required to bind dsDNA and permit selective elution of ssDNA must be determined in a preliminary trial.

How can dsDNA be eluted from HA?

By raising the molarity of the phosphate buffer to 0.4M PB.

VARIATIONS ON THE BASIC METHOD

1) Hybridization at temperatures above T_m - 25°C

The effect of raising the hybridization temperature is to increase the stringency of the hybridization conditions. Sequences with partial homology are less likely to hybridize. Heterologous DNAs from closely related strains will show little change in relative relatedness. More distantly related strains with similar relatedness at optimum temperature may show a marked decrease in relatedness.

Relationships can be expressed independently of the stringency of the hybridization conditions by calculating a thermal binding index (TBI).

$$TBI = \frac{\text{Relative relatedness at } T_m - 10°C}{\text{Relative relatedness at } T_m - 25°C}$$

2) Determination of $T_{m(e)}$ and $\Delta T_{m(e)}$

These are statistics which express the base sequence homology in the fraction of the DNA which hybridizes at a given temperature, usually T_m - 25°C.

At a suitable buffer concentration the ssDNA in a hybridization mixture can be eluted and dsDNA binds to HA. If the temperature is raised 5°C conditions become more stringent and dsDNA with marginal sequence homology will dissociate. More ssDNA can be eluted. Successive 5°C increments will elute all the dsDNA the radioactivity associated with each fraction is measured, the temperature at which half the dsDNA is dissociated can be calculated. This temperature is $T_{m(e)}$. Homologous DNAs will dissociate over a small temperature range and have $T_{m(e)}$ values close to T_m. Heterologous DNAs melt over a wider temperature range and $T_{m(e)}$ will be less than T_m.

The difference between the $T_{m(e)}$ of the homologous DNA and that of a heterologous pair of DNAs is $\Delta T_{m(e)}$. This is an index of base sequence homology in the hybrid DNA. $\Delta T_{m(e)}$ has a range of 0°C to 15 or 20°C and a $\Delta T_{m(e)}$ of 1°C is approximately equal to 1% mismatch in the hybridizing sequences.

ADVANTAGES AND DISADVANTAGES OF SOLUTION HYBRIDIZATION AND HYDROXYAPATITE SEPARATION COMPARED WITH OTHER DNA:DNA HYBRIDIZATION METHODS

Advantages

1) Hybridization approaches completion for a given stringency.
2) The method is quantitative and reproducible within ± 2.5%.
3) Whole genomes are compared.
4) Unaffected by methylation of bases in DNA.
5) Variations on the basic method permit estimation of the average sequence homology in the fraction of the DNA which hybridizes at a given stringency.

Disadvantages

1) A relatively large quantity of DNA is required.
2) DNA of high purity is required.
3) Use of ^{32}P to label DNA.
4) The optimum buffer concentration for elution of ssDNA from HA may vary between batches.
5) Some more or less specialised equipment is required.

A PROTOCOL FOR THE TEACL METHOD OF DNA-DNA HYBRIDIZATION

Adalgisa Caccone
Dipartimento di Biologia
II Università di Roma "Tor Vergata"
00173 Roma, ITALY

Jeffrey R. Powell
Department of Biology
Yale University
P.O. Box 6666
New Haven, CT 06511 USA

The TEACL method of DNA-DNA hybridization was originally developed in Roy Britten's laboratory. The best reference for the original technical aspects is Hunt et al. (1981). Following from these procedures, we have used this technique for about 6 years. During that time we have modified and (we think) improved various aspects. In this document we detail our procedures as they have evolved over this six year period. The attached references give the published procedures, however, this document is more detailed. We have attempted many variations and there may, of course, still be room for improvement. However, we would suggest contacting us if you intend to try a variation and we can let you know if we have tried it and what the result was.

In addition to the two authors on this paper, other members of our laboratory have contributed to the development of these techniques: Clifford Cunningham, Cort Anderson, George Amato, and Kathyrn Goddard.

PREPARATION OF SINGLE-COPY DNA

Sonicated total DNA to an average size of ~1kb. Usually we use a microprobe for a 250ul solution in an Eppendorf tube. Check size on gel.

To ~50-100µg of sonicated DNA add an equal volume of 0.96*M* PB*

---

*0.96 *M* PB for
100ml: 6.62gr NaH_2PO_4
6.81gr Na_2HPO_4

NATO ASI Series, Vol. H 57
Molecular Techniques in Taxonomy
Edited by G. M. Hewitt et al.

(Phosphate Buffer) to a final concentration of 0.48*M* PB. Boil in Eppendorf 5', quickly spin, and incubate at 60°C to reach the desired C_ot.

Example of C_ot calculation:
For C_ot 200:
Concentration of DNA/310# = Moles/liter
C_ot 200 / (Moles/liter) = seconds in 0.12*M* PB
Seconds in 0.12 MPB / 5.65 = Seconds in 0.48*M* PB
5.65 is an acceleration factor, which varies depending on what molarity of phosphate buffer is used.
Other acceleration rates are:

Mol. PB	Rate
0.12	1
0.48	5.65
0.60	6.55
0.85	7.86
0.95	8.24
1	8.4

After the incubation is completed, dilute the sample with H_2O to 0.12*M* PB and loaded on an hydroxylapatite (HAP, BIORAD cat. 130-0420) column, which is kept in a circulating water bath at 60°C. The column is made in a 3ml sterile disposable syringe, plugged with sterile glass wool or another kind of filter. Attached to the syringe is an 18G needle fitted with plastic tubing to control the flow. Wash the column with 0.12*M* PB. Add enough HAP diluted in 0.12*M* PB for a 0.5cc bed. The volume of the HAP bed to use depends on the quantity of DNA you start with. Keep in mind that 1ml of packed HAP can bind 200µg of DNA. When possible use the least amount of HAP so that volumes of the eluted fractions can be kept small and the DNA is more concentrated. We commonly use a 3cc syringe, but for minute amounts of DNA we use 0.1-0.2cc HAP beds in a 1ml syringe. Wash the column with 5-10 bed volumes of 0.12*M* PB. Let the column run dry and immediately add the sample. Wait 5-10'. Collect the first fraction in an Eppendorf. Add slowly 0.12*M* PB to fill the syringe. Wait 5'. Collect 6-7 fractions of 500µl each. Slowly add 0.12*M* PB on top

----

#Average molecular weight of the nucleotides

of the column. Let the column run dry, then add 0.48*M* PB (fill the syringe). Wait 5'. Start collecting 1.5ml fractions (usually 5 or 6).

Measure by spectrophotometer the DNA in the 0.12*M* PB and 0.48*M* PB fractions. The single-copy and single-stranded DNA should be in the 0.12*M* PB fractions. For *Drosophila* an incubation to C_ot 200 usually is enough (50% of the input DNA is in the 0.12*M* PB fractions). Sometimes the spectrophotometer readings are not accurate (the phosphate buffer and/or the HAP may alter readings); you may wish to add some labelled total DNA from the same DNA preparation to follow the reassociation kinetics by scintillation counting.

Dialyze the 0.12*M* PB single-copy DNA fractions against dH_2O for 48 hrs and then concentrate it by speed-VAC. Read concentration with spectrophotometer (remember it is single-stranded). Store in freezer. We tried to desalt by glass milk purification and by using Centricon 30 columns (Amicon), but we lost 80% for the single-stranded DNA.

RANDOM PRIMER LABELLING OF SINGLE-COPY DNA

Dry down 50µl of tritiated thymidine (Amersham TRK .576; in EtOH: water 1:1) in each of 2 Eppendorf tubes. Each 50µl will be used for 1 reaction. We use the Boehringer random primed labeling kit (Cat. 1004760) with some modification to increase the average size of the product. To each vial with the dried hot TTP add:

0.25µg DNA
4 µl of the 4 dNTP's (1µl each dATP, dCTP, dGTP, dTTP
1 µl Reaction mixture
1 µl Reaction Buffer
X µl H_2O
<u>1 µl</u> Klenow enzyme
20 µl final volume

The reaction buffer is: 500m*M* Tris, 100m*M* $MgCl_2$, 1m*M* DTT, 2µg/µl BSA ultrapure pH 7.2. Add the H_2O first and resuspend the hot nucleotides, then add the other ingredients. Do not boil the DNA, it should already be single-stranded. The buffer in the above protocol is added to reduce by 50% the primer concentration of the Boehringer protocol. We add both hot and cold TTP to avoid running out of hot TTP during the reaction. We have also done this reaction with S^{35}, and with double labelling to increase specific activity. Incubate the reaction at 37°C for 4 hrs, combine the 2 reactions, and precipitate by adding NaAcetate to 0.3*M* and 2 volumes of 100% EtOH. Let sit for 1 hr at -20°C. Spin 15'. Wash with 70% EtOH and resuspend in 20µl dH_2O. Read 1µl of a 1:10 dilution in a scintillation counter. Use the remainder of your dilution to size your tracer on an alkaline gel (see appropriate section). The average size is usually 400-600 bp, and total cpm's are $20\text{-}30 \times 10^6$.

TEACL PREPARATION

We use TEACL (Tetraethyl ammonium chloride) from Sigma (Cat. T2265). It is important to have a stock of TEACL solution as concentrated as possible. Usually we start by adding 400gr TEACL slowly to 300ml dH_2O. Since TEACL is highly hygroscopic, the volume of the final solution will be almost tripled. The yellow colored TEACL solution is then mixed with Norite A-activated charcoal twice and millipore filtered (0.45µ). Molarity is checked by measuring its refractive index:

$$\text{Mol. TEACL} = \frac{R^* - 1.3333 + 0.008}{0.033}$$

Usually solutions prepared this way range in molarity from 3.7 to 3.2*M* TEACL. Store this solution at room temperature. Prepare a diluted stock solution of 2.4*M* TEACL, which can be kept in refrigerator. 2.4*M* TEACL should have a refractive index of 1.4042. Acceptable readings are from 1.4041 - 1.4043.

HYBRID PREPARATION

Preliminary preparation:

1. Mix your stock solution of TEACL and check the concentration by

----

*Refractive index.

refractive index. If TEACL stays for a long time on the shelf, H_2O may evaporate or the solution may become layered.

2. Mix your tracer well and check its specific activity by scintillation counting.

3. Check the concentration of the total sonicated DNA's that are going to be your drivers. Concentrations should be around 3-5µg/µl. If they are lower, concentrate the sample using, if possible, speed-vac. Tracers should be kept in H_2O because any buffer containing salt will interfere with the reassociation conditions and subsequent S1 digestions.

General strategy to follow:

Prepare at the same time the hybrids for the homologue reactions (i.e. A*-A) and all the heterologue reactions where A* is the tracer (i.e. A*-B, A*-C, etc.).

Divide the tracer (as measured by cpm's) evenly among all reactions. The quantity of driver in each reaction should be about the same. In this way the concentration of all your hybrids will be comparable as will be their incubation times to reach the desired C_ot value.

How much tracer to use: It mostly depends on how many total cpm's you have, how many comparisons you want to do, and how distant the taxa are (the percent reassociation decreases as the distance between related taxa increases). In ideal conditions use 100,000-200,000 cpm's, in 3-4µl solution. Do not worry about the µg of tracer DNA. If the tracer was prepared by oligolabelling, the actual concentration is very low.

How much driver to use: Again it still depends on how much DNA you have. Generally use a minimum of 10-30µg of DNA.

How much TEACL to use: The reassociation reaction should occur in a 1*M* TEACL solution. Since tracer and driver are in H_2O it is necessary to have a solution of TEACL as concentrated as possible to add to the tracer-driver mix in order to obtain a final solution of 1*M* TEACL. Usually you can make 3.6*M* TEACL. This solution should be kept at room temperature since it is near saturation and will crystallize if refrigerated.

Total volumes of reaction: volumes should be, if possible, between 10-30μl.

TRACER-DRIVER hybrid preparation:

Example:

Mix:	Vol. (μl)	Amount DNA (μg)	Total cpm's
TRACER	4 μl		400,000
DRIVER	10 μl	50	
TEACL (3.32 *M*)	6.04 μl		
Total	20.04 μl	50 μg	400,000 cpm's

Conc. 2.49 μg/μl in 1*M* TEACL

The final concentration of this solution should be around 2μg/μl to ensure that the incubation time does not exceed 7-8 days.

<u>How to mix these solutions and let them reassociate</u>:

Mix tracer-driver-TEACL in the desired proportions in a 0.5ml Eppendorf tube. Close the tube well and boil them for 7'. Spin for few seconds, wrap the top with parafilm, place them in a 50ml Falcon tube filled with 45°C H_2O. Close the Falcon tube tightly and submerge rapidly in a 45°C water bath. Doing the reassociations directly in an Eppendorf tube has the advantages of being faster, both at this stage and also after incubation when the samples have to be digested with S1 nuclease.

<u>How long should the hybrids stay at 45C</u>: It is necessary that the reaction arrives at equilibrium (i.e. no more hybrid duplexes are formed). The C_ot (concentration x time) necessary to reach this plateau depends on the size of the genome of the organisms studied. For insects a C_ot of 12000 is the value we usually use. Remember that the acceleration rate for 1*M* TEACL is 4.5.

<u>Example of C_ot calculation for sample in 1M TEACL</u>:

If the concentration of the solution is 2.2μg/μl and the C_ot value desired is 12,000:

2.2 / 310 = 0.0070967 (Moles per liter)

12000 / 0.0070967 = 1690926.7 (seconds in 0.12 *M* phosphate buffer)

1690926.7 / 4.5 = 375761.48 (seconds in 1*M* TEACL)

375761.48 / 3600 = 104 (hours in 1*M* TEACL)

104 / 24 = 4 days and 8 hours of incubation.

Hybrids should be kept at 45°C for at least this amount of time, or they may be left an extra day or so for convenience (if, for instance, other hybrids are coming out then). Once the incubation time is completed, hybrids should be treated with S1 nuclease, without storing them in fridge or freezer.

FIRST S1 NUCLEASE DIGESTION

At the end of the time necessary to reach the desired C_0t value, the reassociated hybrid duplex must be separated from the unreassociated DNA. This is achieved by lightly digesting the samples with S1 Nuclease to 95% completion and then separating the undigested double-stranded DNA from the free nucleotides (digested single-stranded DNA) by column chromatography.

Amount of S1 Nuclease necessary:

It is necessary to calibrate each batch of S1 Nuclease to make sure that the quantity used is enough to digest single-stranded DNA but not enough to digest double stranded DNA. To calibrate the enzyme, divide a spare homoduplex hybrid into aliquots, each of which is digested with different amounts of S1 (for instance from 0.5U/20µg DNA to 10U/µg DNA). When you graph the amount of DNA digested in each aliquot against the amount of S1 used, there will be a plateau, where, for increasing concentrations of enzyme, you do not have more single stranded DNA digested. Choose an S1 concentration in that plateau.

S1 is usually sold in a very concentrated form (from 400 to 1000U/µl). Since you need only a few units for each digestion, it is necessary to dilute the enzyme in order to accurately aliquot it in small quantities. Dilute enzyme in the buffer suggested by the manufacturer of the enzyme. To avoid wasting enzyme, try to do as many S1 nuclease digestions as you can in a day. The diluted enzyme may not be stored.

Comments: We have done enzyme calibration experiments several times. Usually 5U of S1 for each 20μg of DNA will work fine.

Digestion protocol:

General strategy: The S1 one will not work in presence of 1*M* TEACL, therefore it is necessary to dilute the sample at least 0.6*M* TEACL, or less. In addition, it is necessary to add a low pH ZnS04 buffer for the S1 to work. As a general strategy try to do the S1 digestion for the hybrid duplexes of a single tracer all together, so that the reaction conditions are as similar as possible among the homoduplex and heteroduplexes for each tracer. Try to keep the volume of your digestion between 50μl and 100μl.

Digestion protocol:

For each sample you already know the total volume. To each sample add 2 volumes H_2O and 1/10 final volumes Nuclease buffer (=NB, 0.5*M* Na Acetate, 5 m*M* ZnS04, pH 4.4). Add more water, if necessary, to have the same final volume for each sample but always dilute NB to 0.1. Add 5U of S1 for each 20μg of DNA. Mix lightly, spin for a few seconds in microfuge to bring down the all solution, and incubate in a 37°C water bath for 30'. Stop the reaction by adding 3μl of 1*M* EDTA pH 8 for each 100μl solution. Put the sample on ice. Save an aliquot of this sample (usually 1/10 of it) for the sizing gel (you may freeze this aliquot and run the gel at your convenience).

Example of calculations for S1 digestion:

Let us assume we have a reassociated hybrid which is in 25μl and contains 30μg of DNA, our S1 Nuclease has been diluted from the high concentration stock to a concentration of 2U/μl:

26	μl Volume Sample
50	μl Volume H_2O to add
10	μl Volume NB buffer to add
3.75	μl of S1 (5U/20 μg of DNA, then 7.5U for 30μg DNA)

11.25 µl H_2O (to add to reach the final volume of 100µl)

---

100 µl Final Volume

Place 30' at 37°C, then add 3µl 1*M* EDTA pH 8. Save 10µl for sizing. Place the rest on ice or, if ready, directly put over a G-100 column.

CHROMATOGRAPHY ON G-100 COLUMNS

Preliminary preparation:

1. It is necessary to have a stock solution of exactly 2.4*M* TEACL. Prepare it from the 3.X *M* solution and check the molarity with a refractometer (refractive index, r, should be 1.4042-1.4044, for 2.4*M* TEACL). Store this solution (1-2 liters) in a refrigerator. To avoid contaminations, take small aliquots (15ml) each time you need it.

2. At least one day in advance (usually it is more convenient to prepare a stock to keep in refrigerator), Sephadex G-100 beads must be rehydrated in 2.4*M* TEACL. Place in a screw cap bottle 1-2gr of beads, add 100ml of 2.4*M* TEACL and mix it until all the beads are in solution. If it becomes almost solid and jelly-like, you need to add more TEACL. Store it in a refrigerator. The chromatography must be done in 2.4*M* TEACL because you want to increase the concentration of the TEACL in your samples from less than 0.6*M* TEACL (after S1 digestion) to 2.4*M* TEACL to run them on the temperature gradient. The chromatography, therefore, serves two purposes: separating the reassociated DNA from the unreassociated, and transferring the reassociated DNA to 2.4*M* TEACL.

3. Sterilize some glass wool.

Column preparation:

Place sterile, short Pasteur pipettes (1.5ml) vertically. Put a small quantity of glass wool inside the pipette, pushing it down as much as you can using a long Pasteur pipette. Use one pipette for each sample.

Mix the G-100 beads in 2.4*M* TEACL and start loading the beads onto the column using a Pasteur pipette. Fill the column, wait for the buffer to flow through, check the level of the compact beads, and add more beads until the layer of compact beads is exactly at the constriction of the neck of the column. Fill the column completely with 2.4*M* TEACL, let it pass through. If you are not ready to place the samples on the columns, continue to add TEACL to prevent desiccation (you may recycle the TEACL already passed through, TEACL is expensive!).

Actual chromatography:

Before loading the samples on the column, prepare five labelled 1.5ml Eppendorf tubes for each sample. Place tubes in a rack below the column. When the top of the column is barely dry, add your sample. Wait until it gets into the beads, then fill the column to the top with 2.4*M* TEACL. Immediately after placing the sample on the column start collecting the eluate in the Eppendorf number 1. Collect 400µl in tube 1. Add more 2.4*M* TEACL on top of the column. Collect 800µl in tube 2. At this point add to the column H_2O and not TEACL, you will not need the other fractions for the temperature gradient, therefore it is not necessary to use 2.4*M* TEACL for the successive elutions. Collect 400µl in tube 3. Continue to add water to the top of the column as the column becomes almost dry. Collect 1000µl in tube 4 and 1500µl or more in tube 5. The elution will go much faster for the last 3 fractions because H_2O is less viscous than TEACL. Place fraction 2 on ice, the others may stay at room temperature.

Calculate the total cpm's in each fraction by scintilation counting. Fraction 1 should have almost no counts, fraction 2 should contain most of the reassociated duplex DNA, fraction 3 still contains some hybrids but in very low quantity and small in size; fractions 4 and 5 will contain the unreassociated DNA digested by S1 (free nucleotides).

From this experiment one calculates the percent reassociation (PR), simply by summing up the radioactivity in the first 3 fractions and dividing it by the total radioactivity of the sample (the sum of the radioactivity in all five fractions). Typically homoduplexes will have a PR of 80-60%, heteroduplexes will have lower PR values depending on their genetic relatedness to the homoduplex. From the PR value, one can calculate the NPR (Normalized Percent Reassociation), which is the PR of the heteroduplex divided by the PR of the homoduplex.

Example of percent reassociation calculation:

	Volume (µl)	cpm's in 50µl	Total cpm's
Fraction 1	400	450	3600
Fraction 2	800	1520 (in 10µl)	121600
Fraction 3	400	1260	10080
Fraction 4	1000	1200	24000
Fraction 5	1500	360	10800
			170080

Percent reassociation = 135280/ 170080 = 79.54%

Preparation of fraction 2 for temperature gradient:

For each experiment you need at least 1000 cpm's per tube in 100µl. To do a full gradient one uses at least 10 tubes plus two tubes to use as normalization controls. For each replicate experiment, one needs 4 tubes. You need to replicate the experiment at least 3 times to calculate a standard error. If you have more counts you can do more replicates. Therefore, the minimum volume necessary is 2.4ml of duplex in 2.4*M* TEACL solution with at least 24000 cpm's.

DETERMINATION OF THERMAL STABILITY OF DNA DUPLEXES

Preparation of the samples:

Use fraction 2 from the first S1 digestion. After determining the cpm's in this fraction, add 2.4*M* TEACL to a volume of 2.4-3.5ml. The final volume depends on how many cpm's you actually have, keep in mind you need at least 1000 cpm's in each 100µl aliquot, and you need at least 24 aliquots.

After adding the 2.4*M* TEACL, mix gently, and check the actual molarity by measuring the refractive index in 5µl aliquots. For 2.4*M* TEACL the refractive index should be 1.4042; readings between 1.4041-1.4043 are acceptable. A 1% change in the TEACL concentration at 2.4*M* gives about 0.4°C change in Tm, therefore it is important to be accurate.

The same person should check the refractive index of homoduplex and heteroduplexes for a certain tracer and all should be done at the same time. If the refractive index needs to be changed, adjust with 3.X *M* TEACL or H_2O. Usually you do not need large quantities, 20-50µl should be enough. Once the refractive index is correct, aliquot 100µl of your sample in Eppendorfs. Label the Eppendorf with a permanent ink pen. Be careful in making the 100µl aliquots, the 2.4*M* solution is quite viscous and it tends to remain in the pipette tip. After aliquoting all the sample, spin the Eppendorfs for a few seconds to bring all the sample down, and store at -20°C. The samples are now ready to be placed in the aluminum block. I try to use them within two weeks (less if ^{35}S is used).

Solutions needed:

1. Prepare 0.0001 *M* EDTA solution (it is easy to do from a 1*M* EDTA stock solution). Use distilled H_2O. Prepare large quantities, at least 2 liters. Store at room temperature.

2. Prepare 1-2 liters of 70% EtOH 0.1 *M* NaAc.

3. Prepare 250-500 ml of 10% CTAB (Hexadecyltrimethylammonium Bromide; SIGMA No. H-5882). Mix 10gr of CTAB in enough H_2O for 100ml solution. It takes some time and some warming up (37-45°C) to get into solution. Store at room temperature. If the room temperature gets too low it will precipitate. Place it at 37°C until it goes into solution again.

4. Prepare a 10µg/µl solution of sonicated salmon sperm DNA to use as carrier DNA. Mix the Salmon sperm in H_2O until it gets into solution (2-3 hrs), then place the solution in a 50ml Corning or Falcon tube and sonicate it for 5-10 minutes at a medium setting using the regular tip (not the microtip). Try to do the sonication with the tube immersed in cold EtOH to reduce overheating. After sonication place the solution at -20°C (The day you need it remember to defrost it on time). Place at 37°C if you are in a rush.

5. Prepare the "Second S1 digestion Buffer", a dilution of the "First S1 Digestion Buffer" (NB above). In a screw cap glass bottle add 100ml of "First S1 Digestion Buffer" and 650ml of sterile H_2O. Mix and keep refrigerated.

6. The day you are doing the gradient prepare a fresh dilution of S1. You need 15U for each 100µl aliquot. (Therefore if you have 90 aliquots you need 1350U). Prepare a solution of 3U per µl. You want to do all your samples with the same solution of diluted S1.

Preparations before the experiment:

1. In a scintillation rack prepare a series of scintillation vials labelled SS (single stranded). Add 500µl of 0.0001*M* EDTA to each vial. Close the cap loosely. Prepare another series of vials labelled DS (=double stranded). Do not add any EDTA, cover the vials with the caps and store them. I find it easier to label each duplex I am testing that day with a different color to reduce the possibility of mistakes.

2. Start the water baths connected to the aluminum block at least 2 hours before starting gradients to allow the gradient a chance to stabilize. Place a blank tube with 100µl H_2O in each hole. Choose a hole at each end of the gradient and place thermometer, which is kept permanently in place submerged in water. The temperature in each water bath should be regulated so that temperature ranges between 20-27C in the first hole and 65-70C in the last hole.

THE EXPERIMENT

Choice of range of temperatures to use:

For obtaining a nice melting curve for each duplex studied you want to use at least 10 aliquots in the block. You also need two aliquots to use as controls, one is kept on ice, the other is placed in boiling water for at least 10-14 minutes. They will serve to normalize the results, because they represent the amount that is already single stranded without any heating treatment (ice control), and the amount that will still remain double stranded even after prolonged boiling (the boil control). The 10 aliquots are placed in 10 holes in the aluminum gradient. The choice of holes depends on whether you are doing an homoduplex, or an heteroduplex. In otherwords, it depends on where you expect the tm of that duplex to be (i.e. when 50% of your duplex is going to be single stranded and 50% double stranded). The homoduplex is going to have a tm between 53-59°C (depending on its size), the heteroduplexes will have a lower tm, depending on the percent divergence accumulated between the two taxa compared. Therefore, depending on how distant you think the two taxa may be, you should choose a lower or higher range of temperatures to test. Select a range of temperature which will provide you with some helpful points for making your melting curve. I usually always use the first and the last holes (because if something goes wrong with the controls, you can always use these two), 2 holes at lower temperatures, 1 at higher temperature, and the remaining 5 are sequentially chosen on the basis of where I think the tm is going to be. For instance, in case of a homologue the tm should be around 55-59°C (depending on the size of the duplex), therefore I will use the holes from around 52 to 60°C. If I am doing an heterologue I will select lower temperatures. The difference in temperature between successive holes is around 1.8-2.0°C.

Read the temperature measured by the 2 control thermometers. Assuming the temperature will increase linearly, calculate the temperature in each hole by interpolation. Place vials number 1 to number 10 in the holes, vial number 11 on ice and number 12 in boiling water. Let them stay for exactly 30'. Check the temperature again before taking the vials out of the gradient. This operation serves to insure that you will take into account any change occurring in the gradient while the vials are in there (usually it does not oscillate more than 0.2C°).

<u>Second S1 Digestion</u>:

After 30,' take the samples out of the block, and add to each tube 300μl of "Second S1 digestion buffer". Use the same pipette tip for each, do not touch the vial with it. Add 5μl (=15 units) of diluted S1 using the same pipette tip for each, placing the drop of 5μl on the inner walls of the tubes. Close the caps, mix gently and spin few seconds. Incubate at 37°C for <u>exactly</u> 45 minutes.

At this time you may start a new sample in the aluminum block. Remember to check the temperatures again.

Double stranded DNA precipitation:

At the end of the 45 minutes of the S1 digestion place your vials in a rack, open all the caps, and add in each vial 20µl of 10µg/µl of sonicated salmon DNA (use only one pipette tip, do not touch it to the vials). Add 110µl of 10% CTAB. Mix. Spin in microfuge for 10 minutes. After centrifugation, remove the supernatant with a 1ml Pipetman and add it to the appropriate SS (single stranded) scintillation vial. Be careful, it is a very delicate step. You want to get all the supernatant, but you do not want to touch the tip to the pellet or, worse, dislodge the pellet. Add 8ml of scintillation fluid (Ultima Gold) to the scintillation vial. To the pellett, add 1ml of 70% EtOH 0.1 M NaAc using a 10ml plastic disposable pipette. Do not disturb the pellet. Spin for 3 minutes. Pour off the supernatant. Leave the caps open for 5-10 minutes to let the residues of EtOH evaporate. Add 1ml of 0.0001 *M* EDTA. Resuspend the pellet with a 1ml Pipetman, calibrated around 800µl. Pipette up and down several times. Transfer the solution to the appropriate DS scintillation vials. Add an additional 1ml of 0.0001*M* EDTA to each Eppendorf. Try to remove every little residue of pellet you see. Be careful as the pellet tends to stick to the walls of the vials or to the inner part of the pipette tip. If that occurs you <u>have</u> to get it out and put <u>everything</u> in the scintillation vials, otherwise the ratio between cpm's in the SS and DS scintillation vials is altered. Add 16ml Ultima Gold scintillation fluid and count. Sometimes it may help the resuspension of the pellet if you let the sample sit with EDTA for 30 minutes or so, or if you add warm EDTA. Sometimes the pellet are very easy to resuspend, other times very hard, I do not know why!.

Recently we have eliminated the resuspension of the double-stranded pellet altogether. Just count the total radioactivity in 2 or 3 of the 100µl aliquots and use the average value as the total amount of radioactivity present in each tube. The % single-stranded is then calculated by dividing the radioactivity in each ss vial by the total radioactivity present in each vial.

SIZING GEL

Tracers and hybrid duplexes lengths can be checked by running them on a 3% alkaline gel. To 2.4gr of Agarose (Ultra Pure Reagent,

Electrophoresis Grade, IBI. Cat. 70040, 70042) add 80mls of neutral buffer (30m*M* NaCl, 2m*M* EDTA pH 7.0, for 1 liter of a 10X solution use 17.53gr NaCl and 7.44gr of EDTA). After it is solidified, pre-electrophoresed the gel for 1 hr in an alkaline bridge buffer at 50v, 175mA. The alkaline bridge buffer is 30m*M* NaOH and 2m*M* EDTA (for 1 liter of a 20X solution add 14.89gr of EDTA to 24gr of NAOH).

Before loading your sample, add 1/3 volume of a denaturing loading dye (0.12*M* NaoH, 0.008*M* EDTA, 20% v/v glycerol, 0.1% bromocresol green) kept in the refrigerator. Run gel at 25 V for ~12 hrs. Neutralize gel in 250mls 0.1M Tris, pH 7 with 1μg Ethidiumbromide for 30'. Photograph to record position of your marker. Slice each lane that contains a sample in 0.7cm chunks. For our gels we have 17 pieces. Place each piece in a scintillation vial with 1ml of 0.2 *M* HCl. Melt each set of 17 vials in microwave for 1' and then add 16ml of Scintillation fluid (Opti-fluor from Packard, catalog no. 6013425, or Ultima Gold from Packard).

Using the marker DNA, calculate the average fragment length: using a log/linear scale (size / distance) the average size in each slice is determined. The formula to get the tracer size is the one from Hall et al., 1980: $\Sigma Qi/\Sigma(Qi/Li)$, where Li is the average fragment size in slice i and Qi are the cpm/slice. The marker we find convenient to use is pBr322 cut with the enzyme *Hin*f I. Since the bromocresol green in this loading buffer disappears quite rapidly we fill an empty lane with 20-30μl of our regular loading buffer (0.25% bromophenol blue, 0.25% xylene cyanol, 30% glycerol in H_2O). This dye will have a dark blue band which runs fast and a slow lighter blue band. When the dark blue band has run for 9-9.5 cm from the wells then the gel is ready to be stained. This is usually after 12 hrs, but times may vary depending upon the power pack, the thickness of the gels, the melting conditions etc. It is essential if one wants to compare results from different gels to run them to the same length.

DATA ANALYSIS FOR TEACL DNA-DNA HYBRIDIZATION

Attached are some sheets illustrating "raw" data from our DNA-DNA hybridization experiments on primates. We will explain in some detail just how we go about the experiments and calculate tm's (which in our nomenclature are the median melting temperature uncorrected for duplex size).

1. We first perform what we call our "full gradient". This consists of 12 tubes. Ten are placed in the temperature gradient as indicated in the first ten lines in tables 1 and 2. While our gradient blocks have a total of 20 holes, we concentrate the tubes in the area in which we expect the tm should be while skipping other holes. This is why the temperature increments are not constant. Two tubes serve as controls: one is kept on ice and the other is placed in boiling water. These should represent zero and 100 percent single-stranded respectively.

2. We carry out the melts as indicated previously and obtain cpm's such as those illustrated in columns 3 and 4 in table 1 and 2; these are the counts in the single-stranded and double-stranded fractions respectively. We then simply calculate the % single-stranded (column 5).

3. These numbers are then normalized to the zero and 100% controls to give the normalized % ss. This is done by subtracting the ice control %ss and dividing by the range. For example, in the JRP-JRP1 melt, we subtracted 7.7 from each %ss figure and divided by 70.63. This generates column 6.

4. We then note that the median melting temperature occurred between tubes 7 and 8, i.e. between 55.8°C and 58.0°C (Table 1). We then do a linear extrapolation to find where 50% ss would fall.

   First: 66.1% - 44.1% = 22.0% melting between temperatures.

   Second: 50% - 44.05% = 5.9%   5.9/22.0 = 0.2681, the fraction of 22.0% needed to reach 50%

   Third: 58.0 - 55.8 = 2.2°C between tubes

   Fourth: 2.2 x 0.2681 = 0.59

   Fifth: 55.8 + 0.59 = 56.39

5. Once we know about where the tm is, we do replicates using only the four tubes closest to the tm. These are shown in the blocks of 4 lines. The data are handled identically as before.

6. If, as occasionally happens, the first tube in the full gradient has less ss DNA than the ice control, we then use it as the zero % control for normalizing. This is illustrated in the second data set, BAR-CHI.

7. Average t_m values are then transformed into T_m by adding the size correction factor (T_{cor}) obtained from the sizing gels (Table 3)

8. Δt_m and ΔT_m are calculated by subtracting the t_m and T_m of each heterologue to the respective homologue (Table 4).

9. The SEs of the ΔT_m's are calculated as:
$(VAR1/N1 + VAR2/N2)^{1/2}$
where VAR1 and VAR2 are the respective variances of the Tm's for the homoduplex and heteroduplex and N1 and N2 are the respective number of determinations (Table 4).

10. For the weighted mean and its SE for reciprocal values (both strains used as tracer and driver) we used the following formula (see Table 4 for example):
Weighted mean = $[\Delta Tm1/(SE1^{*})^2 + \Delta Tm2/(SE2)^{2*}]/K$
SE of weighted mean = $(1/K)^{1/2}$
where $K = 1/(SE1)^2 + 1/(SE2)^2$

11. Percent reassociation (PR) are calculated after the first S1 Digestion, Normalized Percent Reassociation (NPR) are the PRs of the heterologues divided by the PR of the respective homologue.

---

*SE1 and SE2 standard errors of Tm1 and Tm2

Table 1: JRP-JRP1

	TEMP	cpm's SS	cpm's DS	% SS	% SS Norm.	tm's
1	30.0	191.0	2152.0	8.10	0.6	
2	37.4	176.5	1924.5	8.40	1.0	
3	44.6	285.5	2123.5	11.80	5.9	
4	48.2	304.0	1988.5	13.30	7.9	
5	52.0	497.0	2179.5	18.60	15.4	
6	54.2	672.0	1858.5	26.60	26.7	
7	55.8	805.5	1270.0	38.80	44.1	
8	58.0	1479.0	1241.0	54.40	66.1	
9	60.0	2340.5	797.5	74.60	94.7	
10	66.0	2211.5	755.5	74.50	94.6	56.39
11	53.4	544.0	2533.5	17.70	14.2	
12	55.2	856.0	1809.5	32.10	34.6	
13	57.0	1270.0	1529.5	45.40	53.3	
14	59.0	2334.0	856.0	73.20	92.7	56.69
15	54.0	615.0	1930.5	24.20	23.3	
16	56.0	864.0	1705.5	33.60	36.7	
17	58.0	1609.5	1254.5	56.20	68.7	
18	59.8	2183.5	748.0	74.50	94.5	56.83
19	52.4	566.0	1961.0	22.40	20.8	
20	54.4	726.5	1846.0	28.20	29.1	
21	56.2	1192.5	1574.0	43.10	50.1	
22	58.0	1937.5	894.5	68.40	85.9	56.19
23	53.2	510.0	1853.5	21.60	19.7	
24	55.2	784.5	1545.5	33.70	36.8	
25	57.0	1241.0	1145.0	52.00	62.7	
26	59.2	2034.5	896.0	69.40	87.4	56.12
27	Ice cont.	196.0	2358.5	7.70		
28	Boil cont.	2111.0	584.0	78.33		
				Norm. range=78.33-7.70=70.63		

Table 2: BAR-CHI

	TEMP	cpm's SS	cpm's DS	% SS	% SS Norm.	tm's
1	32.6	154.5	1654.0	8.50	0.00	
2	35.6	143.0	1474.0	8.80	0.40	
3	45.6	207.0	1327.5	13.50	6.40	
4	49.0	318.5	1650.5	16.20	9.80	
5	50.6	290.5	1243.5	18.90	13.40	
6	52.2	396.0	1582.5	20.00	14.80	
7	54.0	594.5	1385.5	30.00	27.70	
8	55.4	931.0	1053.0	46.90	49.40	
9	57.0	1564.5	633.5	71.20	80.70	
10	61.8	1884.0	430.5	81.40	93.80	55.43
11	50.0	320.0	1716.5	15.70	9.20	
12	51.6	381.5	1770.0	17.70	11.80	
13	53.6	526.0	1365.0	27.80	24.80	
14	55.4	1290.5	807.0	61.50	68.20	54.64
15	50.2	358.0	1440.5	19.90	14.60	
16	52.4	556.0	1459.0	27.60	24.50	
17	54.4	802.0	1240.5	39.30	39.60	
18	56.0	1353.0	870.0	60.90	67.40	55.00
19	50.0	311.5	1737.0	15.20	86.0	
20	51.8	368.0	1737.0	17.50	11.50	
21	53.8	462.5	1574.5	22.70	18.20	
22	55.6	1416.0	764.5	64.90	72.60	54.85
23	50.6	335.0	1649.0	16.90	10.70	
24	52.0	479.0	1656.0	22.40	17.90	
25	53.4	664.0	1211.0	35.40	34.60	
26	55.0	1155.5	1140.0	50.30	53.80	54.68
27	52.2	603.0	1765.5	25.50	21.80	
28	53.8	685.0	1226.5	35.80	35.20	
29	55.2	959.5	1295.0	42.50	43.80	

30	56.6	1547.0	401.5	79.40	91.25	55.38
32	Ice cont.	263.5	1419.5	15.66		
32	Boil cont.	2164.0	347.0	86.18		
				Norm. Range=86.18-8.54=77.64		

Table 3. DNA-DNA Hybridization Studies with Parthenogenetic *Drosophila*

Tracer & Driver	Tm±SE	Tracer Length (bp)	T_{cor}	T_mCOR	% Reassoc.	Normalized % Reassoc.
KP:						
KP	56.99±.07	498	1.00	57.99	73.3	100
KWP	55.61±.06	500	1.00	56.61	69.4	94.8
SP	55.65±.04	484	1.03	56.68	60.3	82.3
KWP:						
KWP	56.63±.19	412	1.21	57.84	80.8	100
KP	55.11±.11	326	1.53	56.64	69.9	86.5
SP	55.51±.15	374	1.33	56.84	64.4	79.7
SP:						
SP	56.65±.25	335	1.49	58.14	77.4	100
KP	56.69±.06	403	1.24	56.93	66.0	85.2
KWP	55.37±.21	446	1.12	56.49	58.7	75.8

Table 4. ΔT_m's for Data in Table 3

Tracer-Driver	Δtm	ΔTmCOR	SE
KP-KWP	1.39	1.38	0.09
KWP-KP	1.52	1.20	0.22
KP-KWP[a]	1.40	1.36	0.08
KP-SP	1.34	1.31	0.07
SP-KP	0.96	1.21	0.26
KP-SP[a]	1.21	1.28	0.03

KSP-SP	1.12	1.00	0.24
SP-KWP	1.28	1.65	0.33
KWP-SP[a]	1.20	1.33	0.20

[a]Weighted mean of the reciprocals, calculated as described in text.

TECHNICAL LITERATURE ON THE SUBJECT:

Britten, R.J., A. Cetta, and E.H. Davidson. 1978. The single-copy sequence polymorphism of the sea urchin Strongylocentrotus purpuratus. Cell 15:1175-1186.

Britten, R.J., D.E. Graham, and B.R. Neufeld. 1974. Analysis of repeating DNA sequences by reassociation. Pp. 363-418 in L. Grossman and K. Moldave, eds. Methods in enzymology. Vol. 29e. Academic Press, New York.

Caccone, A., and J.R. Powell. 1987. Molecular evolutionary divergence among North American cave crickets. II. DNA-DNA hybridization. Evolution 41:1215-1238.

Caccone, A., and J.R. Powell. 1989. DNA divergence among hominoids. Evolution 43:925-942.

Caccone, A., and J.R. Powell. 1990. Extreme rates and heterogeneity in insect DNA evolution. J. Mol. Evol. 30:273-280.

Caccone, A., G.D. Amato, and J.R. Powell. 1987. Intraspecific DNA divergence in Drosophila: a study on parthenogenetic D. mercatorum. Mol. Biol. Evol. 4:343-350.

Caccone, A., R. DeSalle, and J.R. Powell. 1988a. Calibration of the change in thermal stability of DNA duplexes and degree of base pair mismatch. J. Mol. Evol. 27:212-216.

Chang, C., T.C. Hain, J.R. Hutton, and J.G. Wetmur. 1974. Effects of microscopic and macroscopic viscosity on the rate of renaturation of DNA. Biopolymers 13:1847-1858.

Crothers, D.M., N.R. Kallenback, and B.H. Zimm. 1965. The melting transition of low-molecular-weight DNA: theory and experiment. J. Mol. Biol. 11:802-820.

Galau, G.A., B.R. Chamberlin, B.R. Hough, R.J. Britten, and E.H. Davidson. 1976. Evolution of repetitive and nonrepetitive DNA. In: Ayala FR (ed) Molecular evolution. Sinauer Associates, Sunderland MA, pp. 200-224.

Hall, T.J., J.W. Grula, E.H. Davidson, and R.J. Britten. 1980. Evolution of sea urchin non-repetitive DNA. J. Mol. Evol. 16:95-110.

Hayes, F.N., E.H. Lilly, R.L. Ratliff, D.A. Smith, and D.H. Williams. 1970. Thermal transitions in mixtures of polydeoxynucleotides. Biopolymers 9:1105-1117.

Hunt, J.A., T.J. Hall, and R.J. Britten. 1981. Evolutionary distance in Hawaiian Drosophila measured by DNA reassociation. J. Mol. Evol. 17:361-367.

Hutton, J.R., and J.G. Wetmur. 1973. Effect of chemical modification on the rate of renaturation of deoxyribonucleic acid: deamination and glyoxalated deoxyribonucleic acid. Biochemistry 12:558-563.

Kohne, D.E. 1970. Evolution of higher organism DNA. Q. Rev. Biophys. 33:327-375.

Melchior, W.B., and P.H. Von Hippel. 1973. Alteration of the relative stability of dA-dT and dG-dC base pairs in DNA. Proc. Natl. Acad. Sci. USA 70:298-302.

Orosz, J.M., and J.G. Wetmur. 1977. DNA melting temperature and renaturation rates in concentrated alkylammonium salt solutions. Biopolymers 16:1183-1199.

Powell, J.R., and A. Caccone. 1990. The TEACL method of DNA-DNA hybridization: technical considerations. J. Mol. Evol. 30:267-272.

List of Participants

Prof Godfrey Hewitt	Norwich	UK	(Director)
Prof Jean David	Gif	France	(Committee)
Prof Andy Johnston	Norwich	UK	(Committee)
Dr Peter Young	Norwich	UK	(Committee)
Prof Sam Berry	London	UK	
Dr Roy Carter	Nottingham	UK	
Prof Hampton Carson	Hawaii	USA	
Prof Robert Cedergen	Montreal	Canada	
Dr. Jane Doyle	New York	USA	
Prof Jeffrey Doyle	New York	USA	
Dr. Noel Ellis	Norwich	UK	
Prof Antonio Fontdevila	Barcelona	Spain	
Dr Michael Gasson	Norwich	UK	
Prof Martin Kreitman	New Jersey	USA	
Dr. Daniel Lachaise	Gif	France	
Prof Gary Olsen	Illinois	USA	
Dr. David Parkin	Nottingham	UK	
Dr. Rory Post	Manchester	UK	
Prof Jeffrey R. Powell	Conneticut	USA	
Prof Valerio Sbordoni	Rome	Italy	
Prof Chris Simon	Hawaii	USA	
Prof Michel Solignac	Gif	France	
Prof Dieter Sperlich	Tubingen	Germany	
Prof David L. Swofford	Illinois	USA	
Dr. Richard Thomas	London	UK	
Prof Eleftherios Zouros	Iraklion	Greece	
Dr Begona Arano	Madrid	Spain	
Mr Lutz Bachmann	Tubingen	Germany	
Mr F.R. Bakker	Groningen	Netherlands	
Dr John W.O. Ballard	Brisbane	Australia	
Mr Eladio Barrio	Valencia	Spain	
Dr Michel Baylac	Paris	France	
Dr Jose Luis Bella	Madrid	Spain	
Dr Paul Bentzen	Halifax	Canada	
Dr Giorgio Piertro Mario Binelli	Milan	Italy	
Ms Josephine Bowes	Herson	Australia	
Dr Paul D Bridge	Surrey	UK	
Dr John F.Y. Brookfield	Nottingham	UK	
Mr Andrew V.Z. Brower	New York	USA	
Dr Jonathan M. Brown	New York	USA	
Dr Roger K. Butlin	Cardiff	UK	
Dr Adalgisa Caccone	Rome	Italy	
Dr Francois M. Catzeflis	Montpellier	France	
Ms Magda Charalambous	London	UK	
Dr Jan Evelyn Conn	Caracas	Venezuela	
Dr Alastair Culham	Reading	UK	
Dr Danielle De Faye	Paris	France	

Dr Joaquina De La Torre	Madrid	Spain
Dr Ronald W. Debry	Florida	USA
Mr Robert A. Feldman	Hawaii	USA
Dr Enrique Figueroa	Seville	Spain
Dr Maria Grazia Filippucci	Rome	Italy
Dr Jonathan P.A. Gardner	Leicester	UK
Ms Jennifer M. Gleason	Conneticut	USA
Dr George N. Goulielmos	Iraklion	Greece
Mr Sigurdur Greipsson	Reykjavik	Iceland
Ms Alison M. Griffin	London	UK
Mr. Oliver Hannotte	Mons	Belgium
Mr Stephen A. Harris	Fife	UK
Ms Cheryl Y. Hayashi	Conneticut	USA
Dr Jean-Louis Hemptinne	Gembloux	Belgium
Dr Hannelore Hoch	Marburg	Germany
Dr Annelies Hofman	Groningen	Netherlands
Dr Carlos Juan	Palma de Mallorca	Spain
Mr Kemal Kazan	Ankara	Turkey
Dr Elizabeth A. Kellog	Surrey	UK
Dr. Nazy Khadem	Athens	Greece
Dr. Lesley N. Manchester	Leicester	UK
Mr Kevin McCluskey	Oregon	USA
Mr Michael Milinkovitch	Bruxelles	Belgium
Dr Enrique Monte	Salamanca	Spain
Dr Loredana Nigro	Padova	Italy
Dr Svante Paabo	California	USA
Mr Herve Phillipe	Paris	France
Mr Aloysius Phillips	Hawaii	USA
Dr Wayne Powell	Dundee	UK
Ms Carol A. Reeb	Hawaii	USA
Dr Michael G. Ritchie	Leicester	UK
Mr Jose Miquel Rubio	Madrid	Spain
Ms Marta Ruitort	Barcelona	Spain
Dr Richard R. Snell	Otowa	Canada
Dr Pierre Taberlet	Grenoble	France
Dr. William Kelley Thomas	California	USA
Mr Diogo M.P.F. Thomaz	Oporto	Portugal
Dr Fiona Thomson-Carter	Aberdeen	UK
Mr Yves Van De Peer	Wilrijk	Belgium
Mr Peter Van Dijk	Heteren	Netherlands
Ms Patricia Vasquez	Madrid	Spain
Dr Frederica Venanzetti	Rome	Italy
Ms Sonia Virdee	Norwich	UK
Robert G. Wisotzkey	Hawaii	USA
Dr Kirsten Wolff	Haren	Netherlands
Mr Wolfgang Wuster	Aberdeen	UK
Dr Brenda L. Young	Plymouth	UK
Ms Irena Zajc	Ljubeljana	Yugoslavia
Mr Kees Zwanenburg	Nova Scotia	Canada

NATO ASI Series H

NATO ASI Series H

Vol. 21: **Cellular and Molecular Basis of Synaptic Transmission.**
Edited by H. Zimmermann. 547 pages. 1988.

Vol. 22: **Neural Development and Regeneration. Cellular and Molecular Aspects.**
Edited by A. Gorio, J.R. Perez-Polo, J. de Vellis, and B. Haber. 711 pages. 1988.

Vol. 23: **The Semiotics of Cellular Communication in the Immune System.**
Edited by E.E. Sercarz, F. Celada, N.A. Mitchison, and T. Tada. 326 pages. 1988.

Vol. 24: **Bacteria, Complement and the Phagocytic Cell.**
Edited by F.C. Cabello und C. Pruzzo. 372 pages. 1988.

Vol. 25: **Nicotinic Acetylcholine Receptors in the Nervous System.**
Edited by F. Clementi, C. Gotti, and E. Sher. 424 pages. 1988.

Vol. 26: **Cell to Cell Signals in Mammalian Development.**
Edited by S.W. de Laat, J.G. Bluemink, and C.L. Mummery. 322 pages. 1989.

Vol. 27: **Phytotoxins and Plant Pathogenesis.**
Edited by A. Graniti, R.D. Durbin, and A. Ballio. 508 pages. 1989.

Vol. 28: **Vascular Wilt Diseases of Plants. Basic Studies and Control.**
Edited by E.C. Tjamos and C.H. Beckman. 590 pages. 1989.

Vol. 29: **Receptors, Membrane Transport and Signal Transduction.**
Edited by A.E. Evangelopoulos, J.P. Changeux, L. Packer, T.G. Sotiroudis, and K.W.A. Wirtz. 387 pages. 1989.

Vol. 30: **Effects of Mineral Dusts on Cells.**
Edited by B.T. Mossman and R.O. Bégin. 470 pages. 1989.

Vol. 31: **Neurobiology of the Inner Retina.**
Edited by R. Weiler and N.N. Osborne. 529 pages. 1989.

Vol. 32: **Molecular Biology of Neuroreceptors and Ion Channels.**
Edited by A. Maelicke. 675 pages. 1989.

Vol. 33: **Regulatory Mechanisms of Neuron to Vessel Communication in Brain.**
Edited by F. Battaini, S. Govoni, M.S. Magnoni, and M. Trabucchi. 416 pages. 1989.

Vol. 34: **Vectors as Tools for the Study of Normal and Abnormal Growth and Differentiation.**
Edited by H. Lother, R. Dernick, and W. Ostertag. 477 pages. 1989.

Vol. 35: **Cell Separation in Plants: Physiology, Biochemistry and Molecular Biology.**
Edited by D.J. Osborne and M.B. Jackson. 449 pages. 1989.

Vol. 36: **Signal Molecules in Plants and Plant-Microbe Interactions.**
Edited by B.J.J. Lugtenberg. 425 pages. 1989.

Vol. 37: **Tin-Based Antitumour Drugs.** Edited by M. Gielen. 226 pages. 1990.

Vol. 38: **The Molecular Biology of Autoimmune Disease.**
Edited by A.G. Demaine, J-P. Banga, and A.M. McGregor. 404 pages. 1990.

Vol. 39: **Chemosensory Information Processing.** Edited by D. Schild. 403 pages. 1990.

Vol. 40: **Dynamics and Biogenesis of Membranes.**
Edited by J.A.F. Op den Kamp. 367 pages. 1990.

Vol. 41: **Recognition and Response in Plant-Virus Interactions.**
Edited by R.S.S. Fraser. 467 pages. 1990.

NATO ASI Series H

Vol. 42: Biomechanics of Active Movement and Deformation of Cells.
Edited by N. Akkaş. 524 pages. 1990.

Vol. 43: Cellular and Molecular Biology of Myelination.
Edited by G. Jeserich, H. H. Althaus, and T. V. Waehneldt. 565 pages. 1990.

Vol. 44: Activation and Desensitization of Transducing Pathways.
Edited by T. M. Konijn, M. D. Houslay, and P. J. M. Van Haastert. 336 pages. 1990.

Vol. 45: Mechanism of Fertilization: Plants to Humans.
Edited by B. Dale. 710 pages. 1990.

Vol. 46: Parallels in Cell to Cell Junctions in Plants and Animals.
Edited by A. W. Robards, W. J. Lucas, J. D. Pitts, H. J. Jongsma, and D. C. Spray. 296 pages. 1990.

Vol. 47: Signal Perception and Transduction in Higher Plants.
Edited by R. Ranjeva and A. M. Boudet. 357 pages. 1990.

Vol. 48: Calcium Transport and Intracellular Calcium Homeostasis.
Edited by D. Pansu and F. Bronner. 456 pages. 1990.

Vol. 49: Post-Transcriptional Control of Gene Expression.
Edited by J. E. G. McCarthy and M. F. Tuite. 671 pages. 1990.

Vol. 50: Phytochrome Properties and Biological Action.
Edited by B. Thomas and C. B. Johnson. 337 pages. 1991.

Vol. 51: Cell to Cell Signals in Plants and Animals.
Edited by V. Neuhoff and J. Friend. 404 pages. 1991.

Vol. 52: Biological Signal Transduction.
Edited by E. M. Ross and K. W. A. Wirtz. 560 pages. 1991.

Vol. 53: Fungal Cell Wall and Immune Response.
Edited by J. P. Latgé and D. Boucias. 472 pages. 1991.

Vol. 54: The Early Effects of Radiation on DNA.
Edited by E. M. Fielden and P. O'Neill. 448 pages. 1991.

Vol. 55: The Translational Apparatus of Photosynthetic Organelles.
Edited by R. Mache, E. Stutz, and A. R. Subramanian. 260 pages. 1991.

Vol. 56: Cellular Regulation by Protein Phosphorylation.
Edited by L. M. G. Heilmeyer, Jr. 520 pages. 1991.

Vol. 57: Molecular Techniques in Taxonomy.
Edited by G. M. Hewitt, A. W. B. Johnston, and J. P. W. Young. 420 pages. 1991.

Printing: Druckerei Zechner, Speyer
Binding: Buchbinderei Schäffer, Grünstadt